Progress in Mathematics
Volume 70

Analytic Number Theory and Diophantine Problems

Proceedings of a Conference at Oklahoma State University, 1984

Edited by
A.C. Adolphson
J.B. Conrey
A. Ghosh
R.I. Yager

1987

Birkhäuser
Boston · Basel · Stuttgart

Math

sep/ae

A.C. Adolphson
J.B. Conrey
A. Ghosh
Department of Mathematics
Oklahoma State University
Stillwater, OK 74078
U.S.A.

R.I. Yager
Macquarie University
New South Wales 2113
Australia

Library of Congress Cataloging-in-Publication Data
Analytic number theory and diophantine problems.
(Progress in mathematics ; v. 70)
Includes bibliographies.
1. Numbers, Theory of—Congresses. I. Adolphson, A.C.
II. Series: Progress in mathematics (Boston, Mass.) ;
vol. 70
QA241.A487 1987 512′.73 87-14635

CIP-Kurztitelaufnahme der Deutschen Bibliothek
Analytic number theory and diophantine problems:
proceedings of a conference at Oklahoma State Univ.,
1984 / ed. by A.C. Adolphson . . . —Boston ;
Basel ; Stuttgart : Birkhäuser, 1987.
(Progress in mathematics ; Vol. 70)
ISBN 3-7643-3361-8 (Basel . . .)
ISBN 0-8176-3361-8 (Boston)
NE: Adolphson, A.C. [Hrsg.] ; Oklahoma State
University [Stillwater, Okla.]; GT

ISBN 0-8176-3361-8
ISBN 3-7643-3361-8

Text prepared by the editors in camera-ready form.
Printed and bound by Edwards Brothers Incorporated, Ann Arbor, Michigan.
Printed in the U.S.A.

9 8 7 6 5 4 3 2 1

PREFACE

A conference on Analytic Number Theory and Diophantine Problems was held from June 24 to July 3, 1984 at the Oklahoma State University in Stillwater. The conference was funded by the National Science Foundation, the College of Arts and Sciences and the Department of Mathematics at Oklahoma State University.

The papers in this volume represent only a portion of the many talks given at the conference. The principal speakers were Professors E. Bombieri, P. X. Gallagher, D. Goldfeld, S. Graham, R. Greenberg, H. Halberstam, C. Hooley, H. Iwaniec, D. J. Lewis, D. W. Masser, H. L. Montgomery, A. Selberg, and R. C. Vaughan. Of these, Professors Bombieri, Goldfeld, Masser, and Vaughan gave three lectures each, while Professor Hooley gave two. Special sessions were also held and most participants gave talks of at least twenty minutes each. Prof. P. Sarnak was unable to attend but a paper based on his intended talk is included in this volume.

We take this opportunity to thank all participants for their (enthusiastic) support for the conference. Judging from the response, it was deemed a success.

As for this volume, I take responsibility for any typographical errors that may occur in the final print. I also apologize for the delay (which was due to the many problems incurred while retyping all the papers).

A special thanks to Dollee Walker for retyping the papers and to Prof. W. H. Jaco for his support, encouragement and hard work in bringing the idea of the conference to fruition.

A. Ghosh

(on behalf of the Editors).

TABLE OF CONTENTS

PARTICIPANTS

Adolphson, A.
Alladi, K.
Bateman, P.
Beukers, F.
Bombieri, E.
Brownawell, D.
Chakravarty, S.
Chen, W. W. L.
Cisneros, J.
Conrey, J. B.
Cooper, C.
Diamond, H. G.
Friedlander, J.
Gallagher, P. X.
Ghosh, A.
Goldfeld, D.
Goldston, D. A.
Gonek, S. M.
Graham, S.
Greenberg, R.
Gupta, R.
Halberstam, H.
Harman, G.
Hensley, D.
Hildebrand, A.
Hooley, C.
Iwaniec, H.
Jaco, W.
Kano, T.
Kennedy, R. E.
Kolesnik, G.
Kueh, Ka-Lam.
Lewis, D. J.
Maier, H.
Masser, D. W.
McCurley, K.
Montgomery, H. L.
Mueller, J.
Myerson, J.
Nathanson, M.
Ng, E.
Pollington, A.
Schumer, P.
Selberg, A.
Shiokawa, I.
Sieburg, H. B.
Skarda, V.
Spiro, C.
Vaaler, J.
Vaughan, R. C.
Vaughn, J.
Woods, D.
Yildirim, C. Y.
Youngerman, D.
Yager, R.

MULTIPLICATIVE FUNCTIONS AND SMALL DIVISORS

K. Alladi[1], P. Erdős and J.D. Vaaler[2]

1. Introduction[3]

Let S be a set of positive integers and g be a multiplicative function. Consider the problem of estimating the sum

$$S(x,g) = \sum_{\substack{n \leq x \\ n \in S}} g(n). \tag{1.1}$$

A natural way to start is to write

$$g(n) = \sum_{d|n} h(d) \tag{1.2}$$

and reverse the order of summation. This in turn leads to the estimation of the contribution arising from the large divisors d of n, where n S, which often presents difficulties. In this paper we shall characterize in various ways the following idea:

"*Large divisors of a square-free integer have more prime divisors than the small ones.*" (1.3)

When the multiplicative function h is small in size, (1.3) will be useful in several situations to show that the principal contribution is due to the small divisors. The terms 'large' and 'small' will be made precise in the sequel.

An application to Probabilistic Number Theory is discussed in

---

[1]On leave of absence from 'MATSCIENCE', Institute of Mathematical Sciences, Madras, India.

[2]The research of the third author was supported by a grant from the National Science Foundation.

[3]As this paper evolved we had several useful discussions with Amit Ghosh, Roger Heath-Brown and Michael Vose.

Sec.4; indeed, it was this application which motivated the present paper (see [1], [3]). Our discussion in the first two sections is quite general - in Sec.2 the principal result is derived for sets rather than for divisors only and in Sec.3 the main inequality is for submultiplicative functions. This is done in the hope that our elementary methods may have other applications as well, perhaps even outside of Number Theory.

2. A mapping for sets.

If n is not square it is trivial to note that half its divisors are less than $\sqrt{n}$. If n is square-free there is also an interesting one-to-one correspondence, namely: there is a bijective mapping m between the divisors d of n which are less than $\sqrt{n}$ and the divisors d' of n which are greater than $\sqrt{n}$ such that

$$m(d) = d' \equiv 0 \pmod{d} \tag{2.1}$$

(of course the mapping m depends on n). In fact, this mapping is a special case of a rather general one-to-one correspondence that can be set up between subsets of a finite set, as we shall presently see.

Let $\mathbf{S}$ be a finite set and λ a finite measure on the set of all subsets of $\mathbf{S}$. For each $t \geq 0$ define

$$A(t,\mathbf{S}) = \{\mathbf{E} \subseteq \mathbf{S} : \lambda(\mathbf{E}) \leq t\}.$$

We then have

Theorem 1. *For each* $t \geq 0$ *there is a permutation*

$$\pi_{t,\mathbf{S}} : A(t,\mathbf{S}) \to A(t,\mathbf{S})$$

such that for all $\mathbf{E} \subset A(t,\mathbf{S})$ *we have* $\pi_{t,\mathbf{S}}(\mathbf{E}) \cap \mathbf{E} = \emptyset$.

Remark. There are trivial cases here. If $\lambda(\mathbf{S}) \leq t$ then $A(t,\mathbf{S})$ is the power set of $\mathbf{S}$ and so the permutation $\mathbf{E} \to \mathbf{S} - \mathbf{E}$ has the desired property. If $t = 0$ then $A(0,\mathbf{S})$ is the power set of $\mathbf{S}^{(0)}$ where

$\mathbf{S}^{(0)} = \{ s \in \mathbf{S} : \lambda(s) = 0 \}$. Here $\mathbf{E} \to \mathbf{S}^{(0)} - \mathbf{E}$ is an appropriate permutation. So in the proof that follows we assume that $0 < t < \lambda(S)$.

Proof. If $\mathbf{S}$ has cardinality $|\mathbf{S}| = 1$ the result is trivial. We proceed by induction of $|\mathbf{S}|$.

Let $|\mathbf{S}| = N \geq 2$ and assume the result is true for sets with $N - 1$ elements. Pick x in $\mathbf{S}$ with $\lambda(\{x\}) \leq t$. (If such an x does not exist the result is trivially true because $A(t,\mathbf{S}) = \emptyset$.) Let $\mathbf{T} = S - \{x\}$ and note that $|\mathbf{T}| = N - 1$. By our inductive hypothesis there is for each $\tau \geq 0$, a permutation $\pi_{\tau,\mathbf{T}}$ of $A(\tau,\mathbf{T})$ such that

$$\pi_{\tau,\mathbf{T}}(\mathbf{E}) \cap \mathbf{E} = \emptyset \text{ for all } \mathbf{E} \subseteq A(\tau,\mathbf{T}).$$

We partition $A(t,\mathbf{S})$ into three disjoint subsets as follows:

$$A_1(t,\mathbf{S}) = \{\mathbf{E} \subseteq A(t,\mathbf{S}) \mid x \in \mathbf{E}\},$$

$$A_2(t,\mathbf{S}) = \{\mathbf{E} \subseteq A(t,\mathbf{S}) \mid x \notin \mathbf{E},\ t - \lambda(\{x\}) < \lambda(\mathbf{E}) \leq t\},$$

$$A_3(t,\mathbf{S}) = \{\mathbf{E} \subseteq A(t,\mathbf{S}) \mid x \notin \mathbf{E},\ \lambda(\mathbf{E}) \leq t - \lambda(\{x\})\}.$$

Next, define

$$\phi : A_1(t,\mathbf{S}) \cup A_2(t,\mathbf{S}) \to A(t,\mathbf{T})$$

by $\phi(\mathbf{E}) = E - \{x\}$ and

$$\psi : A(t - \lambda(\{x\}),\mathbf{T}) \to A_1(t,\mathbf{S})$$

by $\psi(\mathbf{E}) = E \cup \{x\}$. Clearly both ϕ and ψ are bijective. Also

$$A_2(t,\mathbf{S}) \cup A_3(t,\mathbf{S}) = A(t,\mathbf{T})$$

and

$$A_3(t,\mathbf{S}) = A(t - \lambda(\{x\}), \mathbf{T}).$$

We define $\pi_{t,\mathbf{S}}$ as follows

$$\pi_{t,\mathbf{S}}(\mathbf{E}) = \begin{cases} \pi_{t,\mathbf{T}}(\phi(\mathbf{E})) & \text{if } \mathbf{E} \in A_1(t,\mathbf{S}) \cup A_2(t,\mathbf{S}) \\ \psi(\pi_{t-\lambda(\{x\}),\mathbf{T}}(\mathbf{E})) & \text{if } \mathbf{E} \in A_3(t,\mathbf{S}). \end{cases}$$

It is easy to check that $\pi_{t,\mathbf{S}}$ has the desired properties and this proves Theorem 1.

Corollary. *Let* $\mathbf{S}$, λ *be as above. Define*

$$B(t,\mathbf{S}) = \{\mathbf{E} \subseteq \mathbf{S} : \lambda(\mathbf{S}) - t \leq \lambda(\mathbf{E})\};$$

then there is a bijection $\sigma_{t,\mathbf{S}} : A(t,\mathbf{S}) \to B(t,\mathbf{S})$ *such that* $\mathbf{E} \subseteq \sigma_{t,\mathbf{S}}(\mathbf{E})$ *for all* $\mathbf{E} \in A(t,\mathbf{S})$.

Proof. Define $\sigma_{t,\mathbf{S}}(\mathbf{E}) = \mathbf{S} - \pi_{t,\mathbf{S}}(\mathbf{E})$ and use Theorem 1.

Let $n = p_1 \cdots p_r$ be square-free and $\mathbf{S} = \{p_1, p_2, \ldots p_r\}$ with $\lambda(p_i) = \log p_i$, $i = 1,2,\ldots,r$. We apply the Corollary with this choice of $\mathbf{S}$ and λ (and with t replaced by $\log t$) to obtain the following result, which, in view of its number theoretic form, is given the status of a theorem.

Theorem 2. *Let* n *be square-free and* $t \geq 1$. *Then there is a one-to-one mapping* m_t *between the divisors* d *of* n *which are less than or equal to* t *and those divisors* d' *of* n *which are greater than or equal to* n/t, *such that*

$$m_t(d) = d' \equiv 0 \pmod{d}.$$

Remarks.

1.) In Theorem 2 the parameter t could be greater than $\sqrt{n}$, but only $t \leq \sqrt{n}$ is of interest here. If $t > \sqrt{n}$ then $\tau = n/t < \sqrt{n}$. In this case m_τ produces a correspondence between $d \leq \tau$ and $d' \geq t$. The divisors between τ and t can be mapped onto themselves and m_t for $t > \sqrt{n}$ can be easily constructed from m_τ, where $\tau < \sqrt{n}$.

2.) The case $t = \sqrt{n}$ is of special interest because it shows that for a multiplicative function h satisfying $0 \leq h \leq 1$ we have

$$\sum_{d|n} h(d) \leq 2 \sum_{\substack{d|n \\ d \leq \sqrt{n}}} h(d), \text{ for all square-free } n. \tag{2.2}$$

Note that (2.2) is an immediate consequence of (2.1) (which is Theorem 2 with $t = \sqrt{n}$) because $h(d') \leq h(d)$. Inequality (2.2) can be proved directly without use of (2.1) as was pointed out by Heath-Brown. For this direct proof and applications see [3], [1].

3.) In a private correspondence to one of us (K.A.) R.R. Hall reported that Woodall had arrived at the mapping (2.1) a few years ago. Never-the-less, applications of such mappings or inequalities to Probabilistic Number Theory in [1], [3], appear to be new.

4.) When $h \geq 1$, clearly (2.2) is false. In fact, in this case (2.2) does not even hold if 2 is replaced by an arbitrarily large constant. Note that the constant 2 is best possible in (2.2) by taking $h \equiv 1$.

3. A useful inequality.

In view of (2.2) we may ask as to what sort of conditions one should impose upon h so that for all square-free n,

$$\sum_{d|n} h(d) \ll_k \sum_{\substack{d|n \\ d \leq n^{1/k}}} h(d), \tag{3.1}$$

where $k \geq 2$. Because of (1.3) we may expect (3.1) to hold provided $h(p)$ is quite small.

To get an idea concerning the size of such h we consider the special multiplicative function with $h(p) = c > 0$. Let r be a large integer and $p_1, p_2, \ldots, p_r$ primes such that $p_1 \sim p_2 \sim p_3 \sim \ldots \sim p_r$. Let $n = p_1 p_2 \ldots p_r$. In this situation a divisor d of n satisfies $d \leq n^{1/k}$ provided d has (asymptotically) $\leq r/k$ prime factors. Thus,

$$\Big\{ \sum_{d|n} h(d) \Big\}\Big\{ \sum_{\substack{d|n \\ d \le n^{1/k}}} h(d) \Big\}^{-1} \sim (1+c)^r \Big\{ \sum_{\ell=0}^{r/k} \binom{r}{\ell} c^\ell \Big\}^{-1} . \quad (3.2)$$

The maximum value of $\binom{r}{\ell}c^\ell$ occurs when $\ell \sim rc/(c+1)$, as $r \to \infty$. So the left hand side of (3.2) is unbounded if $c/(c+1) > 1/k$, i.e., if $c > 1/(k-1)$. On the other hand if $c < 1/(k-1)$ then the expressions in (3.2) are ~ 1 as $r \to \infty$. This example led one of us (K.A.) to make the following conjecture, part (i) of which appeared as problem 407 in the *West Coast Number Theory Conference, Asilomar (1983)*:

Conjecture.

(i) *For each* $k \geq 2$, *there exists a constant* c_k *such that* (3.1) *holds for all multiplicative functions* h *satisfying* $0 \leq h(p) \leq c_k$, *for all* p.

(ii) *In part* (i) $c_k = 1/(k-1)$ *is admissible.*

To this end we now prove an inequality for certain submultiplicative functions h, namely, those h for which $h(mn) \leq h(m)h(n)$, if $(m,n) = 1$.

Theorem 3. *Let* $h \geq 0$ *be submultiplicative and satisfy* $0 \leq h(p) \leq c < 1/(k-1)$ *for all primes* p. *Then for all square-free* n *we have*

$$\sum_{d|n} h(d) \leq \Big\{ 1 - \frac{kc}{1+c} \Big\}^{-1} \sum_{\substack{d|n \\ d \le n^{1/k}}} h(d) .$$

Proof. We begin with the familiar decompositon

$$\sum_{d|n} h(d) = \sum_{d|\, n/p} h(d) + \sum_{d|\, n/p} h(pd),$$

where p is any prime divisor of n. Submultiplicativity yields

$$\begin{aligned} h(p) \sum_{d|n} h(d) &= h(p) \sum_{d|\, n/p} h(d) + h(p) \sum_{d|\, n/p} h(pd) \qquad (3.3) \\ &\geq \sum_{d|\, n/p} h(np) + h(p) \sum_{d|\, n/p} h(pd) \end{aligned}$$

$$= \{1 + h(p)\} \sum_{d \mid n/p} h(d) .$$

Next, observe that

$$\sum_{\substack{d \mid n \\ d \leqslant n^{1/k}}} h(d) \geqslant \sum_{d \mid n} h(d) \frac{\log(n^{1/k}/d)}{\log(n^{1/k})} = \sum_{d \mid k} h(d) - \frac{k}{\log n} \sum_{d \mid n} h(d) \log d. \tag{3.4}$$

In addition

$$\sum_{d \mid n} h(d) \log d = \sum_{d \mid n} h(d) \sum_{p \mid d} \log p = \sum_{p \mid n} \log p \sum_{d \mid n/p} h(pd)$$

$$\leqslant \{ \sum_{d \mid n} h(d) \} \{ \sum_{p \mid n} \frac{(\log p)\, h(p)}{1 + h(p)} \} \tag{3.5}$$

because of (3.3). Since $0 \leqslant h(p) \leqslant c$, we have $h(p)/(1 + h(p)) \leqslant c/(1 + c)$. By combining (3.4) and (3.5) we obtain Theorem 3.

Remarks.

1.) Theorem 3 proves Conjecture (i) for any $c_k < 1/(k-1)$. The case $c_k = 1/(k-1)$ (part (ii)) is still open when $k > 2$ (for $k = 2$ this is (2.2)). The analysis underlying (3.2) shows that $c_k > 1/(k-1)$ is not possible.

2.) It would be of interest to see if the constant $\{1 - \frac{kc}{1+c}\}^{-1}$ can be improved. An attempt to deal with the case $c_k = 1/(k-1)$ may throw some light on this question.

3.) R. Balasubramaniam and S. Srinivasan (personal communication to one of us - K.A.) have obtained slightly weaker versions of Theorem 3 in response to our conference query in the course of proving Conjecture (i) for $c_k < 1/(k-1)$.

4.) If h is submultiplicative, then so is $h_T(n)$ which is equal to $h(n)$ when $n \leqslant T$ and is zero for $n > T$. The proof of Theorem 3 shows

$$\sum_{\substack{d \mid n \\ d \leqslant T}} h(d) < \{ \sum_{\substack{d \mid n \\ d \leqslant t}} h(d) \} \{ 1 - \frac{1}{\log t} \sum_{\substack{p \mid n \\ p \leqslant T}} \frac{h(p) \log p}{1 + h(p)} \}^{-1}$$

holds uniformly for all square-free $0 \leqslant t \leqslant T$ and submultiplicative h satisfying $h \geqslant 0$ and $0 \leqslant h(p) \leqslant (\log t)/(\log n/t)$.

5.) Let h be super-multiplicative, that is, $h(mn) \geqslant h(m)h(n)$, for $(m,n) = 1$. Suppose $h(p) \geqslant c > 1/(k-1)$ for all primes p. Then the proof of Theorem 3 can be modified to yield the dual inequality

$$\sum_{d|n} h(d) \geqslant \frac{(1 + c)(k - 1)}{k} \sum_{\substack{d|n \\ d \leqslant n^{1/k}}} h(d)$$

for all square-free n. Here also the situation regarding $c = 1/(k-1)$ is open.

4. An application.

Let **S** be an infinite set of positive integers. Define

$$S_d(x) = \sum_{\substack{s \leqslant x,\ s \in \mathbf{S} \\ s \equiv 0 \pmod{d}}} 1 ,$$

and set $X = S_1(x)$. In addition, let

$$S_d(x) = \frac{X\omega(d)}{d} + R_d(x) ,$$

where ω is multiplicative. First we assume that R_d satisfies the following condition:

(C-1) There exists $\delta > 0$ such that uniformly in x

$$|R_d(x)| \ll \frac{X\omega(d)}{d} \text{ (equivalently } S_d(x) \ll \frac{X\omega(d)}{d}\text{) for } 1 \leqslant d \leqslant x^{\delta} .$$

We also require R_d to satisfy at least one of the following two conditions:

(C-2) $|R_d(x)| \ll \omega(d)$, or

(C-3) There exist $\beta > 0$ such that to each $U > 0$ there is $V > 0$ satisfying

$$\sum_{d \leqslant X^{\beta}/(\log X)^{V}} |R_d(x)| \ll_U X/(\log^U X) .$$

Furthermore, we also require that there exists $c > 0$ such that

$$|R_d(x)| \ll (\frac{X \log X}{d} + 1)\, c^{\nu(d)}, \quad 1 \leqslant d \leqslant x,$$

where $\nu(n) = \sum_{p|n} 1$.

Examples of sets **S** satisfying these conditions include

(E-1) $\mathbf{S} = \{Q(n) \mid n = 1,2,3,\ldots\}$, where $Q(x)$ is a polynomial with positive integer coefficients. Here $\omega(d) = \rho(d)$, the number of solutions of $Q(x) \equiv 0 \pmod d$ and $|R_d| \leqslant \rho(d)$, so (C-2) holds. We may take $\delta = 1/(\deg Q)$ in (C-1).

(E-2) $\mathbf{S} = \{p + a \mid p = \text{prime}\}$, where a is a fixed positive integer. Here $\omega(d) = d/\phi(d)$ where ϕ is Euler's function. By the Brun-Titchmarsh inequality (see Halberstam-Richert [6], p.107) we can take any $\delta \in (0,1)$ in (C-1). By Bombieri's theorem (see [6], p. 111), we see that (C-3) holds with $\beta = 1/2$.

Let f be a (complex valued) strongly additive function, namely, one that satisfies

$$f(n) = \sum_{p|n} f(p).$$

The quantities

$$A(x) = \sum_{p \leqslant x} \frac{f(p)\omega(p)}{p} \quad \text{and} \quad B(x) = \sum_{p \leqslant x} \frac{|f^2(p)|\omega(p)}{p}$$

act like the 'mean' and 'variance' of $f(n)$, for $n \in \mathbf{S}$, $n \leqslant x$. Our problem is to obtain a bound for

$$\sum_{\substack{n \leqslant x \\ n \in \mathbf{S}}} |f(n) - A(x)|^{\ell}, \quad \ell = 1,2,3,\ldots$$

in terms of $B(x)$. In the special case where **S** is the set of all positive integers, Elliott [4] has solved this problem elegantly.

Recently one of us (K.A.) has improved Elliott´s method in order to make it applicable to subsets. In [2] sets **S** with $\delta = 1$ in (C-1) are treated whereas in [3] the situation concerning **S** in (E-2) is investigated. It is this improved method which we shall employ here; we sketch only the main ideas since details may be found in [2], [3].

We start by introducing a simplification: We may assume that $f \geq 0$. This is because the inequality

$$|a + b|^{\ell} \ll_{\ell} |a|^{\ell} + |b|^{\ell} \tag{4.1}$$

is valid for all complex numbers a and b. So a complex function could be decomposed into its real and imaginary parts. If f is real valued we can write $f = f^{+} - f^{-}$, where f^{+}, f^{-} are non-negative strongly additive functions generated by

$$f^{+}(p) = \max(0, f(p)), \quad f^{-}(p) = -\min(0, f(p)).$$

For convenience we introduce the distribution function

$$F_{x}(\upsilon) = \frac{1}{X} \sum_{\substack{n \leq x,\ n \in \mathbf{S} \\ f(n) - A(x) \leq \upsilon\sqrt{B(x)}}} 1.$$

We note that for even ℓ

$$\frac{1}{XB(x)^{\ell/2}} \sum_{\substack{n \leq x \\ n \in \mathbf{S}}} \big(f(n) - A(x)\big)^{\ell} = \int_{-\infty}^{\infty} \upsilon^{\ell} dF_{x}(\upsilon) . \tag{4.2}$$

Our aim is to show that the moments of F_x are bounded (uniformly in x).

To accomplish this we consider the bilateral Laplace transform

$$T_{u}(x) = \int_{-\infty}^{\infty} e^{uv} dF_{x}(v) .$$

If there is $R > 0$ for which $T_u(x) \ll 1$ when $|u| \leq R$, then it follows that the expression in (4.2) is bounded. Note that

$$T_{u}(x) = \frac{1}{X} \sum_{\substack{n \leq x \\ n \in S}} e^{u(f(n) - A(x)/\sqrt{B(x)})} = \frac{e^{-uA(x)/\sqrt{B(x)}}}{X} \sum_{\substack{n \leq x \\ n \in S}} g(n),$$

where

$$g(n) = e^{uf(n)/\sqrt{B(x)}} . \tag{4.3}$$

Of course g is strongly multiplicative (that is $g(n) = \prod_{p|n} g(p)$), because f is strongly additive. Our goal therefore is to bound $S(x,g)$ (see (1.1)) suitably. We have two cases.

Case 1: $u \leq 0 \Rightarrow 0 \leq g \leq 1$.

In a recent paper [2] it was shown by using a sieve method, that in Case 1, for the sets **S** satisfying either (C-2) or (C-3), we have

$$S(x,g) \ll X \prod_{p \leq x} \left(1 + \frac{(g(p) - 1)\omega(p)}{p} \right) . \tag{4.4}$$

Case 2: $u > 0 \Rightarrow g \geq 1$.

Here we let $\delta = 1/k$ (in C-1) and assume that f satisfies

$$\left\{ \max_{p \leq x} f(p) \right\} / \sqrt{B(x)} \ll 1. \tag{4.5}$$

Then we can choose $R > 0$ (sufficiently small) such that

$$1 \leq g(p) \leq 1 + \frac{1}{2(k-1)} .$$

With h as in (1.2) we note that $0 \leq h(p) = g(p) - 1 \leq \frac{1}{2(k-1)}$. Also $h(p^e) = 0$ for all p, $e \geq 2$, because g is strongly multiplicative. So by Theorem 3

$$S(x,g) = \sum_{\substack{n \leq x \\ n \in \mathbf{S}}} \sum_{d|n} h(d) \ll \sum_{\substack{n \leq x \\ n \in \mathbf{S}}} \sum_{\substack{d | n_\delta \\ d \leq n^\delta}} h(d)$$

$$\leq \sum_{d \leq x^\delta} h(d) S_d(x) .$$

By (C-1) we obtain

$$S(x,g) \ll X \sum_{d \leq x^\delta} \frac{h(d)\omega(d)}{d} \leq X \prod_{p \leq x} \left(1 + \frac{h(p)\omega(p)}{p} \right) . \tag{4.6}$$

Inequalities (4.4) and (4.6) combine with (4.3) and (4.5) to yield

$T_u(x) \ll 1$ for $|u| \leqslant R$. For details relating to such calculations see [2], Sec.7 . Therefore by means of this method we obtain the following extension of a result of Elliott [4],

Theorem 4. *Let* f *be as above and* |f| *satisfy* (4.5). *Then*

$$\sum_{\substack{n \leqslant x \\ n \in S}} |f(n) - A(x)|^{\ell} \ll_{\ell} XB(x)^{\ell/2}, \text{ for all } \ell > 0 .$$

Remarks.

1.) Although our discussion was for even ℓ, Theorem 4 is stated for all $\ell > 0$. This is because one can pass from even ℓ to all positive real numbers by a suitable application of the Hölder-Minkowski inequality.

2.) If f satisfies certain additional conditions then one can use the above method more carefully to obtain asymptotic estimates for the moments. In these cases the weak limit of $F_x(v)$ would exist. Such asymptotic estimates are obtained in [2] for S with $\delta = 1$, and in [3] for S in (E-2). For these sets the full strength of Theorem 3 is not required. The inequality (2.2) (which follows from Theorem 2) suffices.

3.) There are certain open problems concerning the behavior of additive functions in polynomial sequences (see Elliott [5], Vol. 2, p. 335). Part of the difficulty in such questions is because we do not fully understand the moments of additive functions in these sequences. Theorem 4 is derived in the hope that it might shed some light on these questions.

4.) We restrict our attention to strongly additive functions for the sake of simplicity. From here the transition to general additive functions is not difficult. This procedure for the case $\delta = 1$ is illustrated in [2], Sec.10.

References.

1. K. Alladi, Moments of additive functions and sieve methods, New York Number Theory Seminar, Springer Lecture Notes **1052** (1982), 1-25.

2. K. Alladi, A study of the moments of additive functions using Laplace transforms and sieve methods, Proceedings Fourth Matscience Conference on Number Theory, Ootacamund, India (1984), Springer Lecture Notes (to appear).

3. K. Alladi, Moments of additive functions and the sequence of shifted primes, *Pacific Journal of Math*. **Ernst Straus Memorial Vol.**, June (1985) (to appear).

4. P.D.T.A. Elliott, High power analogues of the Turán-Kubilius inequality and an application to number theory, *Can. Jour. of Math* **32** (1980), 893-907.

5. P.D.T.A. Elliott, Probabilistic Number Theory, Vol. 1 and 2, Grundelehren **239-240**, Springer-Verlag, Berlin, New York, 1980.

6. H. Halberstam and H.-E. Richert, Sieve Methods, Academic Press, London, New York, 1974.

K. Alladi
University of Hawaii,
Honolulu, Hawaii 96822

P. Erdös
Hungarian Academy of Sciences,
Budapest, Hungary.

J.D. Vaaler
University of Texas,
Austin, Texas 78712, U.S.A.

LECTURES ON THE THUE PRINCIPLE

Enrico Bombieri

I. Introduction.

The aim of these lectures is to give an account of results obtained from the application of Thue's idea of comparing two rational approximations to algebraic numbers in order to show that algebraic numbers cannot be approximated too well by rational numbers. In particular we will give special attention to the problem of obtaining effective measures of irrationality, or types, for various classes of algebraic numbers.

1.1 Notation.

In what follows we shall adhere to the following notation.
k is a number field and K denotes an extension of k of degree $r = [K:k]$, with $r \geqslant 2$.

For each place of k we have an absolute value $|\,|_v$, uniquely defined up to a power. In order to fix this power, let us consider the inclusion of complete fields $\mathbf{Q}_v \subset k_v$ arising from the inclusion $\mathbf{Q} \subset k$; then if v lies over the rational prime p, which we write as $v|p$, we want

$$|p|_v = p^{-[k_v:\mathbf{Q}_v]/[k:\mathbf{Q}]}$$

while if v is archimedean we want

$$|x|_v = |x|^{[k_v:\mathbf{Q}_v]/[k:\mathbf{Q}]}$$

where $|x|$ denotes the usual euclidean absolute value in **R** or **C**. We also write

$$\varepsilon_v = [k_v:\mathbf{Q}_v]/[k:\mathbf{Q}] \quad \text{if } v|\infty$$

$$\varepsilon_v = 0 \quad \text{if v is finite.}$$

If $\alpha \in k$, $\alpha \neq 0$ and if we consider $\alpha \in K$ by means of the inclusion $k \subset K$ then we have

$$\log |\alpha|_v = \sum_{w|v} \log |\alpha|_w \qquad (1)$$

where w runs over the places of K lying over the place v of k. We also write

$$\| \ \|_v = | \ |_v^{[k:\mathbf{Q}]/[k_v:\mathbf{Q}_v]} .$$

Fundamental for us is the product formula in k, which we write as

Product Formula. *If $\alpha \in k$, $\alpha \neq 0$ then*

$$\sum_v \log |\alpha|_v = 0 .$$

1.2 Heights.

Let us abbreviate $\log^+ t = \log t$ if $t > 1$, $\log^+ t = 0$ if $0 < t \leq 1$. As an immediate consequence of the product formula we have

Fundamental Inequality. *Let $\alpha \in k$, $\alpha \neq 0$ and let S be any set of places of k. Then*

$$- \sum_{v \notin S} \log^+ |\alpha|_v \leq \sum_S \log |\alpha|_v \leq \sum_{v \notin S} \log^+ |\alpha|_v .$$

This leads to the definition of height: the *absolute height* of $\alpha \in k$, $\alpha \neq 0$, denoted by $h(\alpha)$, is defined by

$$\log h(\alpha) = \sum_v \log^+ |\alpha|_v \qquad (2)$$

where $\sum_v$ runs over all places of k. The height $h(\alpha)$ has the following properties.

(a) invariance: $h(\alpha)$ does not depend on the field k with $\alpha \in k$ used in the definition (2)

(b) $h(\alpha) = h(\alpha^{-1})$

(c) $h(\alpha\beta) \leqslant h(\alpha)h(\beta)$

(d) $h(\alpha_1 + \ldots + \alpha_n) \leqslant n\, h(\alpha_1)h(\alpha_2)\ldots h(\alpha_n)$.

Of these, (a) follows from (1); (b) follows from the product formula; (c) follows from $\log^+(ab) \leqslant \log^+ a + \log^+ b$; (d) follows from

$$\log^+|\alpha_1 + \ldots + \alpha_n|_v \leqslant \max_i \log^+|\alpha_i|_v \quad \text{if } v \nmid \infty$$

and

$$\log^+|\alpha_1 + \ldots + \alpha_n|_v \leqslant \log^+|n|_v + \max_i \log^+|\alpha_i|_v \quad \text{if } v|\infty .$$

Let $\alpha \in k$, $\alpha \neq 0$ and let

$$f(x) = a_0 x^d + a_1 x^{d-1} + \ldots + a_d = 0, \tag{3}$$

be an irreducible equation for α in $\mathbf{Z}[x]$, with $\mathrm{GCD}(a_0, \ldots, a_d) = 1$. The classical height $H(\alpha)$ of α is given by

$$H(\alpha) = \max_i |a_i|, \tag{4}$$

and the Mahler measure $M(\alpha)$ of α is defined by

$$M(\alpha) = \exp\Big(\frac{1}{2\pi}\int_0^{2\pi} \log|f(e^{i\theta})|\,d\theta\Big). \tag{5}$$

One proves easily, by Jensen's formula or directly, that

$$M(\alpha) = |a_0| \prod_i \max(1, |\alpha_i|) \tag{6}$$

where $\alpha_1, \ldots, \alpha_d$ are the roots of $f(x)$. If $k = \mathbf{Q}(\alpha)$ we get

$$\log M(\alpha) = \log|a_0| + \sum_i \log^+|\alpha_i|$$

and

$$\frac{1}{d}\sum_i \log^+|\alpha_i| = \sum_{v|\infty} \log^+|\alpha|_v$$

$$\frac{1}{d}\log|a_0| = \sum_{v\nmid\infty} \log^+|\alpha|_v ;$$

hence

$$M(\alpha) = h(\alpha)^d \tag{7}$$

where $d = \deg \alpha$.

Also by (5) we have

$$M(\alpha) \leqslant \left(\frac{1}{2\pi}\int_0^{2\pi} |f(e^{i\theta})|^p d\theta\right)^{1/p}$$

for every $p > 0$. The special case $p = 2$ yields

$$M(\alpha) \leqslant \left(\sum_i |a_i|^2\right)^{1/2} \leqslant (d+1)^{1/2} H(\alpha). \tag{8}$$

In the opposite direction, by symmetric functions we have

$$\begin{aligned} |a_0| + \ldots + |a_d| &\leqslant |a_0| \prod_i (1 + |\alpha_i|) \\ &\leqslant 2^d M(\alpha), \end{aligned} \tag{9}$$

so that (8) and (9) prove that $M(\alpha)$ and $H(\alpha)$ have the same order of magnitude.

We may consider $h(\alpha)$ as an intrinsic height on the algebraic group $\mathbf{G}_m$. If P is the point on $\mathbf{G}_m$ corresponding to $\alpha \in k^* = \mathbf{G}_m(k)$ then

$$\log h(\alpha) = \frac{1}{d} \lim_{m\to\infty} \frac{1}{m} \log H(mP) \tag{10}$$

where $mP = \alpha^m$ is the "sum" of P with itself (for the operation in $\mathbf{G}_m$) m times. Formula (10) shows the analogy of $\log h(\alpha)$ with the Tate height on elliptic curves; everything is of course much simpler here.

The definition of height can be carried through in other settings too; of importance to us is the *projective height*, defined as follows.

Let $\mathbf{x} = (x_0, x_1, \ldots, x_N)$ be homogeneous coordinates of a k-rational point in projective space $\mathbf{P}^N$. The projective height of $\mathbf{x}$ is defined by

$$\log h(\mathbf{x}) = \sum_v \log |\mathbf{x}|_v \tag{11}$$

where

$$|\mathbf{x}|_v = \max_i |x_i|_v. \tag{12}$$

By the product formula, $h(\mathbf{x}) = h(\lambda\mathbf{x})$ whenever $\lambda \in k^*$, thus the height $h(\mathbf{x})$ is well-defined on $\mathbf{P}^N(k)$; it is also independent of field extensions. We note that the projective height is compatible with tensor (Kronecker) products:

$$h(\mathbf{x} \otimes \mathbf{y}) = h(\mathbf{x})h(\mathbf{y}). \tag{13}$$

Examples.

(i) $k = \mathbf{Q}$, $\alpha = p/q \in \mathbf{Q}^*$.

In this case

$$h(\alpha) = H(\alpha) = \max(|p|,|q|).$$

(ii) $\alpha = \sqrt{2} - 1$, $k = \mathbf{Q}(\sqrt{2})$.

Here α is integral, thus $|\alpha|_v = 1$ if $v \nmid \infty$. At ∞ we have two inequivalent absolute values v, for which $k_v = \mathbf{R}$; the inclusion $k \subset k_v = \mathbf{R}$ is such that $\sqrt{2}$ is positive in one embedding and negative (the other determination, $-\sqrt{2}$) in the other. Let us call v_+, v_- the corresponding places. Now

$$|\alpha|_{v_+} = (\sqrt{2}-1)^{1/2} < 1,$$

$$|\alpha|_{v_-} = (\sqrt{2}+1)^{1/2}$$

and

$$h(\alpha) = \sqrt{\sqrt{2}+1} = 1.55377\ldots\ .$$

(iii) $\alpha^r - m\alpha^{r-1} + 1 = 0$, $m > 2$.

Let $k = \mathbf{Q}(\alpha)$. Now $[k:\mathbf{Q}] = r$ and α is a unit, thus $|\alpha|_v = 1$ if $v \nmid \infty$. At ∞ we have

(a) one absolute value v_0 with $k_{v_0} = \mathbf{R}$ and α close to m, in fact

$$m - \frac{1}{m^{r-1}} < \alpha < m$$

for the embedding $k \subset k_{v_0} = \mathbf{R}$;

(b) if r is even, one absolute value v_+ with $k_{v_+} = \mathbf{R}$ and such that

$$\alpha \sim m^{-\frac{1}{r-1}}$$

for the embedding $k \subset k_{v_+} = \mathbf{R}$, and $\frac{r}{2} - 1$ absolute values v_j, $j = 1, \ldots, \frac{r}{2} - 1$ with $k_{v_j} = \mathbf{C}$ and such that

$$\alpha \sim m^{-1/(r-1)}\zeta$$

with $\zeta^{r-1} = 1$, $\zeta \neq 1$ for the embedding $k \subset k_{v_j} = \mathbf{C}$ (the conjugate embedding determines the same v_j);

(c) if r is odd, we have a result similar to (b) but with two absolute values v_+, v_- with $k_{v_+} = \mathbf{R}$, $k_{v_-} = \mathbf{R}$ and $\alpha \sim m^{1/(r-1)}$.

If $v|\infty$ and $v \neq v_0$ then $|\alpha|_v < 1$ hence $\log^+|\alpha|_v = 0$ and thus

$$\log h(\alpha) = \log |\alpha|_{v_0} = \log(|\alpha|^{1/r})$$

$$< \log(m^{1/r})$$

$$= \frac{1}{r} \log m.$$

Thus $h(\alpha) < m^{1/r}$ and in fact $h(\alpha)$ is extremely close to $m^{1/r}$.

1.3 General heights.

The above discussion on heights can be extended by introducing different types of local heights. This turns out to be useful in obtaining refined results on roots of special type (for example, roots of unity) of polynomials. Before considering a general construction let us reexamine the height introduced before in the light of different considerations.

Let k be a number field and let $\alpha \in k$. For each place v of k let k_v denote the completion of k with respect to the absolute value $|\,|_v$ determined by v and let Ω_v be the completion of an algebraic closure of k_v with respect to an absolute value, again denoted by

$||_v$, extending the absolute value on k_v.

Lemma 1. *For every* v *we have*

$$\int_{|z|_v=1} \log|z-\alpha|_v \, d_v z = \log^+|\alpha|_v$$

where $d_v z$ *is the normalized Haar measure on the units* $\{z \in \Omega_v : |z|_v = 1\}$ *of* Ω_v.

Proof. If $v|\infty$ this reduces to Jensen's formula

$$\frac{1}{2\pi}\int_0^{2\pi} \log|e^{i\theta}-\alpha|d\theta = \log^+|\alpha|.$$

If instead $v\nmid\infty$ we have

$$|z-\alpha|_v = \max(1,|\alpha|_v)$$

almost everywhere in Ω_v; this is clear if either $|\alpha|_v < 1$ or $|\alpha|_v > 1$ and for $|\alpha|_v = 1$ it reduces to the case in which $\alpha = 1$, where it follows from the fact that the subgroup of units of Ω_v congruent to 1 modulo the maximal ideal $\{|z|_v > 1\}$ of the ring $R_v = \{z \in \Omega_v : |z|_v \leqslant 1\}$ has infinite index in the group of all units of Ω_v.

Corollary. *Let* $f \in \bar{\mathbb{Q}}[x]$ *and let* k *be an algebraic number field containing the coefficients of* f *and all roots of* f. *We have*

$$\sum_\alpha (\mathrm{ord}_\alpha f)\log h(\alpha) = \sum_v \int_{|z|_v=1} \log|f(z)|_v \, d_v z,$$

where $\sum_v$ *runs over all normalized absolute values of* k *and* $d_v z$ *is the normalized Haar measure on the group of units* $|z|_v = 1$ *of* Ω_v.

Let $f(z) = a_0 z^d + \ldots + a_d \in k[z]$ and let us define the local height $H_v(f)$ by means of

$$\log H_v(f) = \max_i \log |a_i|_v \quad \text{if } v \nmid \infty \tag{14}$$

and

$$\log H_v(f) = \frac{\varepsilon_v}{2} \log\left(\sum_i \|a_i\|_v^2\right) \quad \text{if } v|\infty. \tag{15}$$

Lemma 2. *For every* v *we have*

$$\int_{|z|_v=1} \log|f(z)|_v \, d_v z \leq \log H_v(f)$$

and moreover equality holds if $v\nmid\infty$.

Proof. We have

$$\int_{|z|_v=1} \log|f(z)_v \, d_v z = \frac{[k_v:\mathbf{Q}_v]}{2[k:\mathbf{Q}]} \int_{|z|_v=1} \log(\|f(z)\|_v^2 d_v z$$

$$\leq \frac{[k_v:\mathbf{Q}_v]}{2[k:\mathbf{Q}]} \log\left(\int_{|z|_v=1} \|f(z)\|_v^2 \, d_v z\right).$$

Since

$$\int_{|z|_v=1} \|f(z)\|_v^2 \, d_v z_v = \begin{cases} \max_i \|a_i\|_v^2 & \text{if } v\nmid\infty \\ \\ \sum_i \|a_i\|_v^2 & \text{if } v|\infty \end{cases}$$

the first part of Lemma 2 follows from the definition of $H_v(f)$.

In order to prove the second part one may note that the statement is true if f has degree 1 (by Lemma 1) and proceed by induction on deg f using

Gauss' Lemma. *If* $v\nmid\infty$ *then*

$$H_v(fg) = H_v(f)H_v(g).$$

Theorem 1. *For every* $f \in \overline{\mathbf{Q}}[x]$ *we have*

$$\log H(f) - (\log 2)(\deg f) \leq \sum_\alpha (\mathrm{ord}_\alpha f)\log h(\alpha) \leq \log H(f)$$

where $\sum_\alpha$ *runs over all roots of* f.

Proof. The right-hand side inequality is immediate from Lemma 1, Corollary and Lemma 2. The left-hand side inequality can be proved as follows. We may suppose that f is monic, hence

$$f(z) = \prod_{\alpha} (z-\alpha)^{\operatorname{ord}_{\alpha} f}.$$

Now if $v\nmid\infty$ we have

$$\log H_v(f) = \sum_{\alpha} (\operatorname{ord}_{\alpha} f)\log^{+}|\alpha|_v$$

by Gauss' Lemma. If instead $v|\infty$ then

$$\sum_{s=0}^{r} \Big\| \sum_{1\leq i_1<\ldots<i_s\leq r} \alpha_{i_1}\cdots\alpha_{i_s} \Big\|_v^2$$

$$\leq \sum \binom{r}{s} \sum_{1\leq i_1<\ldots<i_s\leq r} \|\alpha_{i_1}\|_v^2 \cdots \|\alpha_{i_s}\|_v^2$$

$$\leq 2^r \prod_{i=1}^{r} (1 + \|\alpha_i\|_v^2)$$

$$\leq 4^r \prod_{i=1}^{r} \max(1,|\alpha_i|_v^2).$$

If we apply this inequality to the roots of the polynomial f we find

$$\log H_v(f) \leq \sum_{\alpha} (\operatorname{ord}_{\alpha} f)\log^{+}|\alpha|_v + \frac{[k_v:\mathbf{Q}_v]}{[k:\mathbf{Q}]} (\log 2)(\deg f)$$

and the left-hand side inequality of Theorem 1 is obtained by summing these local estimates for all v.

For later use we also need bounds for the heights of derivatives of polynomials.

Lemma 3. *Let* $f \in \overline{\mathbf{Q}}[x_1,\ldots,x_N]$ *and let* Δ^I *be the differential operator*

$$\Delta^I = \frac{1}{i_1!\cdots i_N!} \frac{\partial^{i_1+\cdots+i_N}}{\partial x_1^{i_1}\cdots\partial x_N^{i_N}}$$

where $I = (i_1,\dots,i_N)$. *Then we have*

$$\log H(\Delta^I f) \leqslant \log H(f) + \sum_{\nu=1}^{N} \phi\left(\frac{i_\nu}{\deg_{x_\nu} f}\right) \deg_{x_\nu} f$$

where

$$\phi(t) = t \log \frac{1}{t} + (1-t)\log \frac{1}{1-t}.$$

Proof. Clear, because

$$\log\binom{d}{m} \leqslant \phi\left(\frac{m}{d}\right)d$$

for every m, d; this last inequality is most easily proved by noting that

$$\left|\binom{d}{m}\right| = \left|\frac{1}{2\pi i} \int_{|z|=u} \frac{(1+z)^d}{z^m} \frac{dz}{z}\right|$$

$$\leqslant \left(1 + \frac{1}{u}\right)^m (1+u)^{d-m}$$

and choosing $u = \frac{m}{d-m}$.

Now we consider general heights. Let μ_v be a positive measure on Ω_v with total mass $\mu_v(\Omega_v) = 1$ and let us define a height $h(\alpha,\underline{\mu})$ by

$$\log h(\alpha,\underline{\mu}) = \sum_v \int \log|z-\alpha|_v \, d\mu_v. \tag{16}$$

It is clear that

$$\sum_\alpha (\mathrm{ord}_\alpha f)\log h(\alpha,\underline{\mu}) = \sum_v \int \log |f(z)|_v \, d\mu_v. \tag{17}$$

As a special case, suppose that μ_v has support in $|z|_v \leqslant 1$ for every v. Then

$$|f(z)|_v \leqslant \max_i |a_i|_v \quad \text{if } v\nmid\infty \tag{18}$$

and

$$|f(z)|_v \leqslant |d+1|_v \max_i |a_i|_v \quad \text{if } v|\infty . \tag{19}$$

Hence

Theorem 2. *If each* μ_v *has support in* $R_v = \{z \in \Omega_v : |z|_v \leqslant 1\}$ *then*

$$\sum_{\alpha} (\mathrm{ord}_\alpha f) \log h(\alpha, \underline{\mu}) \leqslant \log H(f) + \log(\deg f + 1).$$

Quite often, one uses Theorem 2 for its consequence

$$\mathrm{ord}_\alpha f \leqslant \frac{\log H(f) + \log(\deg f + 1)}{\log h(\alpha, \underline{\mu})}, \qquad (20)$$

which we have wherever $h(\alpha, \underline{\mu}) \geqslant 1$ for all α. In what follows we shall describe one non-trivial application of Theorem 2.

If we use (20) with the height $h(\alpha)$ studied so far we get no result whatsoever in the case in which α is a root of unity, since then $\log h(\alpha)$ vanishes. It is an interesting question in itself to study what is the maximum multiplicity of a root of unity in a polynomial of given degree and given height.

Let p be a rational prime and let us choose

$$\mu_v(z) = \frac{1}{p-1} \sum_{\zeta}{}' \delta_\zeta(z) \quad \text{if } v|\infty$$

where δ_ζ is a Dirac measure at ζ and where $\sum_\zeta'$ runs over the primitive p-th roots of unity; if $v|\infty$ we choose instead μ_v = Haar measure on $\{z \in \Omega_v : |z|_v = 1\}$.

We note that if $v\nmid\infty$ then $\log^+|\alpha|_v = \log^+|\alpha-\zeta|_v$ if α is not a primitive p-th root of unity, and it follows that

$$\begin{aligned}\log h(\alpha, \underline{\mu}) &= \sum_{v\nmid\infty} \log^+|\alpha|_v + \frac{1}{p-1} \sum_{v|\infty} \sum_{\zeta}{}' \log|\alpha-\zeta|_v \\ &= \frac{1}{p-1} \sum_{\zeta}{}' \Big(\sum_{v\nmid\infty} \log^+|\alpha-\zeta|_v + \sum_{v|\infty} \log|\alpha-\zeta|_v \Big) \geqslant 0\end{aligned}$$

by the product formula. Also

$$\log h\big((1, \underline{\mu})\big) = \frac{\log p}{p-1};$$

by Theorem 2 we obtain

Theorem 3. *If* $f(\zeta) \neq 0$ *whenever* ζ *is a primitive* p-th *root of unity,* p *prime, then*

$$\frac{\log p}{p-1}\,\mathrm{ord}_1(f) \leq \log H(f) + \log(\deg f + 1).$$

As a final remark for this section we note that the fundamental theorem of algebra

$$\sum_{\alpha} \mathrm{ord}_\alpha f = \deg f$$

may be considered a limiting case of our considerations, if μ_v becomes a point mass at ∞, for every v.

II. Thue's method.

2.1 As a first application of the estimates of the preceding section we prove the basic Liouville lower bound for the distance of two algebraic numbers.

Let K be an extension of k of degree $r = [K:k]$, let v be a place of k with an extension $\tilde{v}$ to K, with associated absolute values $|\,|_{\tilde{v}}$ and $|\,|_v$. We have

$$|\xi|_v = |\xi|_{\tilde{v}}^{r/[K_{\tilde{v}}:k_v]} \qquad \text{if } \xi \in k \tag{21}$$

and thus we can use (21) to extend the absolute value $|\,|_v$, originally defined in k, to the field K. We can now state

Liouville Bound. *Let* $\alpha \in K$, $\beta \in k$, $\alpha \neq \beta$. *Then*

$$|\alpha-\beta|_v \geq (2h(\alpha)h(\beta))^{-r/\delta}$$

with $\delta = [K_{\tilde{v}}:k_v]$. *In particular, we have*

$$|\alpha-\beta|_v \geq (2h(\alpha)h(\beta))^{-r}.$$

Proof. By the Fundamental Inequality we have

$$\log|\alpha-\beta|_{\tilde{v}} \geq -\log h(\alpha-\beta)$$

and

$$h(\alpha-\beta) \leqslant 2h(\alpha)h(\beta)$$

by property d) of the height. Since $|\alpha-\beta|_{\tilde{v}} = |\alpha-\beta|_v^{\delta/r}$, the result follows.

Definition. *μ is a type of irrationality for α over k with respect to v if*

$$|\alpha-\beta| \geqslant c(\alpha)h(\beta)^{-\mu} > 0$$

for all $\beta \in k$, $\beta \neq 0$. We also say that μ is a measure of irrationality, or type, for α/k, relative to v.

It is clear that it suffices to consider lower bounds for $\alpha-\beta$ only if $h(\beta)$ is larger than a prescribed bound, simply by changing the constant $c(\alpha)$, that is we need to prove

$$|\alpha-\beta|_v \geqslant c_1(\alpha)h(\beta)^{-\mu} > 0$$

for $h(\beta) \geqslant c_2(\alpha)$.

Definition. *μ is an effective type for α over k with respect to v if $c_1(\alpha)$, $c_2(\alpha)$ above can be determined effectively. One then writes*

$$\mu_{eff}(\alpha; k, v) = \inf \mu$$

where the infimum is taken over all admissible μ's for which effectively calculable constants $c_1(\alpha)$, $c_2(\alpha)$ can be found.

It is clear that the Liouville bound implies

$$\mu_{eff}(\alpha; k, v) \leqslant r/\delta. \tag{22}$$

In the other direction, it is known (see [Schmidt 1980]) that if $\delta = 1$, $\alpha \in K$ and $\alpha \notin k$ then

$$|\alpha - \beta|_v \leqslant c_3(\alpha)h(\beta)^{-2}$$

for an effectively computable $c_3(\alpha) > 0$ and infinitely many $\beta \in k$. Thus, if $\delta = 1$,

$$\mu_{eff}(\alpha; k, v) \geq 2 \tag{23}$$

for every $\alpha \in K$, $\alpha \notin k$. The gap between (22) and (23) is considerable and it was only after Baker's work on linear forms in logarithms that the first improvement on (22) was obtained, namely: if $\delta = 1$ then

$$\mu_{eff}(\alpha; \mathbf{Q}, \infty) \leq r - \eta(\alpha) \tag{24}$$

for some very small $\eta(\alpha) > 0$ [Feldman 1971]. Further work showed that $\eta(\alpha)$ can be made to depend only on the field K, and generalized (24) to arbitrary extensions K/k and absolute values v. All these result, although of great theoretical importance, are far away from the celebrated theorom of Roth:

Roth's Theorem. *If* $\delta = 1$ *then* $\mu(\alpha; k, v) = 2$.

On the other hand, Roth's theorem is ineffective and this limits to some extent the range of its applications. In what follows, we shall describe in some detail Thue's method, which is at the origin of all ineffective results such as Roth's, together with some recent effective developments and new applications.

We may summarize the essence of Thue's method in three steps. Let $\alpha_1, \alpha_2 \in K$ and suppose that $\beta_1, \beta_2 \in k$ are approximations to α_1, α_2, for the absolute value $| \, |_v$. For simplicity, we consider the case $k = \mathbf{Q}$ and write $\beta_1 = p_1/q_1$, $\beta_2 = p_2/q_2$; we also write $r = [K:\mathbf{Q}]$.

Step 1. One constructs a polynomial $P(x_1, x_2)$ with rational integral coefficients, vanishing at (α_1, α_2), together with all partial derivatives of order (i_1, i_2), with $\frac{i_1}{d_1} + \frac{i_2}{d_2} < t$, where $d_i = \deg_{x_i} P$, and where t is sufficiently small. The number of coefficients of P at our disposal is asymptotic to $d_1 d_2$, while the number of equations is asymptotic to $(r\, t^2/2) d_1 d_2$. If $t < \sqrt{(2/r)}$ we can solve the corresponding linear system for the coefficients of P, with a height

$$\log h(P) < \frac{r\,t^2/2}{1-rt^2/2}\,(d_1 + d_2)\log C(\alpha_1,\alpha_2),$$

for a suitable $C(\alpha_1,\alpha_2)$. Of course, the construction guarantees that the polynomial P is not identically 0.

Step 2. By modifying P if necessary, and perhaps by imposing a condition of type "q_2 *is much larger than* q_1", one shows that

$$P(p_1/q_1,p_2/q_2) \neq 0.$$

Step 3. By looking at denominators one has the lower bound

$$P(p_1/q_1,p_2/q_2 \geqslant q_1^{-d_1}\,q_2^{-d_2}.$$

Finally one compares this lower bound with an upper bound obtained by using a Taylor series expansion of P at (α_1,α_2), noting that P vanishes to high order at this point:

$$|P(\frac{p_1}{q_1},\frac{p_2}{q_2})| < C_1^{d_1+d_2}\,(|\alpha_1 - \frac{p_1}{q_1}|^{td_1} + |\alpha_2-\frac{p_2}{q_2}|^{td_2}),$$

with a suitable $C_1 = C_1(\alpha_1,\alpha_2,t)$.

Now suppose that the approximations to α_i satisfy

$$|\alpha_i - \frac{p_i}{q_i}| < q_i^{-\eta}, \qquad i=1,2;$$

then from the preceding bounds we obtain

$$q_1^{-d_1}\,q_2^{-d_2} < C_1^{d_1+d_2}(|\alpha_1-\frac{p_1}{q_1}|^{td_1}+|\alpha_2-\frac{p_2}{q_2}|^{td_2})$$

$$< C_1^{d_1+d_2}(q_1^{-t\eta d_1}+q_2^{-t\eta d_2}).$$

The degrees d_1 and d_2 are still at our disposal and we choose them so that $q_1^{d_1}$ and $q_2^{d_2}$ are about of the same magnitude. If q_1 and q_2 are sufficiently large, this implies that $t\eta < 2 + \varepsilon$ for any positive ε. Since any $t < \sqrt{(2/r)}$ is allowed, one deduces

$$\eta < \sqrt{2r} + \varepsilon,$$

which is the Thue-Siegel-Dyson theorem.

It is clear that the preceding argument requires two approximations p_1/q_1 and p_2/q_2, with q_1 and q_2 large (otherwise the presence of the constant C_1 in the estimates becomes too important), while it may very well be that such approximations do not exist. Moreover, all existing arguments for Step 2 require that q_2 be much larger than q_1. This means that if we seek for a bound Q for which

$$|\alpha_2 - \frac{p}{q}| > q^{-\sqrt{2r} - \varepsilon} \quad \text{for } q > Q,$$

then the preceding argument will obtain Q as a function

$$Q = Q(\alpha_1, \alpha_2, \varepsilon, p_1/q_1),$$

but only provided p_1/q_1 *is a sufficiently good approximation to* α_1 *and provided* q_1 *is sufficiently large as a function of* α_1, ε *and the approximation*. Two cases now may occur:

Case 1. α_1 does not admit such a good approximation.In this case, we conclude an effective type of irrationality for α_1.

Case 2. There is at least one good approximation to α_1.In this case, we conclude a type of irrationality for elements α_2 of the field K, which depends on the denominator q_1 of the good approximation to α_1.

No procedure is given to decide between Case 1 and Case 2, and in Case 2 we have no information on the location q_1 of the approximation. The ineffectivity of the method depends on the fact that the statement of Case 2 is an existence statement whose truth is not determined in the course of our arguments.

Until recently, no instance of Case 2 was known. However a refinement of the notion of good approximation led to the first explicit examples in which Case 2 would hold, thus leading to new non-trivial types of approximation for a class of algebraic numbers [Bombieri 1982]. In what follows, we shall carry out the steps in the preceding program, with sufficient accuracy to obtain effective approximation results. We shall proceed using invariant techniques.

Let $P(x_1,x_2) \in k[x_1,x_2]$ denote a polynomial of degree d_1 in x_1 and d_2 in x_2; the totality of such polynomials is a k-vector space $V(d_1,d_2)$, of dimension $(d_1+1)(d_2+1)$. Let θ be positive and let $G(t)$ be the set of pairs (i_1,i_2) such that

$$\theta^{-1}\frac{i_1}{d_1} + \theta\frac{i_2}{d_2} < t.$$

Let $\alpha_1, \alpha_2 \in K$ where $[K:k] = r \geqslant 2$. We want to find $P \in V(d_1,d_2)$ such that

$$\Delta^I P(\alpha_1,\alpha_2) = 0$$

for $I = (i_1,i_2) \in G(t)$ and where

$$\Delta^I = \frac{1}{i_1!i_2!}\left(\frac{\partial}{\partial x_1}\right)^{i_1}\left(\frac{\partial}{\partial x_2}\right)^{i_2}.$$

If we write

$$P = \sum a_{j_1j_2} x_1^{j_1} x_2^{j_2}$$

this means solving the linear system of equations

$$\sum_{j_1=0}^{d_1}\sum_{j_2=0}^{d_2} a_{j_1j_2}\binom{j_1}{i_1}\binom{j_2}{i_2}\alpha_1^{j_1-i_1}\alpha_2^{j_2-i_2} = 0$$

for $(i_1,i_2) \in G(t)$, with $a_{j_1j_2} \in k$ not all zero.

Siegel's Lemma.

Axel Thue was the first to use Dirichlet's Box Princple in order to construct P. This was made explicit by Siegel ([Siegel 1929]), who proved:

Lemma. *Let*

$$\begin{aligned} a_{11}x_1 + \dots + a_{1N}x_N &= 0 \\ a_{21}x_1 + \dots + a_{2N}x_N &= 0 \\ &\cdots\cdots\cdots \\ a_{M1}x_1 + \dots + a_{MN}x_N &= 0 \end{aligned}$$

be a linear system of equations with rational integral coefficients not all 0 *and with* $M < N$. *Then there is a rational integral solution* $(x_1,\ldots,x_N)$ *with not all* x_i's *equal to* 0, *with*

$$\max_i |x_i| \leqslant (N \max_{i,j} |a_{ij}|)^{\frac{M}{N-M}}.$$

Statement of this type are now called Siegel's Lemma. It is a curious fact that the name Siegel's Lemma became associated to weaker statements, replacing the bound given above by

$$c_1\left(c_2 N \max_{ij} |a_{ij}|\right)^{\frac{M}{N-M}}$$

for unspecified constants c_1, c_2, so that we find in the literature "versions of Siegel's Lemma" which are distinctly worse than Siegel's!

The preceding result is sufficient for most applications but for our purposes we need a more sophisticated result. So let us consider more closely the problem of finding solutions in k^N of the linear system

$$A\mathbf{x} = 0,$$

where A is an $M \times N$ matrix with entries in k. The following remarks are useful.

Remark 1. We are dealing with a homogeneous problem, i.e., a problem in projective space. Thus it appears that integrality of coordinates, which is a property in affine space, should be totally irrelevant.

In other words: it is a bad procedure to mix projective and affine points of view.

Remark 2. The system may be supposed of maximal rank. It defines a projective subspace $\Pi \subset \mathbf{P}^{N-1}$ of codimension M. Thus Π is a point defined over k of Grass(N-1,N-1-M), the Grassmannian of (N-1-M)-planes in (N-1)-space. We should regard this point as our basic object and not the individual linear equations defining our system.

In other words: the linear system $A\mathbf{x} = 0$ is not intrinsically defined and therefore it should be replaced by an invariant treatment.

Remark 3. Solutions defined over k correspond to elements of $\Pi(k)$, the points of Π defined over k. Thus we may want to study a basis of solutions, rather than one solution at a time.

We proceed as follows. Let $M \leq N$ and let

$$X = (x_{ij}),\ i = 1, \ldots, M;\ j = 1, \ldots, N$$

be an $M \times N$ matrix with elements in k. For $J \subset \{1, \ldots, N\}$ with $|J| = M$ let X_J denote the $M \times N$ matrix

$$X_J = (x_{ij}),\quad i = 1, \ldots, M,\quad j \in J.$$

We assume that at least one matrix X_J is non-singular, that is X is of maximal rank. Then for each place v of k we define a *local height* by

$$H_v(X) = \max_J |\det X_J|_v \qquad \text{if } v \nmid \infty$$

and

$$H_v(X) = |\det(XX^*)|_v^{1/2} \qquad \text{if } v \mid \infty,$$

and a global height $H(X)$ by

$$\log H(X) = \sum_v \log H_v(X).$$

The height $H(X)$ so defined does not depend on a field of definition k for X and it is *invariant* :

$$H(CX) = H(X)$$

for $C \in GL(M,k)$. We also have the useful property that if $X = \begin{pmatrix} X_1 \\ X_2 \end{pmatrix}$ then

$$H_v(X) \leq H_v(X_1)\, H_v(X_2)$$

for all places v. This is easily seen if $v \nmid \infty$ by using Laplace's expansion, while if $v|\infty$ it is a generalization of Hadamard's inequality due to Fischer in 1908.

The following result is due to Bombieri and Vaaler.

Theorem. *Let* $A\mathbf{x} = 0$ *be a linear system of* M *equations in* N *unknowns, defined over* k *and of maximal rank. There exist* N-M *linearly independent vector solutions* $\mathbf{x}_1, \ldots, \mathbf{x}_{N-M}$ *such that*

$$\prod_{i=1}^{N-M} h(\mathbf{x}_i) \leq |\Delta_k|^{\frac{N-M}{2d}} H(A),$$

where Δ_k *is the absolute discriminant of* k *and where* $d = [k:\mathbf{Q}]$.

If A is defined over a field K with $[K:k] = r$ let $\sigma_i(K)$, $i = 1,\ldots,r$ be the conjugate fields of K/k. Let us suppose that $rM < N$ and let

$$\mathcal{A} = \begin{pmatrix} \sigma_1(A) \\ \sigma_2(A) \\ \vdots \\ \sigma_r(A) \end{pmatrix}.$$

Assume that $\mathcal{A}$ is of rank rM. Then there are $\mathbf{x}_1, \ldots, \mathbf{x}_{N-rM} \in k^N$ such that

$$\prod_{i=1}^{N-rM} h(\mathbf{x}_i) \leq |\Delta_k|^{\frac{N-rM}{2d}} H(\mathcal{A}).$$

Moreover

$$H(\mathcal{A}) \leq H(A)^r.$$

Analogous statements hold for

$$A = \begin{pmatrix} A_1 \\ \vdots \\ A_L \end{pmatrix}$$

where A_ℓ is an $M_\ell \times N$ matrix over a field K_ℓ of degree $[K_\ell:k] = r_\ell$ over k. One defines $\mathcal{A}$ accordingly and replaces rM by $\sum_\ell r_\ell M_\ell$, with the same conclusion.

Suppose

$$A = \left(\binom{j_1}{j_1} \binom{j_2}{j_2} \alpha_{\sigma 1}^{j_1 - i_1} \alpha_{\sigma 2}^{j_2 - i_2} \right)$$

is a matrix with rM rows indexed by (σ, i_1, i_2) and N columns indexed by (j_1, j_2), where: σ denotes conjugation of K over k (there are r such conjugate fields), $(i_1, i_2) \in G$, and $j_1 \leqslant d_1$, $j_2 \leqslant d_2$. Let us assume for simplicity that A is of maximal rank $rM = r|G|$. Then the preceding results on Siegel's Lemma show that there are polynomials $P_\ell(x_1, x_2) \in k[x_1, x_2]$, not identically 0, of degree at most d_i in x_i, such that

(i)
$$\Delta^I P_\ell(\alpha_1, \alpha_2) = 0$$

for $I \in G$;

(ii) $P_1, P_2, \ldots, P_{N-rM}$ are linearly independent over k;

(iii)
$$\prod_{\ell=1}^{N-rM} h(P_\ell) \leqslant |\Delta_k|^{\frac{N-rM}{2d}} H(A).$$

In evaluating $H(A)$ we have to consider the maximal minors of A. A typical determinant is a polynomial of degree $\lesssim rd_1^2 d_2$ in the variables $\alpha_{\sigma 1}$, and of degree $\lesssim rd_1 d_2^2$ in the variables $\alpha_{\sigma 2}$. Since $N \sim d_1 d_2$, we expect an estimate, as d_1, d_2 tend to infinity:

$$\frac{1}{N-rM} \log H(A) \leqslant (A_1 + o(1))d_1 + (A_2 + o(1))d_2$$

where A_1 and A_2 are bounded functions of α_1, α_2. An important but rather difficult problem is the determination of A_1, A_2 as functions of α_1, α_2 and r, Θ, t and $\delta = d_2/d_1$ (the quantities Θ, t appear in the description of G). If $t \leqslant \Theta \leqslant t^{-1}$, which we shall suppose from now on, we have $N \sim d_1 d_2$, $M \sim \frac{1}{2} t^2 d_1 d_2$. With $\delta = d_2/d_1$ we now get

$$\frac{1}{N-rM} \sum_{\ell=1}^{N-rM} \log h(P_\ell) \lesssim (A_1 + A_2 \delta) d_1,$$

so that if $h(P_1) \leqslant h(P_2) \leqslant \ldots$ we obtain

$$\log h(P_1) \lesssim (A_1 + A_2\delta)d_1.$$

The Thue Principle.

Let $d_2/d_1 = \delta$ and let

$$A = \left(\Delta^I \alpha_{\sigma_1}^{j_1} \alpha_{\sigma_2}^{j_2}\right)$$

with rows indexed by (σ, I), $I \in G$ and columns indexed by (j_1, j_2). Let $P_1, \ldots, P_L$, $L = N-rM$, be the polynomials constructed in the preceding section and let (β_1, β_2) be an approximation in k to (α_1, α_2), relative to an absolute value v. By this we mean

$$|\alpha_i - \beta_i|_v \leqslant 1, \quad |\alpha_i|_v \leqslant 1$$

for $i = 1, 2$. Let $I_\ell^* = (i_{1\ell}^*, i_{2\ell}^*)$ be an index such that

$$\Delta^{I_\ell^*} P_\ell(\beta_1, \beta_2) \neq 0$$

and let

$$\tau_\ell^* = \theta^{-1} \frac{i_{1\ell}^*}{d_1} + \theta \frac{i_{2\ell}^*}{d_2}$$

and let τ_ℓ be a real number with $\tau_\ell \geqslant \tau_\ell^*$. Let $\tilde{P}_\ell = \Delta^{I_\ell^*} P_\ell$. Since $\tilde{P}_\ell(\beta_1, \beta_2) \neq 0$, the product formula in k yields

$$\sum_w \log|\tilde{P}_\ell(\beta_1, \beta_2)|_w = 0 .$$

For simplicity of notation we now drop ℓ, ~ and set $\tau_\ell^* = 0$. Thus

$$\sum_w \log |P_\ell(\beta_1, \beta_2)|_w = 0.$$

We estimate separately each $|P(\beta_1, \beta_2)|_w$.

Case (i). $w \neq v$.

In this case

$$|P(\beta_1,\beta_2)|_w \leqslant \max(1, |(d_1+1)(d_2+1)|_w) \times |P|_w \max(1, |\beta_1|_w)^{d_1} \max(1, |\beta_2|_w)^{d_2},$$

hence

$$\log |P(\beta_1,\beta_2)|_w \lesssim d_1 \log^+|\beta_1|_w + d_2 \log^+|\beta_2|_w + \log |P|_w .$$

Case (ii). $w = v$.

In this case

$$P(\beta_1,\beta_2) = \sum \Delta^I P(\alpha_1,\alpha_2)(\beta_1 - \alpha_1)^{i_1}(\beta_2 - \alpha_2)^{i_2}$$

and now $\Delta^I P(\alpha_1,\alpha_2) = 0$ if $\theta^{-1}\frac{i_1}{d_1} + \theta\frac{i_2}{d_2} \leqslant t - \tau_\ell^*$. We write t for $t - \tau_\ell^*$. Let

$$f(x) = x \log \frac{1}{x} + (1-x) \log \frac{1}{1-x},$$

so that

$$\log\binom{d}{i} \leqslant d\, f\left(\frac{i}{d}\right)$$

for every i, d. Let us write

$$a = \|\alpha_1 - \beta_1\|_v , \qquad \beta = \|\alpha_2 - \beta_2\|_v .$$

Subcase I.

We have $\frac{a}{1+a} \leqslant \Theta t$, $\frac{b}{1+b} \leqslant \Theta^{-1} t$. Now

$$|P(\beta_1,\beta_2)|_v \leqslant \max\left(1, (|d_1+1)(d_2+1)|_v\right)^2 \cdot |P|_v \max_{(*)} \left|\binom{d_1}{i_1}\binom{d_2}{i_2}(\beta_1 - \alpha_1)^{i_1}(\beta_2 - \alpha_2)^{i_2}\right|_w$$

where (*) means $\Theta^{-1}\frac{i_1}{d_1} + \Theta\frac{i_2}{d_2} \geqslant t$, thus

$$\log |P(\beta_1,\beta_2)|_v \lesssim d_1 \log^+|\beta_1|_v + d_2 \log^+|\beta_2|_v + \log |P|_v + \varepsilon_v \max_{\theta^{-1}x+\theta y \geqslant t} \left\{d_1(\varepsilon f(x) + x \log a) + d_2(\varepsilon f(y) + y \log b)\right\}$$

where $\varepsilon = 1$ if $v|\infty$ and $\varepsilon = 0$ otherwise, and $\varepsilon_v = \frac{[k_v : \mathbf{Q}_v]}{[k : \mathbf{Q}]}$.
Say $\varepsilon = 1$. The absolute maximum of $f(x) + x \log a$ occurs at $x = a/(1+a)$ and it is $\log(1+a)$. Thus the hypothesis of subcase I implies that the maximum occurs on the line $\theta^{-1}x + \theta y = t$, and thus $x \leqslant \theta t$, $y \leqslant \theta^{-1}t$. Thus the maximum is not more than

$$\begin{aligned} &\varepsilon_v \varepsilon\, f^*(\theta t) d_1 + \varepsilon_v \varepsilon\, f^*(\theta^{-1}t) d_2 \\ &\qquad + \varepsilon_v \max_{\theta^{-1}x+\theta y=t} \left(d_1 x \log a + d_2 y \log b\right) \\ &= - t \min\left(d_1\theta \log \frac{1}{|\alpha_1 - \beta_1|_v},\ d_2\theta^{-1} \log \frac{1}{|\alpha_2 - \beta_2|_v}\right) \\ &\qquad + \varepsilon_v \varepsilon \left(f^*(\theta t)\, d_1 + f^*(\theta^{-1}t)\, d_2\right), \end{aligned}$$

where $f^*(x) = f(x)$ if $0 \leqslant x \leqslant 1/2$ and $f^*(x) = \log 2$ if $1/2 \leqslant x \leqslant 1$.
If we put together the information obtained so far we deduce

$$\begin{aligned} \log |P(\beta_1, \beta_2)|_v \lesssim\ & d_1 \log^+ |\beta_1|_v + d_2 \log^+ |\beta_2|_v + \log |P|_v \\ & + \varepsilon_v \varepsilon\left(f^*(\theta t) d_1 + f^*(\theta^{-1}t) d_2\right) \\ & - t \min\left(d_1 \theta \log \frac{1}{|\alpha_1 - \beta_1|_v},\ d_2 \theta^{-1} \log \frac{1}{|\alpha_2 - \beta_2|_v}\right). \end{aligned}$$

Subcase II. $\theta^{-1}\frac{a}{1+a} + \theta \frac{b}{1+b} \geqslant t$. In this case

$$\begin{aligned} \min &\left(d_1 \theta\, t \log \frac{1}{a},\ d_2 \theta^{-1} t \log \frac{1}{b}\right) \\ &\qquad \leqslant d_1 f^*(\theta t) + d_2 f^*(\theta^{-1}t). \end{aligned}$$

Suppose subcase I holds. We put together all the estimates for $\log |P(\beta_1, \beta_2)|$ with the product formula, and find

$$\begin{aligned} & t \min\left(d_1 \theta \log \frac{1}{|\alpha_1 - \beta_1|_v},\ d_2 \theta^{-1} \log \frac{1}{|\alpha_2 - \beta_2|_v} \right) \\ & \lesssim \left(\varepsilon_v \varepsilon f^*(\theta t) + \log h(\beta_1)\right) d_1 + \left(\varepsilon_v \varepsilon f^*(\theta^{-1}t) + \log h(\beta_2)\right) d_2 \\ & \qquad + \log h(P)\ . \end{aligned}$$

If instead subcase II holds, the above formula still holds.

Finally we replace P by P^m in the above calculation and let $m \to \infty$. We obtain

Theorem. *Let* $P \in k[x_1, x_2]$ *be of bidegree* d_1, d_2 *such that*

$$\Delta^I P(\alpha_1, \alpha_2) = 0$$

for

$$\theta^{-1}\frac{i_1}{d_1} + \theta\frac{i_2}{d_2} \leqslant t,$$

for some θ, $t \leqslant \theta \leqslant t^{-1}$. *Suppose also that* $\beta_1, \beta_2 \in k$ *are such that*

$$P(\beta_1, \beta_2) \neq 0$$

and

$$|\alpha_1|_v \leqslant 1,\ |\alpha_2|_v \leqslant 1,\ |\alpha_1 - \beta_1|_v \leqslant 1,\ |\alpha_2 - \beta_2|_v \leqslant 1 .$$

Then we have

$$t \min\left(d_1\theta \log \frac{1}{|\alpha_1 - \beta_1|_v},\ d_2\theta^{-1}\log \frac{1}{|\alpha_2 - \beta_2|_v} \right)$$

$$\leqslant \log h(P) + d_1\left(\varepsilon_v \varepsilon f^*(\theta t) + \log h(\beta_1)\right)$$

$$+ d_2\left(\varepsilon_v \varepsilon f^*(\theta^{-1} t) + \log h(\beta_2)\right)$$

where

$$\varepsilon_v\varepsilon = 0 \text{ if } v\nmid\infty,\quad \varepsilon_v\varepsilon = \frac{1}{[k:\mathbf{Q}]} \text{ if } k_v = \mathbf{R},$$

$$\varepsilon_v\varepsilon = \frac{2}{[k:\mathbf{Q}]} \text{ if } k_v = \mathbf{C}.$$

As was remarked before, the condition $P(\beta_1, \beta_2) \neq 0$ is the hardest part to verify and usually one replaces P by $\Delta^{I^*}P$ for some I^*. So if τ is real with

$$\tau \geqslant \tau^* = \theta^{-1}\frac{i_1^*}{d_1} + \theta\frac{i_2^*}{d_2}$$

and if $\tau < t$ then we can apply the preceding theorem to $\Delta^{I^*}P$. This means replacing t by $t - \tau$ and using Lemma 3 to estimate $h(\Delta^{I^*}P)$. Then we get

$$(t-\tau)\min\left(d_1\theta \log \frac{1}{|\alpha_1 - \beta_1|_v}, d_2\theta^{-1} \log \frac{1}{|\alpha_2 - \beta_2|_v}\right)$$

$$\leq d_1\left(\log 4h(\beta_1) + A_1\right) + d_2\left(\log 4h(\beta_2) + A_2\right).$$

We now choose

$$d_1 = \left[\frac{D}{\log 4h(\beta_1) + A_1}\right], \quad d_2 = \left[\frac{D}{\log 4h(\beta_2) + A_2}\right]$$

and θ such that

$$d_1\theta \log \frac{1}{|\alpha_1 - \beta_1|_v} = d_2\theta^{-1}\log \frac{1}{|\alpha_2 - \beta_2|_v},$$

and let $D \to \infty$. We have proved

Thue-Siegel Principle. *Let*

$$|\alpha_i - \beta_i|_v \leq \left(4e^{A_i}h(\beta_i)\right)^{-\eta_i}, \quad i = 1, 2.$$

Then we have

$$\eta_1\eta_2 \leq \left(\frac{2}{t-\tau}\right)^2.$$

Two more steps are needed: the estimation of A_i and that of τ.

Application of Dyson's Lemma.

Let us assume that $t \leq \theta \leq t^{-1}$ and let $P(x,y) \in k[x,y]$ be a polynomial of bidegree d_1, d_2 such that

$$\Delta^I P(\alpha_1, \alpha_2) = 0$$

for $\theta^{-1}\frac{i_1}{d_1} + \theta\frac{i_2}{d_2} < t$ and

$$\Delta^I P(\beta_1, \beta_2) = 0$$

for $\theta^{-1}\frac{i_1}{d_1} + \theta\frac{i_2}{d_2} < \tau$. Suppose that $\alpha_1, \alpha_2 \in K$ have degree r over k. We have

Dyson's Lemma.

$$\frac{1}{2}\, rt^2 + \frac{1}{2}\, \tau^2 \leqslant 1 + \frac{r-1}{2}\, \frac{d_2}{d_1} .$$

In terms of $\delta = d_2/d_1$, this yields

$$\tau \leqslant \left(2 - rt^2 + (r-1)\delta\right)^{1/2} \sim \left(2 - rt^2\right)^{1/2}$$

as $\delta \to 0$. If $\sqrt{\frac{2}{r+1}} < t < \sqrt{\frac{2}{r}}$ then $\tau < t$, which is what we need.

Let $P_1, P_2, \ldots, P_L$, $L = N - r|G|$, be the set of linearly independent polynomials constructed in the section on Siegel's Lemma. Let $1 \leqslant \ell \leqslant L$. Since $P_1, P_2, \ldots, P_\ell$, are linearly independent we can find a linear combination of $P_1, \ldots, P_\ell$ which vanishes at (α_1, α_2) to order t_ℓ namely

$$\Delta^I P(\alpha_1, \alpha_2) = 0$$

for

$$\theta^{-1}\, \frac{i_1}{d_1} + \theta\, \frac{i_2}{d_2} < t_\ell$$

where t_ℓ is the largest for which

$$r\left(|G(t_\ell)| - |G(t)|\right) < \ell .$$

By Dyson's Lemma, this linear combination will not vanish at (β_1, β_2) more than

$$\Delta^I P(\beta_1, \beta_2) = 0$$

for

$$\theta^{-1}\, \frac{i_1}{d_1} + \theta\, \frac{i_2}{d_2} < \tau_\ell$$

with

$$\frac{1}{2}\, \tau_\ell^2 \leqslant 1 - \frac{1}{2}\, rt_\ell^2 + \frac{r-1}{2}\, \frac{d_2}{d_1}$$

$$= 1 - \frac{1}{2}\, rt^2 - \frac{\ell}{d_1 d_2} + \frac{r-1}{2}\, \frac{d_2}{d_1} + O\left(\frac{1}{d_2}\right) .$$

Since a linear combination of $P_1,\ldots,P_\ell$ does not vanish at (β_1,β_2) more than calculated before we see that one of them, say $P_{\ell'}$, does not vanish at (β_1,β_2) in the same way. By considering either P_ℓ or $P_\ell + P_{\ell'}$ and replacing P_ℓ by $P_\ell + P_{\ell'}$ we see that we may suppose that P_ℓ itself does not vanish at (β_1,β_2) more than stated before. In doing so, we may have to increase the height of P_ℓ by a factor of 2, or less. In conclusion we find that the polynomials $P_1,\ldots,P_\ell$ satisfy:

(a) $P_1,\ldots,P_L \in k[x_1,x_2]$ have bidegree $\leqslant (d_1,d_2)$ and satisfy

$$\Delta^I P_\ell(\alpha_1,\alpha_2) = 0$$

for $\theta^{-1}\frac{i_1}{d_1} + \theta\frac{i_2}{d_2} < t$;

(b) for each ℓ there is I_ℓ^* such that

$$\Delta^{I_\ell^*} P_\ell(\beta_1,\beta_2) \neq 0$$

and

$$\theta^{-1}\frac{i_{1\ell}^*}{d_1} + \theta\frac{i_{2\ell}^*}{d_2} < \tau_\ell$$

with

$$\frac{1}{2}\tau_\ell^2 \leqslant 1 - \frac{1}{2}rt^2 - \frac{\ell}{d_1 d_2} + \frac{r-1}{2}\frac{d_2}{d_1} + o\left(\frac{1}{d_2}\right);$$

(c) if the $\tilde{P}_\ell$ are the successive minima for (a) then

$$h(\tilde{P}_\ell) \leqslant h(P_\ell) \leqslant Ch(\tilde{P}_\ell)$$

for some bounded C (independent of d_1,d_2). In particular,

$$\frac{1}{L}\sum_{\ell=1}^{L} \log h(P_\ell) = \frac{1}{L}\log H(A) + o(d_1)$$

as d_1, $d_2 \to \infty$, $d_1 \geqslant d_2$.

We apply the preceding result to the case in which $d_1 \to \infty$, $d_2 \to \infty$, $d_2/d_1 \to 0$, $\log h(\beta_2) \to \infty$. We also note that in Dyson's Lemma the condition that α_2 be of degree r over k can be removed and replaced by α_2 of degree $s \geqslant 2$ over k, $\alpha_2 \in k(\alpha_1)$, and

$$t \leqslant \theta \leqslant \frac{s}{2}\, t$$

[Viola 1984]. We obtain

Main Theorem. *Let* $K = k(\alpha_1)$ *be of degree* r *over* k *and let* v *be a place of* k*, extended to* K*. Let* $\beta_1 \in k$ *be an approximation to* α_1 *, in the sense that*

$$|\alpha_1 - \beta_1|_v < 1, \qquad |\alpha_1|_v \leqslant 1.$$

Let $\alpha_2 \in k(\alpha_1)$, $\alpha_2 \notin k$. *Then the effective type of irrationality for* α_2 *over* k *satisfies*

$$\mu_{\mathrm{eff}}(\alpha_2; k, v) \leqslant \frac{4}{(t - \frac{2}{3}\tau)^2}\, \eta_1^{-1}$$

where η_1 *is determined by*

$$|\alpha_1 - \beta_1|_v \leqslant \left(\gamma\, e^{A_1}\, h(\beta_1)\right)^{-\eta_1}.$$

Here $\tau = \sqrt{2 - rt^2}$, *and* $\log \gamma = \theta t \log \frac{2}{\theta t} + (1 - \theta t) \log \frac{1}{1 - \theta t}$ *and* A_1 *is given by*

$$A_1 = \lim_{\substack{d_2 \to \infty \\ d_2/d_1 \to 0}} \frac{1}{(1 - \frac{1}{2} r t^2) d_1^2 d_2} \log H(A)$$

with A *the matrix*

$$A = \left(\binom{j_1}{i_1} \binom{j_2}{i_2} \alpha_{1\sigma}^{j_1 - i_1} \alpha_{2\sigma}^{j_2 - i_2} \right)$$

indexed by rows $(\sigma, i_1, i_2,)$ *and columns* (j_1, j_2), *with* σ *ranging over conjugation of* K/k, *and*

$$\theta^{-1} \frac{i_1}{d_1} + \theta \frac{i_2}{d_2} < t,$$

and j_1, j_2 *ranging over* $j_1 \leqslant d_1$, $j_2 \leqslant d_2$.

Proof. By the preceding theorem we have

$$(t - \tau_\ell)\min\left(d_1,\ \theta \log \frac{1}{|\alpha_1 - \beta_1|_v},\ d_2\,\theta^{-1}\log\frac{1}{|\alpha_2 - \beta_2|_v}\right)$$

$$\leqslant \log h(P_\ell) + d_1 \log(\gamma h(\beta_1)) + d_2 \log(4h(\beta_2))\ .$$

If we take the *average* of this relation with respect to ℓ we get the result, because

$$\int_0^1 \sqrt{1-x}\,dx = \frac{2}{3}.$$

Applications.

If $\alpha_1 = \sqrt[r]{\xi}$, $\xi \in k$ then we can bound A_1 with some precision. There are several ways of doing it and the best one yields

$$A_1 = \frac{1}{2}rt^2 \log h(\alpha_1) + 0\left(\frac{1}{1-{}^1\!/_2\, rt^2}\right);$$

we refer to [B-M 1983] for similar explicit estimates. For example, if

$$\alpha_1 = \sqrt[r]{\frac{a}{b}} \quad \text{and} \quad b > |a| \quad \text{and}$$

$$\lambda = \frac{\log |b - a|}{\log b}\ ,$$

then α_1 has an effective type

$$\mu_{\text{eff}}(\alpha_1;\mathbf{Q}) \leqslant \frac{2}{1-\lambda} + 0\left(\frac{1}{(\log b)^{1/3}}\right)$$

[Bombieri-Mueller,1983]. If b is large and $\lambda < 1 - \frac{2}{r}$ this represents an improvement over the Liouville bound. Previous exponents, obtained with the Pade´ technique yielded

$$1 + \frac{1}{1-2\lambda} + 0\left(\frac{1}{\log b}\right) \qquad \text{(Thue, Baker)}$$

and

$$s + \frac{s}{s-(s+1)\lambda} + 0\left(\frac{1}{\log b}\right) \qquad \text{(Chudnovsky) for } s = 1, 2, \ldots, r-1.$$

In this case, one chooses $\beta_1 = 1$ and $|\alpha_1 - \beta_1|$ is of order $\frac{|b-a|}{b}$. Now $h(\alpha_1) = b^{1/r}$; hence

$$|\alpha_1 - \beta_1| < C(r)h(\alpha_1)^{-r(1-\lambda)}.$$

Since $h(\beta_1) = 1$ we see that η_1 is determined by $(\gamma e^{A_1})^{\eta_1}$ $= h(\alpha_1)^{r(1-\lambda)}/C(r)$, hence

$$\eta_1 = \frac{r(1-\lambda)}{\frac{1}{2} rt^2} + O\left(\frac{1}{\tau^2 \log h(\alpha_1)}\right)$$

and

$$\mu_{eff}(\alpha_2; k, v) \leqslant \frac{2}{1-\lambda} + O(\tau) + \left(\frac{1}{\tau^2 \log h(\alpha_1)}\right).$$

Choosing $\tau = (\log h(\alpha_1))^{-1/3}$ we obtain

$$\mu_{eff}(\alpha_2; k, v) \leqslant \frac{2}{1-\lambda} + O\left(\frac{1}{(\log h(\alpha_1))^{1/3}}\right),$$

as asserted.

At the other hand of calculations of this type we have situations in which $h(\beta_1)$ is large. A typical example is the following. Let α_1 be the root $\sim \frac{1}{m}$ of the equation

$$x^r - mx + 1 = 0,$$

where m is a large integer; here we choose $k = Q$, v the infinite place, so that $| \ |_v$ is the ordinary absolute value. We have already computed the height of α_1, with the bound

$$h(\alpha_1) < \frac{1}{r} \log m.$$

We choose $\beta_1 = \frac{1}{m}$, hence $h(\beta_1) = m$ and note that

$$|\alpha_1 - \beta_1| = \frac{1}{m|\alpha_1|^r}$$

is $\sim m^{-r-1} \sim (h(\alpha_1)h(\beta_1))^{-r}$; thus the pair α_1, β_1 nearly satisfies the Liouville bound.

It remains to estimate $H(A)$. This is a difficult problem. If one uses a Laplace expansion and uses the fact that no term in a row or column may appear twice, then one can prove

$$A_1 \leqslant \frac{r}{2}t^2 \frac{1 - \frac{r}{4}t^2}{1 - \frac{r}{2}t^2} \log h(\alpha_1) + O\left(\frac{1}{1 - \frac{r}{2}t^2}\right).$$

The Main Theorem now yields

$$\mu_{\mathrm{eff}}(\alpha_2;\mathbb{Q},\infty) \leq \frac{4}{(t - \frac{2}{3}\tau)^2}\left(1 + \frac{t^2(1 - \frac{r}{4}t^2)}{\tau^2}\right)\frac{1}{r+1} + O\left((\log h(\alpha_1))^{-1}\right).$$

For large r, this does not exceed

$$\min_a \frac{2}{\left(1 - \frac{2}{3}a\right)^2}\left(1 + \frac{1}{2a^2}\right) < 13.209446$$

with $a = .5674$. Thus if $r \geq r_0$ and $m \geq m(r)$ we have

$$\mu_{\mathrm{eff}}(\alpha_2;\mathbb{Q},\infty) < 13.209446 .$$

It is easy to generalize the last example to equations of the sort

$$x^{r-s}f(x) + mx - 1 = 0$$

where $f(x) = x^s + a_1x^{s-1} + \ldots + a_s$ is a polynomial with bounded coefficients. What appears however of more interest is the fact that for every algebraic α we can find k such that $\mu_{\mathrm{eff}}(\alpha;k,v)$ is small. The following is proved in [Bombieri-Mueller,1986].

Theorem. *Let α be a real algebraic number of degree $r \geq 3$ and let $\eta > 0$ be any positive constant. Then one can find infinitely many real algebraic number fields k of degree $r - 1$ such that*

$$\mu_{\mathrm{eff}}(\alpha;k,\infty) \leq 2 + \eta .$$

In order to apply the Thue-Siegel Principle to such a situation we use Wirsing's result that real numbers admit very good approximations by algebraic numbers of *fixed degree*.

Proposition. *Let α, $|\alpha| \leq 1/2$, be real algebraic of degree r and height $H(\alpha)$. For every $X \geq 2$ there is β, algebraic of degree at most $r - 1$, such that*

$$H(\beta) \leqslant 2^r\big(r(r+1)H(\alpha)\big)^{(r-1)^2/r} X$$

and

$$|\alpha - \beta| \leqslant \frac{r!(r-1)}{X^r}.$$

If we take $k = Q(\beta)$ and $K = k(\alpha)$ one can then show that

$$\mu_{eff}(\alpha; k,\infty) \leqslant 2 + O((\log X)^{-1/3}).$$

Another type of applicaton relates to Thue equations. The Thue-Siegel Principle can be used to obtain bounds for the number of solutions of equations $F(x,y) = c$, since every sufficiently large solution is an *anchor pair* with a root of $F(x,1) = 0$. This allows one to count in an efficient way the solutions to a Thue equation exceeding a certain bound. Coupled with a counting of the remaining "small" solutions, Bombieri and Schmidt proved

Theorem. *Let* $F(x,y)$ *be an irreducible form over* $\mathbf{Z}$ *of degree* $r \geqslant 3$. *Then the number of solutions* $\pm (x,y)$ *to* $|F(x,y)| = 1$ *does not exceed* cr, *for some absolute constant* c. *If* r *is large, one can take* $215r$ *for such a bound.*

This result has been generalized to the so-called Thue-Mahler equation.

Further Applications.

We have not touched in these lectures upon the problem of proving the non-vanishing of the auxiliary polynomial at β_1,β_2, i.e. Dyson's Lemma or Roth's Lemma.

The classical argument goes by induction on the number of variables of P and is roughly as follows. Suppose we know that

$$\Delta^I P(\alpha_1,\alpha_2,\ldots,\alpha_n) = 0$$

for $I = (i_1, i_2, \ldots, i_n) \in G$ and want to show that P cannot vanish too much at some other point $(\beta_1, \ldots, \beta_n)$. If $n = 1$, we have discussed the situation in great detail:

a) The fundamental theorem of algebra

b) Gauss´ Lemma

c) vanishing at 1 or roots of unity.

Let us decompose P as

$$P(x_1, \ldots, x_n) = \sum f_j(\mathbf{x}') \, g_j(x_n)$$

where $\mathbf{x}' = (x_1, \ldots, x_{n-1})$; we may assume that the f_j are linearly independent, and so are the g_j. Because of linear independence, some generalized Wronskian of the f_j is not identically 0, and so is some Wronskian of the g_j, say $W(\mathbf{f})$ and $W(\mathbf{g})$. But now this means that some generalized Wronskian of P, say W(P), is *non-zero and factorizes as a polynomial in* $\mathbf{x}'$ *and a polynomial in* x_n:

$$W(P) = W(\mathbf{f}) W(\mathbf{g}) \; ,$$

as one sees by

$$\det\left(\delta^{(i)} f_j\right) \det\left(\Delta^k g_j\right) = \det\left(\delta^{(i)} \Delta^k \sum f_j g_j\right) = \det\left(\delta^{(i)} \Delta^k P\right).$$

Now the vanishing of P determines the vanishing of W(P), which in turn determines the vanishing of $W(\mathbf{f})$ and $W(\mathbf{g})$, which in turn determines the vanishing of $W(\mathbf{f})$ and $W(\mathbf{g})$, which are polynomials in a lower number of variables. Thus, by induction, we obtain a control on the amount of vanishing of $W(\mathbf{f})$, $W(\mathbf{g})$, hence of W(P) and finally P itself. The final result now depends on how one wants to control the start of the induction, namely a) or b). The technique in b) leads to the famous Roth´s Lemma, which shows that if the heights of $\beta_1, \ldots, \beta_n$ go to ∞ sufficiently rapidly then P has very limited vanishing at $\beta_1, \ldots, \beta_n$. The technique in a) leads for the case $n = 2$ to Dyson´s Lemma ([Dyson 1947], [Bombieri 1982]). The main advantage in b) is the fact that no conditions on the height of β_1, β_2 are needed (recall that in applications, such as $\alpha_1 = \sqrt[r]{\frac{a}{b}}$, one may want to take $\beta_1 = 1$). On the other hand, the result one obtains in the case $n \geq 3$ is much weaker and it is not directly usable in applicatons; in particular, one could not obtain a new proof of Roth's theorem using Dyson's technique.

The new ideas needed in this direction were provided independently by C. Viola [Viola 1984] and H. Esnault and E. Viehweg [Esnault and Viehweg 1984], using methods from algebraic geometry. Viola's idea, so far carried out completely only in the case $n = 2$, relates the multiplicity of zeros to local contributions to the calculation of invariants of the curve $P = 0$, such as the genus. A global control of the genus (for example, genus ≥ 0 if the curve is irreducible) yields an inequality which implies a sharp form of Dyson's Lemma. A nice feature of Viola's result is that it allows the case in which α_2, which is an element of $k(\alpha_1)$, may have degree over k strictly less than the degree of α_1 over k. This means that the measure of effectivity obtained is valid for all elements of $k(\alpha_1)$ not in k and not just for generators of $k(\alpha_1)$ over k. It is conceivable that examples may be found in which an irrationality type for α_2 is obtained by *constructing* α_1 with $\alpha_2 \in k(\alpha_1)$ and with exceptionally good approximations $\beta_1 \in k$; the degree of α_1 could very well be much larger than the degree of α_2 over k.

The approach of Esnault and Viehweg is based on algebraic geometry and it works in any dimension. It is not possible to describe here their technique, which uses very deep results such as variation of Hodge structures and Kawamata's vanishing theorems. Their result however is easily described. Let us say that P has a *zero of type* $(\mathbf{a},t)$ *at* ζ if

$$\Delta^I P(\zeta) = 0$$

whenever

$$\sum_{\nu=1}^{n} i_\nu a_\nu \leq t .$$

Let $I(\mathbf{d},\mathbf{a},t)$ be the set

$$I(\mathbf{d},\mathbf{a},t) = \{(\xi_\nu) \in I^n : \sum_{\nu=1}^{n} d_\nu \xi_\nu a_\nu \leq t\}$$

where I^n is the unit cube $0 \leq \xi_\nu \leq 1$. We have [Esnault and Viehweg, 1984]:

Dyson's Lemma. *Assume that*

a) $\zeta_\mu = (\zeta_{\mu 1},\ldots,\zeta_{\mu n})$ *for* $\mu = 1,\ldots,M$ *are* M *points in* $\mathbf{C}^n$ *such that* $\zeta_{\mu,\nu} \neq \zeta_{\gamma,\nu}$ *for* $\mu \neq \gamma$ *and* $\nu = 1,\ldots,n$;

b) $\mathbf{a} = (a_1,\ldots,a_n)$ *has* $a_i \geq 0$ *and* $t_\mu \geq 0$ *for* $\mu = 1,\ldots,M$;

c) $d_1 \geq d_2 \geq \ldots \geq d_n$.

Then if P *is a polynomial in* $\mathbf{C}[x_1,\ldots,x_n]$ *of multidegree* $d_1,\ldots,d_n$, *not identically* 0, *and if* P *has a zero of type* $(\mathbf{a},t_\mu)$ *at* ζ_μ *for* $\mu = 1,\ldots,M$ *we have*

$$\sum_{\mu=1}^{M} \mathrm{Vol}(I(\mathbf{d},\mathbf{a},t_\mu)) \leq \prod_{j=1}^{n} \Big(1 + (M'-2)\Big) \sum_{i=j+1}^{n} \frac{d_i}{d_j} \Big)$$

with $M' = \max(M,2)$.

Roughly speaking, this result shows that if $\frac{d_{i+1}}{d_i} \to 0$ for $i = 1,\ldots, n-1$ then the vanishing of P at different points implies "almost independent" conditions on the coefficients of P, as long as we require vanishing of the same "type" and the technical condition a). As proved by Esnault and Viehweg, this implies the Roth theorem. There is another application of this result, which is worth mentioning here. Let α be real algebraic of degree ≥ 3 and let us consider the problem of solutions to

$$\left|\alpha - \frac{p}{q}\right| < q^{-2 - \varepsilon(q)}$$

where $\varepsilon(q) \to 0$ as $q \to \infty$. The so-called Cugiani-Mahler theorem asserts that if $\varepsilon(q) = c_1(\alpha)(\mathrm{logloglog}\ q)^{-1/2}$, for a suitable $c_1(\alpha)$, then the sequence $\{q_i\}$ of solutions to the above inequality satisfies

$$\limsup \frac{\log q_{i+1}}{\log q_i} = +\infty .$$

It is now possible to show that using Dyson's Lemma in place of Roth's Lemma in the proof of the Cugiani-Mahler theorem one reaches the same conclusion with the better value

$$\varepsilon(q) = c_2(\alpha)\left(\frac{\mathrm{loglog}\ q}{\mathrm{logloglog}\ q}\right)^{-1/4};$$

the gain is thus almost of one logarithm, from a triple log decay to a double log.

It is conceivable that the several variable generalization of Dyson's Lemma can be used to improve our knowledge about effective approximations to algebraic numbers. Here the main obstacle appears to be that of estimating in an efficient way the height of the polynomials in the auxiliary construction. It would be of great interest to produce new examples, say from a three variables construction, which could not be treated equally well with the two variable construction of Thue and Siegel.

References.

E. Bombieri, On the Thue-Siegel-Dyson theorem, *Acta Mathematica* **148** (1982), 255-296.

E. Bombieri and J. Mueller, On effective measures of irrationality for $\sqrt[r]{\frac{a}{b}}$ and related numbers, *J. Reine Angew. Math* **342** (1983), 173-196.

E. Bombieri and J. Mueller, Remarks on the approximation to an algebraic number by algebraic numbers, *Michigan Math J.* **33** (1986), 83-93.

E. Bombieri and J. Vaaler, On Siegel's Lemma, *Invent. Math.* **73** (1983),11-32. Addendum to "On Siegel's Lemma", *Invent. Math.* **75** (1984), 377.

F. Dyson, The approximation to algebraic numbers by rationals, *Acta Mathematica* **79** (1947), 225-240.

H. Esnault and E. Viehweg, Dyson's Lemma for polynomials in several variables (and the Theorem of Roth), *Invent. Math.* **78** (1984),445-490.

N. I. Feldman, An effective refinement of the exponent in Liouville's theorem (Russian), *Izv. Akad. Nauk SSSR Ser. Mat.* **35** (1971), 973-990; *Math. USSR Izv.* **5** (1971), 985-1002.

W. M. Schmidt, Diophantine Approximation, Lecture Notes in Math., **785**, Springer, Berlin 1980.

C. L. Siegel, Über einige Anwendungen diophantischer Approximationen, *Abh. der Preuß. Akad. der Wissenschaften. Phys.-Math. Kl.* 1929, Nr. **1** (= Ges. Ahb. I, 209-266).

C. Viola, On Dyson's Lemma, *Annali Scuola Norm. Sup Pisa*, **12** (1985), 105-135.

Enrico Bombieri,
Institute for Advanced Study,
Princeton, N.J., 08540, U.S.A.

POLYNOMIALS WITH LOW HEIGHT AND PRESCRIBED VANISHING

Enrico Bombieri and Jeffrey D. Vaaler*

1. Introduction.

In a recent paper [2] we obtained an improved formulation of Siegel's classical result([9],Bd. I,p. 213, Hilfssatz) on small solutions of systems of linear equations. Our purpose here is to illustrate the use of this new version of Siegel's lemma in the problem of constructing a simple type of auxiliary polynomial. More precisely, let k be an algebraic number field, O_k its ring of integers, $\alpha_1,\alpha_2,\ldots,\alpha_J$ distinct, nonzero algebraic numbers (which are not necesarily in k), and $m_1,m_2,\ldots,m_J$ positive integers. We will be interested in determining nontrivial polynomials P(X) in $O_k[X]$ which have degree less than N, vanish at each α_j with multiplicity at least m_j and have low height. In particular, the height of such plynomials will be bounded from above by a simple function of the degrees and heights of the algebraic numbers α_j and the remaining data in the problem: $m_1,m_2,\ldots,m_J$, N and the field constants associated with k.

This type of construction has been used recently by Mignotte [6],[7] (see also [8, pp.281-288]) and by Dobrowolski [4]. Our bounds provide a simpler and somewhat sharper form of their results. In section 5 we consider the special case of polynomials in Z[X] which vanish at 1 with high multiplicity and yet have relatively low height.

If N is sufficiently large, the set S of polynomials in $k[X]$ which have degree less than N and vanish at each α_j with multiplicity at least m_j, forms a vector space over k of positive dimension L. An interesting feature of the Siegel lemma obtained in

---

* The research of the second author was supported by a grant from the National Science Foundation.

[2] is that it allows us to determine L polynomials $P_1(X), P_2(X), \ldots, P_L(X)$ in $O_k[X] \cap S$ which form a basis for S and are such that the average height of these polynomials is small.

Acknowledgement. We wish to thank Dr. D. Bump for calling our attention to references [5] and [10] on Schur polynomials.

2. Statement of results.

We summarize, briefly, our notation for heights of algebraic numbers, vectors and matrices. This is identical to the notation used in [2]. We suppose that the number field k has degree d over $\mathbf{Q}$ and write v for a place of k. Then k_v is the completion of k at v and $[k_v, \mathbf{Q}_v] = d_v$ is the local degree. At each place v we normalize an absolute value $|\ |_v$ as follows. If $v|\infty$ we set $|x|_v = |x|^{d_v/d}$ where $||$ is the ordinary absolute value on $\mathbf{R}$ or $\mathbf{C}$. If v is a finite place then $v|p$ for some rational prime p. In this case we require that $|p|_v = p^{-d_v/d}$. Because of our normalizations the product formula

$$\prod_v |\alpha|_v = 1$$

holds for $\alpha \in k$ and $\alpha \neq 0$. Also, it will be convenient to use a second nomalized absolute value $\|\ \|_v$ at each place v. These are related by $\|\ \|_v = |\ |_v^{d/d_v}$.

We extend the definition of $||_v$ to (column) vectors $\mathbf{x}$ in k^N with

$$\mathbf{x} = \begin{pmatrix} x_1 \\ x_2 \\ \vdots \\ x_N \end{pmatrix}$$

by $|\mathbf{x}|_v = \max_n |x_n|_v$. The homogenous height of a vector $\mathbf{x}$ in k^N is given by

$$h(\mathbf{x}) = \prod_v |\mathbf{x}|_v .$$

In view of the product formula we have $h(\alpha\mathbf{x}) = h(\mathbf{x})$ for all scalars $\alpha \neq 0$. Thus h is a height on the projective space $\mathbf{P}_k^{N-1}$. If f(X) is a polynomial in $k[X]$ we write h(f) for the height of the vector of coefficients of f.

Let $Y = (y_{mn})$ be an M×N matrix over k with rank(Y) = M < N. For each subset $I \subset \{1,2,\ldots,N\}$ of cardinality $|I| = M$ we write

$$Y_I = (y_{mn}), \quad m \in \{1,2,\ldots,M\}, \quad n \in I,$$

for the corresponding M×M submatrix. At each place v we define a local height H_v on matrices by

$$H_v(Y) = \Big(\sum_{|I|=M} |\det Y_I|_v^2 \Big)^{d_v/2d} \quad \text{if } v|\infty ,$$

$$H_v(Y) = \max_{|I|=M} |\det Y_I|_v \quad \text{if } v\nmid\infty .$$

We then obtain a global height by setting

$$H(Y) = \prod_v H_v(Y) .$$

For elements γ in k we use the inhomogeneous height

$$h_1(\gamma) = \prod_v \max\{1, |\gamma|_v\} .$$

If γ has degree d over **Q** then the quantity $(h_1(\gamma))^d$ is also the Mahler measure of the algebraic number γ, as defined, for example, in [8]. A basic property of each of our heights is that they do not depend on the field containing γ or the entries in **x** or Y.

As before we assume that $\alpha_1, \alpha_2, \ldots, \alpha_J$ are distinct nonzero algebraic numbers with degrees $r_j = [k(\alpha_j):k]$ over k. We also assume that for $i \neq j$ the minimal polynomials for α_i and α_j over k have no common zeros. This allows us to avoid some trivial complications. We write

$$M = \sum_{j=1}^{J} m_j r_j$$

and let N be a positive integer such that N - M = L is positive. It follows easily that the vector space

$$S = \{ f \in k[X]: \deg(f) < N, f^{(\mu)}(\alpha_j) = 0$$

$$\text{for } \mu = 0,1,\ldots,m_j-1 \text{ and } j = 1,2,\ldots,J\} \tag{2.1}$$

has dimension L over k. If $q_j(X)$ is the minimal polynomial of α_j over k and

$$Q(X) = \prod_{j=1}^{J} \{q_j(X)\}^{m_j},$$

then the poynomials $F_\ell(X) = X^{\ell-1}Q(X)$, $\ell = 1,2,\ldots,L$, clearly form a basis for S. Now by a basic result on heights ([1], section 2) we have

$$\sum_{\ell=1}^{L} \log h(F_\ell) = L \log h(Q) \tag{2.2}$$

$$\leq (N - M) \sum_{j=1}^{J} m_j r_j \log h_1(\alpha_j) + N^2 t(\tfrac{M}{N}) ,$$

where $t(\theta) = \theta(1-\theta)\log 2$ for $0 \leq \theta \leq 1$. If the ratio M/N is close to 1 we cannot expect to do much better than (2.2). On the other hand, when M/N is near zero it is possible to determine L polynomials in S for which this bound can be substantially improved.

Let $u(\theta)$ be defined for $0 \leq \theta \leq 1$ by $u(0) = u(1) = 0$ and, if $0 < \theta < 1$, by

$$u(\theta) = \frac{1}{4}\theta^2\log\left(\frac{1 - \theta^2}{16\theta^2}\right) + \frac{1}{2}\theta\log\left(\frac{1 + \theta}{1 - \theta}\right) + \frac{1}{4}\log(1 - \theta^2) .$$

The function $u(\theta)$ is continuous and satisfies the inequalities

$$u(\theta) < t(\theta) \tag{2.3}$$

and

$$u(\theta) < \frac{1}{2}\theta^2\log\left(\frac{1}{4\theta}\right) + \frac{3}{4}\theta^2 , \tag{2.4}$$

for $0 < \theta < 1$. To establish (2.3) on the interval $0 < \theta \leq 2^{-1/2}$ we note that $t(0) - u(0) = 0$, $t(2^{-1/2}) - u(2^{-1/2}) > 0$, and

$$t''(\theta) - u''(\theta) = \frac{1}{2}\log\left(\frac{1 - \theta^2}{\theta^2}\right) \leq 0 .$$

On the remaining interval $2^{-1/2} < \theta < 1$ we have $t''(\theta) - u''(\theta) > 0$. Now we use $t(1) - u(1) = t'(1-) - u'(1-) = 0$ to show that

$$t(\theta) - u(\theta) = \int_\theta^1 (\xi - \theta)\{t''(\xi) - u''(\xi)\}\, d\xi .$$

This proves (2.3). The second inequality follows from the series expansion

$$u(\theta) = \frac{1}{2}\theta^2 \log\left(\frac{1}{4\theta}\right) + \frac{3}{4}\theta^2 - \sum_{n=2}^{\infty} \{4(2n-1)(n^2 - 2n + 2)\}^{-1}\, \theta^{2n} .$$

Of course (2.3) is sharp when θ is near one while (2.4) is most useful for values of θ near zero.

Theorem 1. *There exist polynomials* $P_1(X), \ldots, P_L(X)$ *in* $O_k[X]$ *which form a basis for the space S, defined by* (2.1), *and satisfy*

$$\sum_{\ell=1}^{L} \log h(P_\ell) \leqslant (N - M) \sum_{j=1}^{J} m_j r_j \log h_1(\alpha_j)$$

$$+ N^2 u\left(\frac{M}{N}\right) + (N - M) \log c_k . \tag{2.5}$$

Here $c_k = \left(\frac{2}{\pi}\right)^{s/d} |\Delta_k|^{1/2d}$, *where* s *is the number of complex places of* k *and* Δ_k *is the discriminant of* k.

There are alternative bounds which can be obtained from our method and in some situations these may be sharper than (2.5)

Corollary 2. *The polynomials* $P_1(X), \ldots, P_L(X)$ *in Theorem* 1 *also satisfy*

$$\sum_{\ell=1}^{L} \log h(P_\ell) \leqslant \sum_{j=1}^{J} (N - m_j r_j)\, m_j r_j \log h_1(\alpha_j)$$

$$+ N^2 \sum_{j=1}^{J} u\left(\frac{m_j r_j}{N}\right) + (N - M) \log c_k , \tag{2.6}$$

and

$$\sum_{\ell=1}^{L} \log h(P_\ell) \leqslant \sum_{j=1}^{J} (N - m_j) m_j r_j \log h_1(\alpha_j) \tag{2.7}$$

$$+ N^2 \sum_{j=1}^{J} r_j u(\frac{m_j}{N}) + (N - M)\log c_k .$$

For example, $u(\theta)$ is convex on $[0,(17)^{-1/2}]$ and $u(0) = 0$. It follows that

$$\sum_{j=1}^{J} r_j u(\frac{m_j}{N}) \leq \sum_{j=1}^{J} u(\frac{m_j r_j}{N}) \leq u(\frac{M}{N})$$

whenever $M \leq (17)^{-1/2} N$. Thus the bounds (2.6) and (2.7) are most useful when $\log h_1(\alpha_j)$ is small on average. In particular, if each α_j is a root of unity then (2.7) is clearly sharper than (2.5).

3. Preliminary lemmas.

Let $I \subset \{0,1,2,\ldots,N-1\}$ with $|I| = M$ and $I = \{n_1 < n_2 < \ldots < n_M\}$. We define polynomials $Q_I(\mathbf{x})$ and $P_I(\mathbf{x})$ in M variables as follows. We set

$$Q_I(\mathbf{x}) = \det \left(x_i^{n_j} \right) ,$$

where $i = 1,2,\ldots,M$ indexes rows and $j = 1,2,\ldots,M$ indexes columns, and

$$Q_I(\mathbf{x}) = V(\mathbf{x})P_I(\mathbf{x}) , \qquad (3.1)$$

where

$$V(\mathbf{x}) = \prod_{1\leq i<j\leq M} (x_j - x_i)$$

is the Vandermonde determinant. The polynomials $P_I(\mathbf{x})$ are the Schur polynomials (or S-functions) whose basic properties are given in Macdonald [5] and Stanley [10]. Clearly Q_I and P_I each have integer coefficients. In fact, P_I has nonnegative integer coefficients. This will be useful for our purposes and is contained in [5, p.42, equation (5.12)] and [10, p. 181, Theorem 10.1]. If we evaluate P_I at the vector all of whose coordinates are 1 we find that

$$(1!)(2!) \ldots((M - 1)!)\, P_I(1,1,\ldots,1) = \prod_{1\leq i<j\leq M} (n_j - n_i) , \qquad (3.2)$$

([5, pp. 27-28]).

Lemma 3. *If* M *and* N *are integers* $1 \leqslant M \leqslant N$, *then*

$$\log\{\sum_{|I|=M} P_I(1,1,\ldots,1)^2\} = \sum_{m=-(M-1)}^{M-1} (M - |m|)\log\left(\frac{N+m}{M+m} \right) \tag{3.3}$$

$$\leqslant 2N^2 u(\tfrac{M}{N}) \ .$$

Proof. Let $a(\mu) = \sum_{n=0}^{N-1} n^{\mu}$ (with $a(0) = N$) and let

$$D_{m-1} = \det\{a(\mu + \nu)\},$$

$\mu = 0,1,2,\ldots,m-1$, and $\nu = 0,1,2,\ldots,m-1$, be the corresponding Hankel determinant. Here we assume that $0 \leqslant m \leqslant N-1$ with $D_{-1} = 1$. If A denotes the N×M matrix

$$A = \left((i-1)^{j-1}\right),\ i = 1,2\ldots,N\ ,\ j = 1,2,\ldots,M\ ,$$

then $\det\{A^T A\} = D_{M-1}$. When we expand $\det\{A^T A\}$ using the Cauchy-Binet formula we find that

$$D_{M-1} = \sum_{|I|=M} \{ \prod_{1\leqslant i<j\leqslant M} (n_j - n_i)^2\} \tag{3.4}$$

$$= \{(1!)(2!)\ldots((M-1)!)\}^2 \sum_{|I|=M} P_I(1,1,\ldots,1)^2 \ .$$

The determinants D_{m-1} occur in the construction of orthonormal polynomials on the set $\{0,1,2,\ldots,N-1\}$. Specifically, the polynomials

$$p_m(x) = (D_{m-1}D_m)^{-1/2} \det \begin{pmatrix} a(0) & a(0) & \cdots & a(m) \\ a(1) & a(2) & \cdots & a(m+1) \\ \cdots & & & \cdots \\ a(m-1) & a(m) & \cdots & a(2m-1) \\ 1 & x & \cdots & x^m \end{pmatrix} \tag{3.5}$$

have degree m, where $m = 0,1,2,\ldots,N-1$, and are orthonormal on $\{0,1,2,\ldots,N-1\}$ (see Szego [11, p. 27, equation (22.6)]). That is,

$$\sum_{\xi=0}^{N-1} p_m(\xi)p_n(\xi) = 0 \tag{3.6}$$

if $0 \leq m < n \leq N-1$, and

$$\sum_{\xi=0}^{N-1} \{p_m(\xi)\}^2 = 1 . \tag{3.7}$$

From (3.5) we see that

$$p_m(x) = \left(\frac{D_{m-1}}{D_m} \right)^{1/2} x^m + \dots \tag{3.8}$$

for each m.

A second representation for the polynomials $p_m(x)$, $m = 0, 1, \dots, N-1$, occurs in a paper of Chebyshev [3]. More precisely, Chebyshev showed that the polynomials

$$t_m(x) = m! \Delta^m \binom{x}{m} \binom{x-N}{m} , \tag{3.9}$$

where Δ is the finite difference operator, satisfy (3.6) and

$$\sum_{\xi=0}^{N-1} \{t_m(\xi)\}^2 = (2m+1)^{-1} \prod_{\ell=-m}^{m} (N+\ell) , \tag{3.10}$$

(see [3,p. 552, equation 10] or [11, p. 34, equation (2.3.4)]). Let $T(m,N)$ denote the function on the right of (3.10). Polynomials p_m having degree m, positive leading coefficient and satisfying (3.6) and (3.7) are unique. It follows that $p_m(x) = T(m,N)^{-1/2} t_m(x)$ for each $m = 0,1,2,\dots,N-1$. From (3.9) we have

$$t_m(x) = \binom{2m}{m} x^m + \dots ,$$

and therefore

$$\log D_m - \log D_{m-1} = \log \left\{ T(m,N) \binom{2m}{m}^{-2} \right\} . \tag{3.11}$$

Finally, we sum (3.11) over m in the set $\{0,1,2,\dots,M-1\}$ to obtain

$$\log D_{M-1} = \sum_{m=0}^{M-1} \log \left\{ T(m,N) \binom{2m}{m}^{-2} \right\} . \tag{3.12}$$

Of course the right hand side of (3.12) is known, and when combined with (3.4) leads to the identity in (3.3).

To establish the upper bound in (3.3) we set

$$F(x) = \int_{-x}^{x} \log\left(\frac{N+y}{M+y}\right) dy$$

and note that $\log\left(\frac{N+y}{M+y}\right)$ is positive, decreasing and convex for $-M < y < M$. For $0 \leqslant m \leqslant M-1$ it follows that

$$\sum_{\ell=-m}^{m} \log\left(\frac{N+\ell}{M+\ell}\right) \leqslant \sum_{\ell=-m}^{m} \int_{\ell-1/2}^{\ell+1/2} \log\left(\frac{N+y}{M+y}\right) dy$$

$$= F\left(m+\frac{1}{2}\right) \leqslant \int_{m}^{m+1} F(x)dx .$$

Therefore we have

$$\sum_{m=-(M-1)}^{M-1} (M-|m|)\log\left(\frac{N+m}{M+m}\right) = \sum_{m=0}^{M-1} \sum_{\ell=-m}^{m} \log\left(\frac{N+\ell}{M+\ell}\right)$$

$$\leqslant \sum_{m=0}^{M-1} \int_{m}^{m+1} F(x)dx$$

$$= \int_{-M}^{M} (M-|x|) \log\left(\frac{N+x}{M+x}\right)dx$$

$$= 2N^2 u\left(\frac{M}{N}\right) .$$

Next we suppose that $\beta_1, \beta_2, \ldots, \beta_J$ are distinct nonzero algebraic numbers and $m_1, m_2, \ldots, m_2$ are positive integers with $M = \Sigma_{j=1}^{J} m_j$. Throughout the remainder of this section we work in the number field $K = \mathbf{Q}(\beta_1, \beta_2, \ldots, \beta_J)$. We associate an $m_j \times N$ matrix B_j with each β_j by setting

$$B_j = \left(\binom{n}{\mu}\beta_j^{n-\mu}\right)$$

where $\mu = 0,1,2,\ldots,m_j-1$ indexes rows and $n = 0,1,2,\ldots,N-1$ indexes columns. Then we assemble these into an $M \times N$ matrix

$$\mathcal{B} = \begin{pmatrix} B_1 \\ B_2 \\ \cdots \\ B_J \end{pmatrix} .$$

Theorem 4. *If* $1 \leqslant M < N$ *then the matrix* $\mathcal{B}$ *has rank* M *and satisfies*

$$\log H(\mathcal{B}) \leqslant (N-M)\sum_{j=1}^{J} m_j \log h_1(\beta_j) + N^2 u(\tfrac{M}{N}) \ . \tag{3.13}$$

Proof. If ν is a lattice point in $\mathbf{Z}^M$ with nonnegative coordinates ν_m, $m = 1,2,\ldots,M$, we define the partial differential operator D^{ν} by

$$D^{\nu} = \{(\nu_1!)\ \ldots\ (\nu_M!)\}^{-1} \left(\frac{\partial}{\partial x_1}\right)^{\nu_1} \cdots \left(\frac{\partial}{\partial x_M}\right)^{\nu_M} .$$

Then we fix a lattice point λ in $\mathbf{Z}^M$ by setting

$$\lambda^T = (0,1,2,\ldots,m_1-1,\ 0,1,2,\ldots m_2-1,\ 0,1,\ldots,\ 0,1,2,\ldots,m_J-1).$$

For each subset $I \subset \{0,1,2,\ldots,N-1\}$ with $I = \{n_1 < n_2 < \ldots < n_M\}$ we find that

$$D^{\lambda}Q_I(\mathbf{x}) = \det\left\{\binom{n_j}{\lambda_i} x_i^{n_j - \lambda_i}\right\} . \tag{3.14}$$

Now let $\mathbf{b}$ denote the vector

$$\mathbf{b}^T = (\beta_1,\beta_1,\ldots,\beta_1,\beta_2,\beta_2,\ldots,\beta_2,\ldots,\beta_J,\beta_J,\ldots,\beta_J) . \tag{3.15}$$

In (3.15) there are m_1 coordinates equal to β_1, followed by m_2 coordinates equal to β_2, and so forth, ending with m_J coordinates equal to β_J. When (3.14) is evaluated at $\mathbf{x} = \mathbf{b}$ we obtain

$$D^{\lambda}Q_I(\mathbf{x})\big|_{\mathbf{x}=\mathbf{b}} = \det(\mathcal{B}_I) .$$

By using (3.1) and the product rule for derivatives we have

$$D^{\lambda}Q_I(\mathbf{x}) = \sum_{0\leqslant\nu\leqslant\lambda} \{D^{\nu}V(\mathbf{x})\}\{D^{\lambda-\nu}P_I(\mathbf{x})\} . \tag{3.16}$$

When the right side of (3.16) is evaluated at $\mathbf{x} = \mathbf{b}$, each term in the sum with $0 \leqslant \nu < \lambda$ is zero. This can be seen as follows. Let ν be fixed with $0 \leqslant \nu < \lambda$ and let $s_j = \Sigma_{\ell=1}^{j} m_\ell$, with $s_0 = 0$. Then for some integer j, $0 \leqslant j \leqslant J$, we have

$$\sum_{i=s_{j-1}+1}^{s_j} \nu_i < \sum_{i=s_{j-1}+1}^{s_j} \lambda_i = \tfrac{1}{2}\, m_j(m_j-1) .$$

Thus there must be two distinct integers t_1 and t_2 such that $s_{j-1}+1 \leqslant t_1 < t_2 \leqslant s_j$ and $\nu_{t_1} = \nu_{t_2}$. It follows that

$$D^{\nu}V(\mathbf{x})\big|_{\mathbf{x}=\mathbf{b}} = 0 ,$$

since the determinant has two identical rows. This establishes our assertion and the identity (the confluent case of (3.1))

$$\det(\mathcal{B}_I) = \{D^{\lambda}V(\mathbf{x})\}\ P_I(\mathbf{x})\big|_{\mathbf{x}=\mathbf{b}} . \tag{3.17}$$

The first factor on the right of (3.17) can be explicitly given as

$$D^{\lambda}V(\mathbf{x})\big|_{\mathbf{x}=\mathbf{b}} = \prod_{1\leqslant i<j\leqslant J} (\beta_j - \beta_i)^{m_i m_j} = \gamma .$$

Clearly $\gamma \neq 0$ and in particular $\det(\mathcal{B}_I) \neq 0$ when $I = \{0,1,2,\ldots,M-1\}$. This shows that $\mathcal{B}$ has rank M.

Next we write

$$P_I(\mathbf{x})\big|_{\mathbf{x}=\mathbf{b}} = \tilde{P}_I(\beta_1,\beta_2,\ldots,\beta_J)$$

and note that from (3.1),

$$\deg_{\beta_j}(\tilde{P}_I) \leqslant m_j(N-M) .$$

By the product formula

$$H(\mathcal{B}) = \prod_{v\nmid\infty} \{ \max_I |\tilde{P}_I(\beta_1,\ldots,\beta_J)|_v\} \prod_{v|\infty} \{ \sum_I \| \tilde{P}_I(\beta_1,\ldots,\beta_J) \|_v^2 \}^{d_v/2d} , \tag{3.18}$$

where $d = [K:\mathbf{Q}]$. If $v \nmid \infty$ then

$$\max_I |\tilde{P}_I(\beta_1,\ldots,\beta_J)|_v \leqslant \prod_{j=1}^{J} (\max\{1,|\beta_j|_v\})^{m_j(N-M)} .$$

If $v \mid \infty$ we use the fact that P_I, and hence $\tilde{P}_I$, has nonnegative coefficients, so that

$$\sum_I \|\tilde{P}_I(\beta_1,\ldots,\beta_J)\|_v^2 \leqslant \sum_I \tilde{P}_I(\|\beta_1\|_v,\ldots,\|\beta_J\|_v)^2$$

$$\leq \sum_{I} \tilde{P}_I(1,1,\dots,1)^2 \prod_{j=1}^{J} (\max\{1, \| \beta_j \|_v\})^{2m_j(N-M)} .$$

Combining these estimates we find that

$$\log H(\mathcal{B}) \leq \sum_{j=1}^{J} m_j(N-M) \sum_{v \nmid \infty} \log^+ |\beta_j|_v$$

$$+ \sum_{j=1}^{J} m_j(N-M) \sum_{v|\infty} \frac{d_v}{d} \log^+ \| \beta_j \|_v + \sum_{v|\infty} \frac{d_v}{2d} \log\{ \sum_{I} \tilde{P}_I(1,1,\dots,1)^2 \}$$

$$= (N-M) \sum_{j=1}^{J} m_j \log h_1(\beta_j) + \frac{1}{2} \log\{ \sum_{I} \tilde{P}_I(1,1,\dots,1)^2 \} .$$

Of course $\tilde{P}_I\ (1,1,\dots,1) = P_I(1,1,\dots,1)$ and therefore the proof is completed by appealing to Lemma 3.

4. Proof of Theorem 1.

With each algebraic number α_j we associate an $m_j \times N$ matrix A_j defined by

$$A_j = \left(\binom{n}{\mu} \alpha_j^{n-\mu} \right),$$

where $\mu = 0,1,2,\dots,m_j-1$ indexes rows and $n = 0,1,2,\dots,N-1$ indexes columns. Now let F be a number field which is a Galois extension of k and a Galois extension of each of the fields $k(\alpha_j)$, $j = 1,2,\dots,J$. If $G(F/k)$ is the Galois group of F over k, and $G(F/k(\alpha_j))$ is the Galois group of F over $k(\alpha_j)$, then $G(F/k(\alpha_j))$ is a subgroup of $G(F/k)$ having index $[k(\alpha_j) : k] = r_j$. Let $\sigma_1^{(j)}, \sigma_2^{(j)}, \dots, \sigma_{r_j}^{(j)}$ be a set of distinct representatives of the cosets of $G(F/k(\alpha_j))$in $G(F/k)$. For each $\sigma_i^{(j)}$ we write

$$\sigma_i^{(j)}(A_j) = \left(\binom{n}{\mu} \{\sigma_i^{(j)}(\alpha_j)\}^{n-\mu} \right)$$

and assemble these matrices into an $m_j r_j \times N$ matrix

$$\mathcal{A}_j = \begin{pmatrix} \sigma_1^{(j)}(A_j) \\ \sigma_2^{(j)}(A_j) \\ \cdots \\ \sigma_{r_j}^{(j)}(A_j) \end{pmatrix} .$$

Finally, we assemble the matrices A_j, $j = 1,2,\ldots,J$, into a $M \times N$ matrix (where $M = \Sigma_{j=1}^{J} m_j r_j$)

$$Z = \begin{pmatrix} A_1 \\ A_2 \\ \cdots \\ A_J \end{pmatrix} .$$

Now suppose that $\mathbf{x}$ is a nonzero vector in $(O_k)^N$ such that

$$A_j \mathbf{x} = 0, \; j = 1,2,\ldots,J . \tag{4.1}$$

Then form the polynomial

$$P(Y) = \sum_{n=0}^{N-1} x_n Y^n$$

having $\mathbf{x}$ as its vector of coefficients. Of course the equations (4.1) are equivalent to the vanishing conditions which we wish to impose on $P(Y)$, namely,

$$P^{(\mu)}(\alpha_j) = 0, \quad \mu = 0,1,2,\ldots,m_j-1 ,$$

for each j, $j = 1,2,\ldots,J$. Therefore we apply the general form of Siegel's Lemma given in [2] as Theorem 14. By that result there exist $N - M$ linearly independent vectors $\mathbf{x}_1,\mathbf{x}_2,\ldots,\mathbf{x}_{N-M}$ in $(O_k)^N$ which satisfy (4.1) and

$$\sum_{\ell=1}^{N-M} \log h(\mathbf{x}_\ell) \leq (N - M) \log c_k + \log H(Z) . \tag{4.2}$$

The matrix Z has precisely the same structure as the matrix $\mathcal{B}$ of Theorem 3, but now the set $\{\beta_1,\ldots,\beta_J\}$ used to construct $\mathcal{B}$ consists of the $\Sigma_{j=1}^{J} r_j$ distinct nonzero algebraic numbers in the set

$$\{\sigma_i^{(j)}(\alpha_j) : i = 1,2,\ldots,r_j \text{ and } j = 1,2,\ldots,J\} .$$

In the matrix Z the integers m_j correspond to $\sigma_i^{(j)}(\alpha_j)$ for each value of the index i, $i = 1,2,\ldots,r_j$, and so $M = \Sigma_{j=1}^{J} m_j r_j$. Thus by Theorem 3 we have

$$\log H(Z) \leq (N-M)\sum_{j=1}^{J}\sum_{i=1}^{r_j} m_j \log h_1(\sigma_i^{(j)}(\alpha_j)) + N^2 u(\tfrac{M}{N})$$

$$= (N-M)\sum_{j=1}^{J} m_j r_j \log h_1(\alpha_j) + N^2 u(\tfrac{M}{N}) .$$

This completes our proof of Theorem 1.

To establish Corollary 2 we use the inequality

$$\log H(Z) \leq \sum_{j=1}^{J} \log H(A_j)$$

$$\leq \sum_{j=1}^{J}\sum_{i=1}^{r_j} \log H(\sigma_i^{(j)}(A_i))$$

$$= \sum_{j=1}^{J} r_j \log H(A_j) .$$

This follows from [2, equation (2.6)]. We apply Theorem 3 to obtain an upper bound for $\log H(A_j)$. When this upper bound is combined with (4.2) and (4.3) we find that the inequality (2.6) holds. In a similar manner we deduce (2.7) from (4.3) by using Theorem 3 to bound $\log H(A_j)$.

5. Polynomials which vanish at 1.

We apply Theorem 1 in the special case $k = \mathbf{Q}$, $J = 1$, $\alpha_1 = 1$ and so $r_1 = 1$. If follows that for $1 \leq m_1 < N$ there exist $L = N - m_1$ linearly independent polynomials $P_1, P_2, \ldots, P_L$ in $\mathbf{Z}[X]$ having degree less than N, vanishing at 1 with multiplicity at least m_1 and satisfying

$$\sum_{\ell=1}^{L} \log h(P_\ell) \leq N^2 u(\tfrac{m_1}{N})$$

If we arrange the polynomials P_ℓ in order of increasing height we find that

$$N^{-1}\log h(P_1) \leq \left(1 - \frac{m_1}{N}\right)^{-1} u(\tfrac{m_1}{N}) . \tag{5.1}$$

It will be convenient to combine (5.1) and (2.4) as follows.

Corollary 5. *Let* m_1 *and* N *be integers with* $1 \leq m_1 < N$. *Then there exists a nontrivial polynomial* $P_1(X)$ *in* **Z**[X] *having degree less than* N, *vanishing at* 1 *with multiplicity at least* m_1 *and satisfying*

$$N^{-1}\log h(P_1) \leq \tfrac{1}{2}\left(\frac{m_1}{N}\right)^2 \log\left(\frac{N}{m_1}\right)(1 + o(1)). \tag{5.2}$$

Here o(1) *denotes a function of* m_1 *and* N *which tends to zero as* $N \to \infty$ *in such a way that* $m_1/N \to 0$.

We note that the left and right hand sides of (3.13) are asymptotically equal as $N \to \infty$ in the special case $k = \mathbf{Q}$, $J = 1$, and $\alpha_1 = 1$, which leads to Corollary 5. For this reason we expect the bound (5.2) to be rather sharp. In fact, under somewhat more restrictive conditions, we will show that a polynomial of low height cannot vanish at 1 with too high a multiplicity.

Theorem 6. *Let* G(X) *be a nontrivial polynomial in* **Q**[X] *having degree less than* N *and vanishing at* 1 *with multiplicity* e_1. *If* $N \to \infty$ *and* $e_1 \to \infty$ *in such a way that*

$$\frac{e_1}{N} \to 0 \quad \text{and} \quad \frac{(N \log N)^{1/2}}{e_1} \to 0, \tag{5.3}$$

then

$$\frac{1}{4}\left(\frac{e_1}{N}\right)^2 (1 + o(1)) \leq N^{-1}\log h(G). \tag{5.4}$$

Proof. We will work over the field $k = \mathbf{Q}$. Let F_m denote the m-th cyclotomic polynomial. If g(X) is a nontrivial polynomial in **Q**[X] with $F_m \nmid g$ we define

$$L_m(g) = \sum_{v \nmid \infty} \log|g|_v + \phi(m)^{-1}\log|\mathrm{Res}\,\{F_m, g\}|_\infty .$$

Here Res $\{F_m, g\}$ is the resultant of F_m and g, ϕ is Euler's ϕ-function, and $|g|_v$ is the absolute value $|\ |_v$ applied to the vector of coefficients of g. It is clear from the product formula that

$$L_m(\beta g) = L_m(g)$$

for each $\beta \neq 0$ in **Q** . Also, we have

$$L_m(g_1 g_2) = L_m(g_1) + L(g_2) \ . \tag{5.5}$$

If $g \in$ **Z**[X] is irreducible in **Z**[X] then

$$\sum_{v \nmid \infty} \log |g|_v = 0,$$

the resultant of F_m and g is a nonzero integer, and therefore

$$L_m(g) \geqslant 0 \ . \tag{5.6}$$

Since an arbitrary polynomial g(x) in **Q**[X] can be factored into a rational number times a finite product of irreducible polynomials in **Z**[X] it follows that (5.6) holds generally.

Now suppose that G(X) is a nontrivial polynomial in **Q**[X] having degree less than N and vanishing at 1 with multiplicity e_1. Then we may write

$$G(X) = \prod_{n=1}^{\infty} \{F_n(X)\}^{e_n} Q(X) \tag{5.7}$$

where each e_n is a nonnegative integer and Q is not divisible by any cyclotomic polynomials. Let

$$G_m(X) = (e_m!)^{-1} \left(\frac{d}{dx} \right)^{e_m} G(X)$$

so that $F_m \nmid G_m$ and G_m is divisible by (X-1) with multiplicity greater than or equal to $(e_1 - e_m)^+$. Applying (5.5) and (5.6) we find that

$$\begin{aligned} L_m(G_m) &\geqslant (e_1 - e_m)^+ L_m((X - 1)) \\ &\geqslant (e_1 - e_m) \phi(m)^{-1} \log |F_m(1)|_\infty \ . \end{aligned}$$

For $m \geqslant 2$ we have

$$F_m(1) = \lim_{x \to 1} \prod_{d | m} (X^d - 1)^{\mu(m/d)}$$

$$= \lim_{x\to 1} \prod_{d|m} \left(\frac{X^d - 1}{X - 1} \right)^{\mu(m/d)}$$

$$= \prod_{d|m} (d)^{\mu(m/d)}$$

$$= \exp \{\Lambda(m)\} .$$

In this way we obtain the lower bound

$$L_m(G_m) \geq (e_1 - e_m)\, \phi(m)^{-1} \Lambda(m) \tag{5.8}$$

for $m \geq 2$.

Let $G(X) = \Sigma_{n=0}^{N-1} a_n X^n$ so that

$$G_m(X) = \sum_{n=e_m}^{N-1} a_n \binom{n}{e_m} X^{n-e_m} .$$

If $v \nmid \infty$ then if follows that

$$|G_m|_v \leq |G|_v . \tag{5.9}$$

At the infinite place (extended to **C**) we have

$$\begin{aligned} |G_m(\zeta_m)|_\infty &\leq \max_n |\alpha_n|_\infty \sum_{n=e_m}^{N-1} \binom{n}{e_m} \\ &= |G|_\infty \binom{N}{e_m + 1} , \end{aligned} \tag{5.10}$$

where ζ_m is a primitive m-th root of unity. Combining (5.8), (5.9) and (5.10), we obtain the inequality

$$e_1 \Lambda(m) \leq \phi(m) \log h(G) + \phi(m) \log\binom{N}{e_m + 1} + e_m \Lambda(m) . \tag{5.11}$$

Finally, we set m equal to the prime number p and use the bound

$$\log\binom{N}{M} \leq N\psi\left(\frac{M}{N}\right) , \tag{5.12}$$

for binomial coefficients, where $\psi(\theta) = -\theta \log \theta - (1-\theta) \log(1-\theta)$, $0 < \theta < 1$. The inequality (5.12) is most easily proved by noting that

$$\binom{N}{M} = \left| \frac{1}{2\pi i} \int_{|z|=\rho} \frac{(1+z)^N}{z^{M+1}} dz \right|$$

$$\leq (1 + \rho^{-1})^M (1 + \rho)^{N-M}$$

and then choosing $\rho = M/(N-M)$. Thus the polynomial $G(X)$ having the form (5.7) must satisfy

$$e_1 \log p \leq (p - 1)\log h(G) + (p - 1)N\psi\left(\frac{e_p + 1}{N}\right) + e_p \log p \tag{5.13}$$

for each prime number p.

To complete the proof we set

$$x = \left(\frac{4N}{e_1}\right) \log\left(\frac{4N}{e_1}\right),$$

$$\theta(x) = \sum_{p \leq x} \log p ,$$

$$s(x) = \sum_{p \leq x} (p - 1) .$$

We then sum both sides of (5.13) over the set of primes p less than or equal to x. We also use the fact that ψ is concave and increasing on $(0, 1/2]$, and the obvious inequality

$$\sum_p e_p(p-1) \leq N .$$

It follows that

$$e_1 \theta(x) \leq s(x) \log h(G) + N \sum_{p \leq x} (p - 1)\, \psi\left(\frac{e_p + 1}{N}\right) + N$$

$$\leq s(x)\log h(G) + N\, s(x)\psi\left\{(N\, s(x))^{-1} \sum_{p \leq x} e_p(p - 1) + N^{-1}\right\} + N$$

$$\leq s(x) \log h(G) + N\, s(x)\, \psi\{s(x)^{-1} + N^{-1}\} + N ,$$

and therefore

$$\left(\frac{e_1}{N}\right)\left(\frac{\theta(x)}{s(x)}\right) \leq N^{-1}\log h(G) + \psi\{s(x)^{-1} + N^{-1}\} + s(x)^{-1} . \tag{5.14}$$

By the prime number theorem we have

$$\frac{\theta(x)}{x} \to 1 \quad \text{and} \quad \frac{2s(x)\log x}{x^2} \to 1$$

as $x \to \infty$. It follows easily that

$$\frac{\theta(x)}{s(x)} = \frac{1}{2}\left(\frac{e_1}{N}\right)(1 + o(1)) \tag{5.15}$$

and

$$s(x) = 8\left(\frac{N}{e_1}\right)^2 \log\left(\frac{N}{e_1}\right)(1 + o(1)) \tag{5.16}$$

as $e_1/N \to 0$. We also find that

$$\frac{s(x)}{N} \to 0 \text{ as } \frac{(N \log N)^{1/2}}{e_1} \to 0 . \tag{5.17}$$

Using (5.16) and (5.17) we conclude that

$$\psi\{s(x)^{-1} + N^{-1}\} + s(x)^{-1} = \frac{1}{4}\left(\frac{e_1}{N}\right)^2 (1 + o(1)) . \tag{5.18}$$

When (5.14), (5.15) and (5.18) are combined we obtain exactly the statement of the Theorem.

The argument used to prove Theorem 6 suggests that a polynomial of low height which vanishes at 1 with high multiplicity must also vanish at primitive p-th roots of unity, at least for primes p which are not too large. This type of *automatic vanishing* can be made explicit in various ways. Here we provide a simple result which follows easily from (5.11).

Theorem 7. *Let* m_1 *and* N *be integers with* $1 \leq m_1 < N$ *and let* $P_1(X)$ *be a nontrivial polynomial in* $\mathbf{Z}[X]$ *which satisfies the conclusion of Corollary* 5. *If* $N \to \infty$ *and* $m_1 \to \infty$ *in such a way that*

$$\frac{m_1}{N} \to 0 \text{ and } \frac{(N \log N)^{1/2}}{m_1} \to 0 , \tag{5.19}$$

then $F_p \mid P_1$ *for all primes* p *such that*

$$p \leq \left(\frac{2N}{m_1}\right)(1 + o(1)) . \tag{5.20}$$

Proof. Let P_1 vanish at 1 with exactly the multiplicty e_1, so that $m_1 \leq e_1$. If $F_p \nmid P_1$ then we may apply (5.11) with $P_1 = G$, $p = m$ and $e_p = 0$. It follows that

$$\left(\frac{m_1}{N}\right)\left(\frac{\log p}{p-1}\right) \leq N^{-1}\log h(P_1) + N^{-1}\log N .$$

Using the hypothsis on the right of (5.19) and (5.2) we find that

$$\left(\frac{m_1}{N}\right)\left(\frac{\log p}{p-1}\right) \leq \frac{1}{2}\left(\frac{m_1}{N}\right)^2 \log\left(\frac{N}{m_1}\right)(1 + o(1)) . \qquad (5.21)$$

But (5.21) implies that

$$p \geq \left(\frac{2N}{m_1}\right)(1 + o(1)) .$$

Hence we must have $F_p \mid P_1$ for those primes p satisfying (5.20). This proves the Theorem.

References.

[1]. E. Bombieri, Lectures on the Thue Principle, these proceedings.

[2]. E. Bombieri and J.D. Vaaler, On Siegel's Lemma, *Invent. Math.* **73**,(1983),11-32.

[3]. P.L. Chebyshev, Sur l'interpolation, *Zapiski Akademii Nauk*, vol.**4**, Supplement no. 5, (1864). *Oeuvres*, vol. **1**, pp. 539-560.

[4]. E. Dobrowolski, On a question of Lehmer and the number of irreducible factors of a polynomial, *Acta. Arith.*, **34**, (1979),391-401.

[5]. I.G. Macdonald, Symmetric Functions and Hall Polynomials, (1979),Oxford U. Press.

[6]. M. Mignotte, Approximation des nombres algébriques par des nombres algébriques de grande degré, *Ann. Fac. Sci. Toulouse Math.*(5) 1 (1979), no. 2, 165-170.

[7]. M. Mignotte, Estimations élémentaires effectives sur les nombres algébriques, Journées Arithemétiques, 1980; (ed. J.V. Armitage) London Math. Soc. Lecture Note Ser. 56, Cambridge U. Press, (1982).

[8]. W.M. Schmidt, Diophantine Approximation, Lecture Notes in Math. 785, Springer-Verlag, New York, 1980.

[9]. C.L. Siegel, Uber einige Anwendungen diophantisher Approximationen, *Abh. der Preuss. Akad. der Wissenschaften. Phys.-math. Kl.* (1929), Nr. **1** (=Ges. Abh., I, pp. 209-226).

[10]. R.P. Stanley, Theory and application of plane partitions I, II, *Studies Appl Math.* **50**, (1971), 167-188 and 259-279.

[11]. G. Szego, Orthogonal Polynomials, AMS Colloq. Pub. **23**, 4-th ed., Providence, (1975).

E. Bombieri,
Institute for Advanced Study,
Princeton, N.J. 08540, U.S.A.

J. D. Vaaler,
University of Texas,
Austin, TX. 78712, U.S.A.

ON IRREGULARITIES OF DISTRIBUTION AND APPROXIMATE EVALUATION OF CERTAIN FUNCTIONS II

W.W.L. Chen

1. Introduction.

Let $\mathcal{U} = [0,1]$. Suppose that g is a Lebesgue-integrable function, not necessarily bounded, in $\mathcal{U}^2$, and that h is any function in $\mathcal{U}^2$. Let $\mathcal{P} = \mathcal{P}(N)$ be a distribution of N points in $\mathcal{U}^2$ such that $h(\mathbf{y})$ is finite for every $\mathbf{y} \in \mathcal{P}$. For $\mathbf{x} = (x_1,x_2)$ in $\mathcal{U}^2$, let $\mathcal{B}(\mathbf{x})$ denote the rectangle consisting of all $\mathbf{y} = (y_1,y_2)$ in U^2 satisfying $0 \leqslant y_1 < x_1$ and $0 \leqslant y_2 < x_2$, and write

$$Z[\mathcal{P};h:\mathcal{B}(\mathbf{x})] = \sum_{\mathbf{y} \in \mathcal{P} \cap \mathcal{B}(\mathbf{x})} h(\mathbf{y}). \tag{1}$$

Let μ denote the Lebesgue measure in U^2, and write

$$D[\mathcal{P};h;g;\mathcal{B}(\mathbf{x})] = Z[\mathcal{P};h;\mathcal{B}(\mathbf{x})] - N \int_{\mathcal{B}(\mathbf{x})} g(\mathbf{y})d\mu . \tag{2}$$

The aim of this paper is to use a variation of the ideas in Chen [1] on Halász's method in [2] to prove

Theorem 1. *Suppose that* g *is a Lebesgue-integrable function in* $\mathcal{U}^2$. *Suppose further that there exists a measurable subset* S *of* $\mathcal{U}^2$ *such that* $\mu(S) > 0$ *and* $g(\mathbf{y}) \neq 0$ *for every* $\mathbf{y} \in S$. *Then there exists a positive constant* $c_1 = c_1 = c_1(g)$ *such that for every distribution* $\mathcal{P}$ *of* N *points in* $\mathcal{U}^2$ *and for every function* h *bounded in* $\mathcal{U}^2$,

$$\sup_{\mathbf{x} \in \mathcal{U}^2} | D[\mathcal{P};h;g;\mathcal{B}(\mathbf{x})]| > c_1(g)(\log N).$$

Note that this is an improvement of the case K = 2 of Corollary 1 of Chen [1] as well as a generalization of Theorem 2 of Schmidt [4] and Theorem 2 of Halász [2].

As an application of Theorem 1, we shall consider functions in $\mathcal{U}^2$ of the following type.

Definition. We denote by $\mathcal{F}$ the class of all functions of type

$$C + \int_{\mathcal{B}(\mathbf{x})} g(\mathbf{y})d\mu$$

in $\mathcal{U}^2$, where C is a real constant, and where g satisfies the hypotheses of Theorem 1.

We can show that functions in $\mathcal{F}$ cannot be approximated very well by certain simple functions.

Definition. By an M-simple function in $\mathcal{U}^2$, we mean a function ϕ, defined by

$$\phi(\mathbf{x}) = \sum_{i=1}^{M} m_i \chi_{\mathcal{B}_i}(\mathbf{x})$$

for all $\mathbf{x} \in \mathcal{U}^2$, where, for each $i = 1,\ldots, M$, $\mathcal{B}_i$ denotes a rectangle in $\mathcal{U}^2$ of the type $(u_1^{(i)}, u_1^{(i)} + v_1^{(i)}] \times (u_2^{(i)}, u_2^{(i)} + v_2^{(i)}]$, $\chi_{\mathcal{B}_i}$ denotes the characteristic function of the rectangle $\mathcal{B}_i$, and the coëfficients m_i are real.

As in Sec.2 of [1], the following is an easy consequence of Theorem 1.

Theorem 2. *Suppose that* $f \in \mathcal{F}$. *Then there exists a positive constant* $c_2 = c_2(f)$ *such that for every* M-*simple function* ϕ *in* $\mathcal{U}^2$,

$$\sup_{\mathbf{x} \in \mathcal{U}^2} | \phi(\mathbf{x}) - f(\mathbf{x}) | > c_2(f)M^{-1}(\log M).$$

2. An outline of the method of Halász.

Following the method of Halász [2], corresponding to every function of the type $D(\mathbf{x}) = D[\mathcal{P};h;g;\mathcal{B}(\mathbf{x})]$, where $\mathcal{P}$ is a distribution of N points in $\mathcal{U}^2$ (N being sufficiently large), we construct an auxiliary function $F(\mathbf{x}) = F[\mathcal{P};h;g;\mathbf{x}]$ such that

$$\int_{\mathcal{U}^2} |F(\mathbf{x})| \, d\mu \leqslant 2 \; ; \tag{3}$$

Also, there exists a positive constant $c_3 = c_3(g)$ such that

$$\int_{\mathcal{U}^2} F(\mathbf{x})D(\mathbf{x}) \, d\mu \;\geqslant\; c_3(g)(\log N). \tag{4}$$

Theorem 1 follows, on combining (3) and (4) and noting that

$$\int_{\mathcal{U}^2} F(\mathbf{x})D(\mathbf{x}) \, d\mu \;\leqslant\; \sup_{\mathbf{x} \in \mathcal{U}^2} |D(\mathbf{x})| \int_{\mathcal{U}^2} |F(\mathbf{x})| \, d\mu.$$

It remains to establish the existence of such a constant $c_3(g)$ and function $F[\mathcal{P};h;g;\mathbf{x}]$.

Some difficulty arises, as in [1], from the assumption that g can take different signs in any region. We therefore have to look for regions in $\mathcal{U}^2$ where g is "predominantly positive" or "predominantly negative". We deal with the remaining "undesirable" regions by letting F vanish there. On the other hand, the function F is more complicated than the one used by Roth in [3]. For Halász's method to succeed, we also need to make sure that in the regions we have chosen, the value of g is not "too large" "too often". We discuss this in the next section.

3. Preparation for the proof of Theorem 1.

Let g be a Lebesgue-integrable function in $\mathcal{U}^2$. Suppose that $\mathcal{S}$ is a measurable subset of $\mathcal{U}^2$ satisfying $\mu(\mathcal{S}) > 0$ and $g(\mathbf{y}) \neq 0$ for every $\mathbf{y} \in \mathcal{S}$. Then, replacing g by −g if necessary, we may assume, without loss of generality, that there exist three positive constants $c_4 = c_4(g)$, $c_5 = c_5(g)$ and $c_6 = c_6(g)$ and a subset $\mathcal{S}_1 \subset \mathcal{S}$ such that

$$\mu(\mathcal{S}_1) = c_4(g) \tag{5}$$

and

$$c_5(g) \leqslant g(\mathbf{y}) \leqslant c_6(g) \qquad \text{for every } \mathbf{y} \in \mathcal{S}_1 . \tag{6}$$

Consider the function

$$g^{-}(\mathbf{y}) = \max\{-g(\mathbf{y}), 0\}. \tag{7}$$

Then g^{-} is Lebesgue-integrable in U^2. Let

$$c_7(g) = 2^{-5}c_4(g)c_5(g). \tag{8}$$

Consider also the function $|g|$. Then $|g|$ is Lebesgue-integrable in U^2. Let

$$c_8(g) = 2^{-2}c_6(g). \tag{9}$$

Then there exists a positive constant $c_9 = c_9(g)$ such that for every measurable set $E \subset U^2$ satisfying

$$\mu(E) \leqslant c_9(g) \leqslant 2^{-6}c_4(g), \tag{10}$$

we have

$$\int_E g^{-}(\mathbf{y})d\mu \leqslant c_7(g) \tag{11}$$

and

$$\int_E |g(\mathbf{y})| \, d\mu \leqslant c_8(g). \tag{12}$$

By an elementary box in U^2, we mean a set in U^2 of the type

$$[m_1 2^{-t_1}, (m_1+1)2^{-t_1}) \times [m_2 2^{-t_2}, (m_2+1)2^{-t_2}), \tag{13}$$

where m_1, m_2, t_1, t_2 are integers.

Consider the set S_1. Since S_1 is a measurable, there exists a finite union T^* of elementary boxes in U^2 such that

$$\mu(T^* \Delta S_1) \leqslant c_9(g),$$

where $T^* \Delta S_1$ denotes the symmetric difference of T^* and S_1. Hence if

$$E = T^* \setminus S_1, \tag{14}$$

then

$$\mu(E) \leqslant c_9(g). \tag{15}$$

Also, noting (10), we have that

$$\mu(T^*) \geqslant \tfrac{1}{2} c_4(g). \tag{16}$$

Since T^* is a finite union of elementary boxes of the type (13), there is one such elementary box with maximal t_1, and one with maximal t_2. Let T_1 and T_2 denote these maximal values of t_1 and t_2 respectively, and let

$$T = T_1 + T_2. \tag{17}$$

We can now introduce the auxiliary function $F[P;h;g;\mathbf{x}]$.

Any $x \in [0,1)$ can be written in the form

$$x = \sum_{i=0}^{\infty} \beta_i(x) 2^{-i-1},$$

where $\beta_i(x) = 0$ or 1 such that the sequence $\beta_i(x)$ does not end with 1,1,... . For $r = 0, 1, 2, \ldots$, let

$$R_r(x) = (-1)^{\beta_r(x)}.$$

Definition. By an r-interval, we mean an interval of the form $[m2^{-r},(m+1)2^{-r})$, where the integer m satisfies $0 \leqslant m < 2^r$.

Suppose that $\mathbf{r} = (r_1,r_2)$ is an ordered-pair of non-negative integers. Let

$$|\mathbf{r}| = r_1 + r_2.$$

For any $\mathbf{x} \in [0,1]^2$, let

$$R_{\mathbf{r}}(\mathbf{x}) = R_{r_1}(x_1)R_{r_2}(x_2).$$

Definition. By an $\mathbf{r}$-box in U^2, we mean a set of the form $I_1 \times I_2$, where I_1 is an r_1-interval and I_2 is an r_2-interval.

We shall consider a function of the type

$$F(\mathbf{x}) = \prod_{\substack{|\mathbf{r}|=n \\ r_1 \geqslant T_1 \\ r_2 \geqslant T_2}} \left(1 + \alpha f_{\mathbf{r}}(x)\right) - 1 , \tag{18}$$

where n is chosen in terms of N and where $\alpha = \alpha(g) < 1/2$ is a suitably chosen positive constant. In any $\mathbf{r}$-box $\mathcal{B}$, the function $f_{\mathbf{r}}$ is defined by

$$f_{\mathbf{r}}(\mathbf{x}) = \begin{cases} 0 & (\mathcal{B} \cap T^* = \phi \text{ or } \mathcal{B} \cap P \neq \phi); \\ -R_{\mathbf{r}}(\mathbf{x}) & (\mathcal{B} \subset T^* \text{ and } \mathcal{B} \cap P = \phi). \end{cases} \tag{19}$$

It is not difficult to prove

Lemma 1. *Suppose, for* $j = 1,\ldots, k$, *that* $\mathbf{r}^{(j)} = (r_1^{(j)}, r_2^{(j)})$ *satisfies* $|\mathbf{r}^{(j)}| = n$, $r_1^{(j)} \geqslant T_1$ *and* $r_2^{(j)} \geqslant T_2$. *Suppose further that* $\mathbf{r}^{(1)},\ldots,\mathbf{r}^{(k)}$ *are all different. Then if* $\mathbf{S} = (s_1, s_2)$, where

$$s_1 = \max_{1 \leqslant j \leqslant k} r_1^{(j)} \quad \text{and} \quad s_2 = \max_{1 \leqslant j \leqslant k} r_2^{(j)}, \tag{20}$$

then for any $\mathbf{S}$*-box* $\mathcal{B}$, *exactly one of the following three conditions hold:*

(i) $f_{\mathbf{r}^{(1)}} \cdots f_{\mathbf{r}^{(k)}} = R_{\mathbf{S}}$; *or*

(ii) $f_{\mathbf{r}^{(1)}} \cdots f_{\mathbf{r}^{(k)}} = -R_{\mathbf{S}}$; *or*

(iii) $f_{\mathbf{r}^{(1)}} \cdots f_{\mathbf{r}^{(k)}} = 0$.

Furthermore, (iii) holds in any $\mathbf{S}$*-box* $\mathcal{B}$ *where* $\mathcal{B} \cap P \neq \phi$.

4. Completion of the proof of Theorem 1.

We shall only prove (3) and (4) for

$$N > \frac{1}{8} c_4(g) 2^{2T}. \tag{21}$$

Let N satisfying (21) be given. Let n be a positive integer such that

$$\frac{1}{4}\, c_4(g)2^{n-1} < N \leqslant \frac{1}{4}\, c_4(g)2^n. \tag{22}$$

Then we have, in particular, that

$$n \geqslant 2T. \tag{23}$$

Note, first of all, that

$$\prod_{\substack{|\mathbf{r}| = n \\ r_1 \geqslant T_1 \\ r_2 \geqslant T_2}} \left(1 + \alpha f_{\mathbf{r}}(\mathbf{x})\right) = 1 + \alpha F_1(\mathbf{x}) + \sum_{k=2}^{n+1} \alpha^k F_k(\mathbf{x}), \tag{24}$$

where

$$F_1(\mathbf{x}) = \sum_{\substack{|\mathbf{r}| = n \\ r_1 \geqslant T_1 \\ r_2 \geqslant T_2}} f_{\mathbf{r}}(\mathbf{x}), \tag{25}$$

and where, for k = 2, . . ., n + 1,

$$F_k(\mathbf{x}) = \sum_{\substack{|\mathbf{r}^{(1)}|=\dots=|\mathbf{r}^{(k)}| = n \\ r_1^{(1)},\dots,\, r_1^{(k)} \geqslant T_1 \\ r_2^{(1)},\dots,\, r_2^{(k)} \geqslant T_2 \\ \mathbf{r}^{(i)}\neq \mathbf{r}^{(j)} \text{ if } i\neq j}} f_{\mathbf{r}^{(1)}}(\mathbf{x})\dots f_{\mathbf{r}^{(k)}}(\mathbf{x}) \tag{26}$$

In view of Lemma 1, for each k = 2,...,n+1,

$$\int_{U^2} F_k(\mathbf{x})d\mu = 0. \tag{27}$$

Furthermore, for each $f_{\mathbf{r}}$ in (19),

$$\int_{U^2} f_{\mathbf{r}}(\mathbf{x})d\mu = 0. \tag{28}$$

Since

$$|F(\mathbf{x})| \leqslant \prod_{\substack{|\mathbf{r}| = n \\ r_1 \geqslant T_1 \\ r_2 \geqslant T_2}} \left(1+\alpha f_{\mathbf{r}}(\mathbf{x})\right) + 1$$

for all $\mathbf{x} \in \mathcal{U}^2$, (3) follows easily from (24), (25), (26), (27), and (28).

On the other hand, from (18) and (24), we have

$$F(\mathbf{x}) = \alpha F_1(\mathbf{x}) + \sum_{k=2}^{n+1} \alpha^k F_k(\mathbf{x}). \tag{29}$$

Lemma 2. *Suppose that* $|\mathbf{r}| = n$, $r_1 \geq T_1$ *and* $r_2 \geq T_2$. *Then*

$$\int_{\mathcal{U}^2} f_{\mathbf{r}}(\mathbf{x})D(\mathbf{x})d\mu \geq 2^{-n-8} c_4(g)c_5(g)N. \tag{30}$$

Lemma 3. *We have, for* $k = 2,\ldots, n+1$, *that*

$$\left|\int_{\mathcal{U}^2} F_k(\mathbf{x})D(\mathbf{x})d\mu\right| \leq \sum_{r=0}^{n-k+1} \sum_{h=1}^{n-r} c_6(g)2^{-n-h-3} N\binom{h-1}{k-2}. \tag{31}$$

We can deduce (4) from Lemmas 2 and 3 as follows. There are exactly (n−T+1) choices of $\mathbf{r}$ satisfying the hypotheses of Lemma 2. It follows, from (25), (30) and (23), that

$$\begin{aligned}\int_{\mathcal{U}^2} F_1(\mathbf{x})D(\mathbf{x})d\mu &\geq (n-T+1)2^{-n-8}c_4(g)c_5(g)N \\ &\geq 2^{-n-9}c_4(g)c_5(g)Nn. \end{aligned}\tag{32}$$

On the other hand,

$$\begin{aligned}\left|\sum_{k=2}^{n+1} \alpha^k \int_{\mathcal{U}^2} F_k(\mathbf{x})D(\mathbf{x})d\mu\right| &\leq \sum_{k=2}^{n+1} \sum_{r=0}^{n-k+1} \sum_{h=1}^{n-r} \alpha^k c_6(g)2^{-n-h-3} N\binom{h-1}{k-2} \\ &= \sum_{r=0}^{n-1} \sum_{h=1}^{n-r} \sum_{k=2}^{h+1} \alpha^k c_6(g)2^{-n-h-3}N\binom{h-1}{k-2} \\ &\leq \alpha^2 c_6(g)Nn \sum_{h=1}^{\infty} 2^{-n-h-3} \sum_{k=0}^{h-1}\binom{h-1}{k}\alpha^k \\ &\leq \alpha^2 c_6(g)Nn2^{-n-3} \sum_{h=0}^{\infty} \left(\frac{1+\alpha}{2}\right)^h \\ &\leq \alpha^2 2^{-n-1} c_6(g)Nn. \end{aligned}\tag{33}$$

Let

$$\alpha = \frac{2^{-9}c_4(g)c_5(g)}{c_6(g)}.$$

Then clearly $\alpha < 1/2$. By (29), (31), (32) and (33),

$$\int_{U^2} F(\mathbf{x})D(\mathbf{x})d\mu \geqslant \alpha 2^{-n-9}c_4(g)c_5(g)Nn - \alpha^2 2^{-n-1}c_6(g)Nn$$

$$= \alpha 2^{-n-10}c_4(g)c_5(g)$$

$$= c_{10}(g)2^{-n}Nn,$$

where $c_{10}(g)$ is a positive constant. This proves (4), in view of (22).

It remains to prove Lemmas 2 and 3, the proofs of which are based on

Lemma 4. *Suppose that* B *is an* $\mathbf{S}$-*box in* U^2. *If* $B \cap P = \phi$, *then*

$$\int_B R_{\mathbf{S}}(\mathbf{x})D(\mathbf{x})d\mu = -N\int_B K_B(\mathbf{y})g(\mathbf{y})d\mu\ , \tag{34}$$

where, writing $\mathbf{S} = (s_1,s_2)$ *and*

$$B = [m_1 2^{-s_1},\ (m_1+1)2^{-s_1}) \times [m_2 2^{-s_2},(m_2+1)2^{-s_2}), \tag{35}$$

we have

$$K_B(\mathbf{y}) = \left(2^{-s_1-1} - \left|y_1-(m_1+\tfrac{1}{2})2^{-s_1}\right|\right)\left(2^{-s_2-1} - \left|y_2-(m_2+\tfrac{1}{2})2^{-s_2}\right|\right). \tag{36}$$

Proof. The proof is essentially a slight modification of part of the proof of Lemma 2 of [1]. Let

$$B' = [m_1 2^{-s_1},(m_1+1/2)2^{-s_1}) \times [m_2 2^{-s_2},\ (m_2+1/2)2^{-s_2}).$$

Then

$$\int_B R_{\mathbf{S}}(\mathbf{x})D(\mathbf{x})d\mu =$$

$$\int_{\mathcal{B}'} \sum_{\alpha_1=0}^{1} \sum_{\alpha_2=0}^{1} (-1)^{\alpha_1+\alpha_2} D\big((x_1+\alpha_1 2^{-s_1-1}, x_2+\alpha_2 2^{-s_2-1})\big)d\mu \,. \qquad (37)$$

In view of (1), the sum

$$\sum_{\alpha_1=0}^{1} \sum_{\alpha_2=0}^{1} (-1)^{\alpha_1+\alpha_2} Z[\mathcal{P};h;\mathcal{B}((x_1+\alpha_1 2^{-s_1-1}, x_2+\alpha_2 2^{-s_2-1}))]$$

$$= \sum_{\mathbf{y} \in \mathcal{P} \cap \mathcal{B}^*(\mathbf{x})} h(\mathbf{y}), \qquad (38)$$

where $\mathcal{B}^*(\mathbf{x}) = [x_1, x_1+2^{-s_1-1}) \times [x_2, x_2+2^{-s_2-1}) \subset \mathcal{B}$ for every $\mathbf{x} \in \mathcal{B}'$. Hence the sum (38) vanishes, and so, in view of (2), we have that (37) is equal to

$$-N \int_{\mathcal{B}'} \left(\sum_{\alpha_1=0}^{1} \sum_{\alpha_2=0}^{1} (-1)^{\alpha_1+\alpha_2} \int_0^{x_1+\alpha_1 2^{-s_1-1}} \int_0^{x_2+\alpha_2 2^{-s_2-1}} g(\mathbf{y})d\mu \right) d\mu$$

$$= -N \int_{\mathcal{B}'} \left(\int_{x_1}^{x_1+2^{-s_1-1}} \int_{x_2}^{x_2+2^{-s_2-1}} g(\mathbf{y})d\mu \right) d\mu$$

$$= -N \int_{\mathcal{B}} K_{\mathcal{B}}(\mathbf{y})g(\mathbf{y})d\mu$$

on interchanging the order of integration. This completes the proof of Lemma 4.

Proof of Lemma 2. We decompose the integral (30) into integrals over $\mathbf{r}$ -boxes. We shall say the an $\mathbf{r}$- box $\mathcal{B}$ is "good" if it is contained in $\mathcal{T}^*$ and does not contain any point of $\mathcal{P}$. By (19), $f_{\mathbf{r}} = 0$ in any $\mathbf{r}$- box that is not "good". Hence by (34),

$$\int_{\mathcal{U}^2} f_{\mathbf{r}}(\mathbf{x})D(\mathbf{x})d\mu = - \sum_{\mathcal{B}\ \text{"good"}} \int_{\mathcal{B}} R_{\mathbf{r}}(\mathbf{x})D(\mathbf{x})d\mu$$

$$= N \sum_{\mathcal{B}\ \text{"good"}} \int_{\mathcal{B}} K_{\mathcal{B}}(\mathbf{y})g(\mathbf{y})d\mu \,.$$

For any $\mathbf{r}$ satisfying the hypotheses of Lemma 2, there are at least $(\frac{1}{2} c_4(g)2^n - N)$ "good" $\mathbf{r}$-boxes. It follows, by (6), (14), (7), (11), (22), (8) and (10), that

$$\int_{U^2} f_{\mathbf{r}}(\mathbf{x})D(\mathbf{x})d\mu$$

$$\geq Nc_5(g) \sum_{B \text{ "good"}} \int_B K_B(\mathbf{y})d\mu - N2^{-n-2}c_5(g)\mu(E) - N2^{-n-2}\int_E g^-(\mathbf{y})d\mu$$

$$\geq Nc_5(g)2^{-2n-4}\left(\frac{1}{2} c_4(g)2^n - N\right) - N2^{-n-2}c_5(g)\mu(E) - N2^{-n-2}c_7(g)$$

$$\geq N2^{-n-2}(2^{-4}c_4(g)c_5(g) - c_5(g)\mu(E) - c_7(g))$$

$$= N2^{-n-2}\left(2^{-5}c_4(g)c_5(g) - c_5(g)\mu(E)\right)$$

$$\geq N2^{-n-8}c_4(g)c_5(g).$$

This completes the proof of Lemma 2.

Proof of Lemma 3. Consider

$$\int_{U^2} f_{\mathbf{r}(1)}(\mathbf{x})\ldots f_{\mathbf{r}(k)}(\mathbf{x})D(\mathbf{x})d\mu.$$

Let $\mathbf{S} = (s_1, s_2)$ be defined by (20). Let B by an $\mathbf{S}$-box in U^2. Then by Lemmas 1 and 4,

$$\left|\int_B f_{\mathbf{r}(1)}(\mathbf{x})\ldots f_{\mathbf{r}(k)}(\mathbf{x})D(\mathbf{x})d\mu\right| \leq N\int_B K_B(\mathbf{y})|g(\mathbf{y})|d\mu.$$

Hence by (19), (14), (6), (35), (36), (12), (9) and noting that there are $2^{|\mathbf{S}|}$ S-boxes in U^2,

$$\left|\int_{U^2} f_{\mathbf{r}(1)}(\mathbf{x})\ldots f_{\mathbf{r}(k)}(\mathbf{x})D(\mathbf{x})d\mu\right|$$

$$\leq Nc_6(g)\sum_{B\ T^*}\int_B K_B(\mathbf{y})d\mu + N2^{-|\mathbf{S}|-1}\int_E |g(\mathbf{y})|d\mu$$

$$\leq Nc_6(g)2^{-|\mathbf{S}|-4} + Nc_8(g)2^{-|\mathbf{S}|-2}$$

$$= Nc_6(g)2^{-|\mathbf{S}|-3}.$$

If we use the convention $r_1^{(1)} < \dots < r_1^{(k)}$, we have, writing $h = r_1^{(k)} - r_1^{(1)}$,

$$\left|\int_{U^2} f_{\mathbf{r}^{(1)}}(\mathbf{x})\dots f_{\mathbf{r}^{(k)}}(\mathbf{x})D(\mathbf{x})d\mu\right| \leqslant Nc_6(g)2^{-n-h-3}.$$

Now (31) follows on noting that once $r_1^{(1)}$ and h are chosen, there are exactly $\binom{h-1}{k-2}$ ways of choosing (k-2) integers in the interval $(r_1^{(1)}, r_1^{(1)} + h)$. This completes the proof of Lemma 3.

References.

[1] W. W. L. Chen, On irregularities of distribution and approximate evaluation of certain functions, to appear in *Quarterly Journal of Mathematics (Oxford)* 1985.

[2] G. Halász, On Roth´s method in the theory of irregularities of point distributions, Recent progress in analytic number theory, vol. 2, pp. 79-94 (Academic Press, London, 1981).

[3] K. F. Roth, On irregularities of distribution, *Mathematika*, **1** (1954), 73-79.

[4] W. M. Schmidt, Irregularities of distribution VII, *Acta Arith.*, **21** (1972), 45-50.

W. Chen
Huxley Building,
Imperial College,
London SW7, U.K.

SIMPLE ZEROS OF THE ZETA-FUNCTION OF A QUADRATIC NUMBER FIELD, II

J.B. Conrey, A. Ghosh and S.M. Gonek

1. Introduction.

Let K be a fixed quadratic extension of Q and write $\zeta_K(s)$ for the Dedekind zeta-function of K, where $s = \sigma + it$. It is well-known, and easy to prove, that the number $N_K(T)$ of zeros of $\zeta_K(s)$ in the region $0 < \sigma < 1$, $0 < t \leq T$ satisfies

$$N_K(T) \sim \frac{T}{\pi} \log T \tag{1.1}$$

as $T \to \infty$. On the other hand, not much is known about the number of these zeros, $N_K^*(T)$, that are simple. Indeed, it was only recently that the authors [2] showed that

$$N_K^*(T) \gg T^{6/11}$$

and, if the Lindelof hypothesis is true, that

$$N_K^*(T) \gg_\varepsilon T^{1-\varepsilon}$$

for any $\varepsilon > 0$. Before this, it was not even known whether $\zeta_K(s)$ has infinitely many simple zeros in $0 < \sigma < 1$. In this paper we shall prove that if the Riemann hypothesis (RH) is true for $\zeta(s)$, the Riemann zeta-function, then a positive proportion of the zeros of $\zeta_K(s)$ are simple. More precisely we have

Theorem 1. *Assume that RH is true for* $\zeta(s)$. *Then*

$$N_K^*(T) \geq \left(\frac{1}{54} + o(1)\right) N_K(T)$$

Research supported in part by NSF grants.

as $T \to \infty$.

Remark. As we shall see below, if we assume the Riemann hypothesis for $\zeta_K(s)$, the constant 1/54 can be replaced by 1/27.

In the case of the Riemann zeta-function, there are three known methods for proving that a positive proportion of the zeros are simple. They are the pair correlation method of Montgomery [12], the modification of Levinson's method (due to Heath-Brown and Selberg) and the method of Conrey, Ghosh and Gonek [1].

An application of Montgomery's method shows that on GRH a positive proportion of the zeros of $\zeta_K(s)$ have multiplicity less than or equal to two but does not furnish any information on simple zeros (in fact, this statement also holds for L-functions associated with certain cusp-forms on the modular group, if one assumes the appropriate Riemann hypothesis).

The method of Levinson (which is unconditional) may work if one had mean-value theorems of "mollified" L-functions, on the critical line, with mollifiers of long length. Such results as are available at present do not suffice.

The present method (which is a variation of that in C-G-G [1].) overcomes these difficulties by exploiting the factorization

$$\zeta_K(s) = \zeta(s)L(s,\chi); \tag{1.2}$$

here χ is the quadratic (Kronecker) character of the field K and $L(s,\chi)$ is the associated Dirichlet L-function. Unfortunately, our approach has the drawback that it will not apply to functions like the Dirichlet series associated with cusp-forms, for although these functions also have a $\Gamma(s)$ term in their functional equations, they do not factor as a product of two "natural" Dirichlet series.

To establish Theorem 1 we shall require the following result which is of interest in its own right.

Theorem 2. *Assume RH for* $\zeta(s)$ *and let* $\rho = 1/2 + i\gamma$ *denote the typical nontrivial zero of* $\zeta(s)$. *Then if* χ *is any nonprincipal character (not neccessarily quadratic), we have*

$$|\{ 0<\gamma<T : L(\rho,\chi) = 0 \}| \leq (2/3 + o(1))N(T)$$

as $T \to \infty$, *where* $N(T)$ *is the number of zeros of* $\zeta(s)$ *with* $0 < \gamma < T$. *That is, at most two-thirds of the zeros of* $\zeta(s)$ *are also zeros of* $L(s,\chi)$.

With a lot more work, we could actually show that any two L-functions with inequivalent characters have at most two-thirds of their zeros in common, provided the Riemann hypothesis holds for one of them. A result of this type has also been given by A. Fujii [5] using a method different from ours. While his result is unconditional, his constant (which was not evaluated) is presumably quite small and would therefore not serve to prove Theorem 1.

To prove Theorem 1 we first observe from (1.2) that a zero of $\zeta_K(s)$ is simple if and only if it is either

(i) a simple zero of $\zeta(s)$ and not a zero of $L(s,\chi)$

or

(ii) a simple zero of $L(s,\chi)$ and not a zero of $\zeta(s)$.

Furthermore, these two conditions are mutually exclusive. Now it is known that, on RH at least 19/27 of the zeros of $\zeta(s)$ are simple (see C-G-G [1]). Then, by Theorem 2, the number of zeros satisfying (i) is at least $(19/27 - 2/3 + o(1))N(T) = (1/27 + o(1))N(T)$. But as is well-known $N(T) \sim \frac{1}{2} N_K(T)$. Hence, Theorem 1 follows.

We could have appealed to the result of Montgomery and Taylor [11] where 19/27 is replaced by 0.6725 with some loss in the constant in Theorem 1. Also notice that we have assumed RH only for $\zeta(s)$ and not for $L(s,\chi)$. If one assumes it for both functions (or, equiva- lently, for $\zeta_K(s)$), it can be shown by the method in [1] that 19/27ths of the zeros of $L(s,\chi)$ are simple, and by the method in this paper that at most 2/3rds of the zeros of $L(s,\chi)$ are zeros of $\zeta(s)$. In this way one can count the simple zeros of $\zeta_K(s)$ of type (ii) above, thereby doubling the constant in Theorem 1.

Theorem 2 also has an application to the Hurwitz zeta-function $\zeta(s,\alpha)$, namely

Theorem 3. *On RH, if $\alpha = 1/3, 2/3, 1/4, 3/4, 1/6$ and $5/6$, then at least one-third of the zeros of $\zeta(s,\alpha)$ lie off the line $\sigma = 1/2$.*

A proof and discussion of a result of this kind may be found in Gonek [7].

2. Preamble to the proof of Theorem 2.

Throughout, T is large, $L = \log T$, and ε is an arbitrarily small positive number though not necessarily the same one at each occurence. Estimates depending implicitly on ε will be denoted by O_ε or $\ll_\varepsilon$.

It suffices to prove Theorem 2 for a primitive character χ and its modulus q will be fixed from now on. Consequently, the constants implied by the symbols O and $\ll$ may depend on q and χ. Let

$$A(s,\chi) = \sum_{k \leq y} a(k)k^{-s} ,$$

where

$$a(k) = \mu(k)\chi(k)\left(1 - \frac{\log k}{\log y}\right) \quad \text{and} \quad y = T^\eta$$

with μ the Mobius function and $0 < \eta < 1/2$ to be selected later in the proof. Let

$$N = \sum_{0<\gamma\leq T} L(1/2 + i\gamma,\chi)A(1/2 + i\gamma,\chi) \tag{2.1}$$

and

$$D = \sum_{0<\gamma\leq T} |L(1/2 + i\gamma,\chi)A(1/2 + i\gamma,\chi)|^2 , \tag{2.2}$$

with γ running through the ordinates of the zeros of $\zeta(s)$. Then, by the Cauchy-Schwartz inequality we have

$$|\{ 0<\gamma\leq T : L(1/2 + i\gamma,\chi) \neq 0 \}| \geq \frac{|N|^2}{D} . \tag{2.3}$$

The purpose of $A(s,\chi)$ here is to mollify $L(s,\chi)$ and thereby sharpen the inequality. The remainder of the paper is concerned with the evaluation of N and D. We shall show that on RH, if

$y = T^{1/2-\varepsilon}$, then

$$\mathcal{N} \sim \frac{TL}{2\pi} \quad \text{and} \quad \mathcal{D} \sim 3\,\frac{TL}{2\pi} \quad \text{as } T \to \infty . \tag{2.4}$$

Combined with (2.1) and (1.2), these estimates imply the result.

The first step in treating $\mathcal{N}$ and $\mathcal{D}$ is to express them as contour integrals by Cauchy's residue theorem. To this end let Π denote a sequence of numbers T_n such that

$$n < T_n \le n + 1 \qquad (n = 3, 4, \ldots\ldots)$$

and

$$\frac{\zeta'}{\zeta}(\sigma + iT_n) \ll (\log T_n)^2 \tag{2.5}$$

uniformly for $-1 \le \sigma \le 2$ (see Davenport [4; p.108]). In particular, T_n is not the ordinate of any zero of $\zeta(s)$.

Until the very end of the paper, we shall always assume that $T \quad \Pi$.

Next set (once and for all)

$$a = 1 + L^{-1}.$$

Let $\mathcal{R}$ be a positively oriented rectangle with vertices at $a + i$, $a + iT$, $1-a + iT$, and $1-a + i$. Then, on RH, we have

$$\mathcal{N} = \frac{1}{2\pi i}\int_{\mathcal{R}} \frac{\zeta'}{\zeta}(s)L(s,\chi)A(s,\chi)\,ds \tag{2.6}$$

and

$$\mathcal{D} = \frac{1}{2\pi i}\int_{\mathcal{R}} \frac{\zeta'}{\zeta}(s)L(s,\chi)L(1-s,\bar{\chi})A(s,\chi)A(1-s,\bar{\chi})\,ds . \tag{2.7}$$

Let us consider $\mathcal{N}$ first. As it happens, it is easier to work with

$$\mathcal{N} = \frac{-1}{2\pi i}\int_{\mathcal{R}} \frac{\zeta'}{\zeta}(1-s)L(s,\chi)A(s,\chi)\,ds .$$

This is equivalent to (2.6) because $\frac{\zeta'}{\zeta}(s)$ and $-\frac{\zeta'}{\zeta}(1-s)$ have the same poles and residues inside $\mathcal{R}$. Now for s inside or on $\mathcal{R}$,

$$A(s,\chi) \ll_\varepsilon y^{1-\sigma+\varepsilon} \tag{2.8}$$

and

$$L(s,\chi) \ll_\varepsilon T^{\frac{1}{2}(1-\sigma)+\varepsilon} . \tag{2.9}$$

These bounds and (2.5) imply that the top and bottom edges of R contribute $O(yT^{1/2+\varepsilon})$ to N.

For the left edge of R we replace s by 1-s and find that

$$\frac{-1}{2\pi i} \int_{1-a+iT}^{1-iT} \frac{\zeta'}{\zeta}(1-s)L(s,\chi)A(s,\chi)\, ds$$

$$= \frac{-1}{2\pi i} \int_{a-i}^{a-iT} \frac{\zeta'}{\zeta}(s)L(1-s,\chi)A(1-s,\chi)\, ds$$

$$= \overline{\frac{1}{2\pi i} \int_{a+i}^{a+iT} \frac{\zeta'}{\zeta}(s)L(1-s,\bar{\chi})A(1-s,\bar{\chi})\, ds}.$$

For the right-hand side of R we use the identities

$$-\frac{\zeta'}{\zeta}(1-s) = \frac{\zeta'}{\zeta}(s) - \frac{X'}{X}(1-s) ; \tag{2.10}$$

$$\zeta(1-s) = X(1-s)\zeta(s) , \tag{2.11}$$

where

$$X(1-s) = \pi^{\frac{1}{2}-s}\Gamma(s/2)/\Gamma(\tfrac{1-s}{2}) . \tag{2.12}$$

Thus, on substitution, we may write

$$N = \overline{N_1} + N_2 - N_3 + O_\varepsilon(yT^{\frac{1}{2}+\varepsilon}) \tag{2.13}$$

with

$$N_1 = \frac{1}{2\pi i} \int_{a+i}^{a+iT} \frac{\zeta'}{\zeta}(s)L(1-s,\bar{\chi})A(1-s,\bar{\chi})\, ds , \tag{2.14}$$

$$N_2 = \frac{1}{2\pi i} \int_{a+i}^{a+iT} \frac{\zeta'}{\zeta}(s)L(s,\chi)A(s,\chi)\, ds \tag{2.15}$$

and

$$N_3 = \frac{1}{2\pi i} \int_{a+i}^{a+iT} \frac{X'}{X}(1-s)L(s,\chi)A(s,\chi)\, ds\ . \tag{2.16}$$

We now come to $\mathcal{D}$.

The top and bottom edges of $\mathcal{R}$ contribute $O_\varepsilon(yT^{1/2+\varepsilon})$ to $\mathcal{D}$ by (2.5), (2.8), and (2.9). Replacing s by 1-s and using (2.10), we find that the contribution of the left edge of $\mathcal{R}$ equals

$$\frac{1}{2\pi i} \int_{a-i}^{a-iT} \frac{\zeta'}{\zeta}(1-s)L(1-s,\chi)L(s,\bar{\chi})A(1-s,\chi)A(s,\bar{\chi})\, ds$$

$$= \frac{1}{2\pi i} \int_{a-i}^{a-iT} \left(-\frac{\zeta'}{\zeta}(s) + \frac{X'}{X}(1-s)\right) L(1-s,\chi)L(s,\bar{\chi})A(1-s,\chi)A(s,\bar{\chi})\, ds\ .$$

We will write

$$\mathcal{D}_1 = \frac{1}{2\pi i} \int_{a+i}^{a+iT} \frac{\zeta'}{\zeta}(s)L(s,\chi)L(1-s,\chi)A(s,\chi)A(1-s,\bar{\chi})\, ds \tag{2.17}$$

and

$$\mathcal{D}_2 = \frac{1}{2\pi i} \int_{a+i}^{a+iT} \frac{X'}{X}(1-s)L(s,\chi)L(1-s,\bar{\chi})A(s,\chi)A(1-s,\bar{\chi})\, ds\ , \tag{2.18}$$

so that the integral above equals $\overline{\mathcal{D}}_1 - \overline{\mathcal{D}}_2$.

Notice that $\mathcal{D}_1$ is also the contribution of the right-hand side of $\mathcal{R}$ to $\mathcal{D}$. Hence, on combining these results, we obtain

$$\mathcal{D} = 2\ \mathrm{Re}.\ \mathcal{D}_1 - \overline{\mathcal{D}}_2 + O_\varepsilon(yT^{1/2+\varepsilon}).$$

We conclude this section by introducing some useful notation and formulae. As usual we write e(x) in place of exp($2\pi i$x).

We let $\sum_{m=1}^{r}{}'$ to denote a sum with (m,r)=1. Ramanujan's sum is

$$c_r(a) = \sum_{m=1}^{r}{}' e\left(\frac{ma}{r}\right) = \sum_{d|(r,a)} d\mu\left(\frac{r}{d}\right)\ . \tag{2.20}$$

Similarly, we define

$$c_\chi(a) = \sum_{m=1}^{q} \chi(m)\, e\left(\frac{ma}{q}\right) .$$

It is known (see [12; p.358]) that

$$c_\chi(a) = \begin{cases} \bar{\chi}(a)\tau(\chi) & \text{if } (a,q) = 1, \\ 0 & \text{if } (a,q) > 1. \end{cases} \tag{2.21}$$

where $\tau(\chi) = c_\chi(1)$ is the Gauss sum.

We shall write the functional equation for $L(s,\chi)$ in the form

$$L(1-s,\bar{\chi}) = X(1-s,\chi)L(s,\chi), \tag{2.22}$$

where

$$X(1-s,\chi) = \frac{i^{\mathfrak{a}}}{\tau(\chi)} q^s \pi^{1/2-s} \Gamma\left(\frac{s+\mathfrak{a}}{2}\right)/\Gamma\left(\frac{1-s+\mathfrak{a}}{2}\right) \tag{2.23}$$

with

$$\mathfrak{a} = \begin{cases} 0 & \text{if } \chi(-1) = 1, \\ 1 & \text{if } \chi(-1) = -1. \end{cases}$$

Observe that if $q = 1$, χ is principal and $X(1-s,\chi) = X(1-s)$, where $X(1-s)$ is the factor in the functional equation for $\zeta(s)$ (see (2.11) and (2.12)).

Finally, define

$$F_n(s) = \prod_{p|n} (1-p^{-s}) ;$$

$$F_n(s,\chi) = \prod_{p|n} (1 - \chi(p)p^{-s}).$$

3. Auxilliary lemmas.

Lemma 1. *Let* r *be a positive real number and suppose that* $X(1-s,\chi)$ *is given by* (2.23). *Then for* $a = 1 + L^{-1}$ *and* T *large, we have*

$$\frac{1}{2\pi i} \int_{a+i}^{a+iT} X(1-s,\chi)\, r^{-s}\, ds$$

$$= \begin{cases} \dfrac{\chi(-1)}{\tau(\chi)} e\left(\dfrac{-r}{q}\right) + \dfrac{(q/r)^a}{\tau(\chi)} E(r/q,T) & \text{if } r \le \dfrac{qT}{2\pi}, \\[2ex] \dfrac{(q/r)^a}{\tau(\chi)} E(r/q,T) & \text{if } r > \dfrac{qT}{2\pi}, \end{cases}$$

where

$$E(r/q,T) \ll T^{1/2} + \frac{T^{3/2}}{|T-2\pi r/q| + T^{1/2}}.$$

Proof. When q=1, $X(1-s,\chi) = X(1-s)$ and, except for minor modifications, a proof can be found in Gonek [8]. If $q > 1$, then as is easily shown,

$$X(1-s,\chi) = \frac{\chi(-1)}{\tau(\chi)} q^s X(1-s)\left(1 + O(e^{-\pi t})\right). \tag{3.1}$$

Using this and the case q = 1 of the lemma, we obtain the result.

Lemma 2. *Let* $\alpha(n)$, $\beta(n)$ *be arithmetic functions such that* $\alpha(n) = O(1)$ *and* $\beta(n) = O(d_r(n) \log^{\ell} n)$, *where* $d_r(n)$ *is the coefficient of* n^{-s} *in* $\zeta^r(s)$ *and* ℓ *is a non-negative integer. Also let* $a = 1 + L^{-1}$. *Then if* $1 < x \leqslant T$,

$$\frac{1}{2\pi i} \int_{a+i}^{a+iT} X(1-s,\chi)\Big(\sum_{k\leq x} \alpha(k)k^{s-1}\Big)\Big(\sum_{j=1}^{\infty} \beta(j)j^{-s}\Big)\, ds$$

$$= \frac{\chi(-1)}{\tau(\chi)} \sum_{k\leqslant x} \frac{\alpha(k)}{k} \sum_{j\leqslant qkT/2\pi} \beta(j)e\Big(\frac{-j}{qk}\Big) + O_{\varepsilon}(xT^{1/2+\varepsilon}).$$

Proof. This follows from Lemma 1 and Lemma 2 of C-G-G [1].

Lemma 3. *Let* χ mod q *be a primitive character and set*

$$D\Big(s,\chi,\frac{-H}{qK}\Big) = \sum_{n=1}^{\infty} d(n)\chi(n)e\Big(\frac{-nH}{qK}\Big)n^{-s} \qquad (\sigma > 1).$$

If K *is square-free and* $(H,K) = (K,q) = 1$, *then* D *has an analytic continuation to the whole plane except for a pole at* s=1. *At this point it has the same principal part as*

$$\bar{\chi}(H)\chi(K)\tau(\chi)(qK)^{-s}\zeta^2(s)\left(F_q(s)(1+\chi(-1)K^{1-s}) - \phi(q)q^{-s}\right). \tag{3.2}$$

Proof. This is a straightforward generalization of a well-known result of Estermann.

$\Lambda(s,a,k)$ has the same principal part at s=1 as

$-\delta((a,k))\frac{1}{\phi(k)}\frac{\zeta'}{\zeta}(s)$ where $\delta(\ )$ is the Dirac delta-function. Thus, if we call the expression in (3.2) $P(s,\chi,\frac{-H}{qK})$, then by (3.5)

$$A(s,\chi,\frac{-H}{qK}) = -\sum_{d|K}\ \sideset{}{'}\sum_{b=1}^{qK/d}\left(D(s,\chi,\frac{-bH}{qK/d}) - P(s,\chi,\frac{-bH}{qK/d})\right)$$

$$\cdot\left(\Lambda(s,bd,qK) + \zeta(d)\frac{1}{\phi(qK)}\frac{\zeta'}{\zeta}(s)\right)$$

$$-\sum_{d|K}\ \sideset{}{'}\sum_{b=1}^{qK/d} P(s,\chi,\frac{-bH}{qK/d})\Lambda(s,bd,qK)$$

$$+\sum_{d|K}\ \sideset{}{'}\sum_{b=1}^{qK/d}\frac{\delta(d)}{\phi(qK)}D(s,\chi,\frac{-bH}{\phi(qK)})\frac{\zeta'}{\zeta}(s)$$

$$-\sum_{d|K}\ \sideset{}{'}\sum_{b=1}^{qK/d}\frac{\delta(d)}{\phi(qK)}P(s,\chi,\frac{-bH}{qK/d})\frac{\zeta'}{\zeta}(s)$$

$$= -\Sigma_1 - \Sigma_2 + \Sigma_3 - \Sigma_4,$$

say, with Σ_1 regular at s=1. The principal part of Σ_2 is the same as that of

$$\bar{\chi}(H)\tau(\chi)(qK)^{-s}\zeta^2(s)\sum_{d|K}\chi(\frac{K}{d})\,d^s\left(F_q(s)(1+\chi(-1)(\frac{K}{d})^{1-s}) - \phi(q)q^{-s}\right)$$

$$\cdot\ \sideset{}{'}\sum_{b=1}^{qK/d}\bar{\chi}(b)\Lambda(s,bd,qK).$$

The sum over b equals

$$\sideset{}{'}\sum_{b=1}^{qK/d}\bar{\chi}(b)\sum_{n\equiv b(\mathrm{mod}\ qK/d)}\frac{\Lambda(dn)}{(dn)^s} = d^{-s}\sum_{\substack{n=1\\(n,K/d)=1}}^{\infty}\frac{\bar{\chi}(n)\Lambda(nd)}{n^s},$$

and the last sum is zero unless d=1 or p (recall that K is square-free). In any case we may write it as

$$\delta(d)\left(-\frac{L'}{L}(s,\bar{\chi}) - \frac{F_K'}{F_K}(s,\bar{\chi})\right) + \frac{\Lambda(d)}{d^s - \bar{\chi}(d)}.$$

Thus,

$$\Sigma_2 = \bar{\chi}(H)\chi(K)\tau(\chi)(qK)^{-s}\zeta^2(s)\ \left\{\left(-\frac{L'}{L}(s,\bar{\chi}) - \frac{F_K'}{F_K}(s,\bar{\chi})\right)\right.$$

$$\times \left(F_q(s)(1 + \chi(-1)K^{1-s}) - \phi(q)q^{-s}\right)$$

$$+ \sum_{p|K} \frac{\log p\, \bar{\chi}(p)p^s}{p^s - \bar{\chi}(p)} \left(F_q(s)(1 + \chi(-1)(\tfrac{K}{p})^{1-s}) - \phi(q)q^{-s}\right)\}$$

$$= - \bar{\chi}(H)\chi(K)\tau(\chi)(qK)^{-s}\zeta^2(s)G_K(s,\chi).$$

Next,

$$\Sigma_3 = \frac{1}{\phi(qK)} \frac{\zeta'}{\zeta}(s) \sum_{b=1}^{qK}{}' D(s,x,\frac{-bH}{qK})$$

$$= \frac{1}{\phi(qK)} \frac{\zeta'}{\zeta}(s) \sum_{n=1}^{\infty} \frac{d(n)\chi(n)}{n^s} \Big(\sum_{b=1}^{qK}{}' e(\frac{-bnH}{qK})\Big)$$

for $\sigma > 1$. Since $(H,qK) = 1$, the sum over b equals

$$\sum_{a=1}^{qK}{}' e(\frac{an}{qK}) = c_{qK}(n) .$$

Thus by (2.20) we have

$$\sum_{n=1}^{\infty} \frac{d(n)\chi(n)}{n^s} c_{qK}(n) = \sum_{m,n=1}^{\infty}\sum \frac{\chi(mn)}{(mn)^s} \sum_{d|(qK,mn)} d\mu(\frac{qK}{d})$$

$$= L(s,\chi)^2 \sum_{d|qK} d^{1-s}\mu(\frac{qK}{d})\chi(d) \sum_{e|d} F_{d/e}(s,\chi) ,$$

after some simplifications. Since this function is regular at $s=1$, Σ_3 has the same principal part as

$$\frac{1}{\phi(qK)} \frac{\zeta'}{\zeta}(s)L(1,\chi)^2 \sum_{d|qK} \mu(\frac{qK}{d})\chi(d) \sum_{e|d} F_e(1,\chi) .$$

Evidently we may restrict the first sum to one over $d|K$. Also, since $(q,K) = 1$ and K is square-free, the double sum equals

$$\mu(qK) \sum_{d|K} \mu(d)\chi(d) \prod_{p|d} (1 + F_p(1,\chi)) = \mu(qK)g(K) .$$

Thus, the principal part of Σ_3 is identical to that of

$$\frac{\mu(qK)}{\phi(qK)} g(K)L(1,\chi)^2 \frac{\zeta'}{\zeta}(s) .$$

Lemma 4. *Let* χ mod q *be a primitive character and set*

$$Q(s,\chi,\frac{-H}{qK}) = -\sum_{m,n=1}^{\infty}\sum \frac{\Lambda(m)d(n)\chi(n)}{(mn)^s} e(-\frac{mnH}{qK}) \qquad (\sigma > 1).$$

Then if H, K *and* q *are pairwise coprime and* K *is squarefree,* Q *has a meromorphic continuation to the whole plane. The only pole for* $\sigma \geq 1$ *is at* s=1 *where it has a pole whose principal part is the same as that of*

$$\bar{\chi}(H)\chi(K)\tau(\chi)(qK)^{-s}\zeta^2(s)G_K(s,\chi) + \frac{\mu(qK)g(K)}{\phi(qK)} L^2(1,\chi)\frac{\zeta'}{\zeta}(s),$$

where

$$g(K) = \prod_{p|K} (1 - 2\chi(p) + \frac{\chi^2(p)}{p}) \tag{3.3}$$

and

$$G_K(s,\chi) = \Big(\frac{L'}{L}(s,\bar{\chi}) + \frac{F_K'}{F_K}(s,\bar{\chi})\Big)\Big(F_q(s)(1 + \chi(-1)K^{1-s}) - \phi(q)q^{-s}\Big)$$

$$- \prod_{p|K} \frac{\bar{\chi}(p)\log p}{1-\bar{\chi}(p)/p^s}\Big(F_q(s)(1 + \chi(-1)(\frac{K}{p})^{1-s}) - \phi(q)q^{-s}\Big). \tag{3.4}$$

Proof. For $\sigma > 1$ we have

$$\begin{aligned} Q(s,\chi,\frac{-H}{qK}) &= -\sum_{a=1}^{qK} D(s,\chi,\frac{-aH}{qK})\Lambda(s,a,qK) \\ &= -\prod_{d|K} \sum_{b=1}^{qK/d} D(s,\chi,\frac{-bH}{qK/d})\Lambda(s,bd,qK), \end{aligned} \tag{3.5}$$

where D(s,1,1) is as in Lemma 3, and

$$\Lambda(s,a,k) = \sum_{n\equiv a(\mathrm{mod}\ k)} \Lambda(n)n^{-s} \qquad (\sigma > 1).$$

It is well known that $\Lambda(s,a,k)$ has a meromorphic continuation to the whole plane with a simple pole at s=1 if and only if (a,k) = 1. Also by Lemma 3, $D(s,\chi,.)$ is regular everywhere except for a possible double pole at s=1. Thus, $Q(s,\chi,.)$ is meromorphic in the complex plane and has no poles in $\sigma \geq 1$ except possibly at s=1.

To find the principal part at this point, first note that

Finally,

$$\Sigma_4 = \frac{1}{\phi(qK)} \frac{\zeta'}{\zeta}(s)\ \bar{\chi}(H)\chi(K)\tau(\chi)(qK)^{-s}\zeta(s)^2$$

$$\times \left(F_q(s)\left(1 + \chi(-1)K^{1-s}\right) - \phi(q)q^{-s}\right) \sum_{b=1}^{qK}{}' \ \bar{\chi}(b)$$

$$= 0.$$

Collecting these results, we find that $Q(s,\chi,\frac{-H}{qK})$ has the same principal part as

$$\bar{\chi}(H)\chi(K)\tau(\chi)(qK)^{-s}\zeta(s)^2 G_K(s,\chi) + \frac{\mu(qK)g(K)}{\phi(qK)} L(1,\chi)^2 \frac{\zeta'}{\zeta}(s) .$$

this completes the proof.

Lemma 5. *Let* χ mod q *be a primitive character and suppose that* $(H,K) = (K,q) = 1$. *Set*

$$L(s,\chi,\frac{-H}{qK}) = \sum_{n=1}^{\infty} \chi(n)e(\frac{-nH}{qK})n^{-s} \qquad (\sigma > 1) .$$

Then L *has an analytic continuation to the whole plane except for a possible pole at* $s=1$. *At this point it has the same principal part as*

$$\delta(K)\bar{\chi}(-H)\tau(\chi)q^{-s}\zeta(s),$$

where $\delta(K) = 1$ *if* $K=1$ *and is zero otherwise.*

Proof. This follows on relating the Dirichlet series to a Hurwitz zeta-function in an obvious manner.

Lemma 6. *Let* χ mod q *be a primitive character and write*

$$R(s,\chi,\frac{-1}{qk}) = -\sum_{m,n=1}^{\infty}\sum \frac{\Lambda(m)\chi(n)}{(mn)^s} e(\frac{-mn}{qK}) \qquad (\sigma > 1).$$

If $(K,q) = 1$ *and* K *is squarefree, then* R *has a meromorphic continuation to the entire complex plane. Its only pole in* $\sigma \geqslant 1$ *is at* $s=1$ *where it has a pole with the same principal part as*

$$\frac{\mu(qK)}{\phi(qK)}L(1,\chi)F_K(0,\chi)\frac{\zeta'}{\zeta}(s) + \frac{\chi(-1)\tau(\chi)}{qK}\left(\delta(K)\frac{L'}{L}(1,\bar{\chi}) - \frac{\Lambda(K)}{1-\bar{\chi}(K)/K}\right)\zeta(s),$$

where $\delta(K) = 1$ *if* $K=1$ *and is* 0 *otherwise.*

Proof. This is similar to the proof of Lemma 4.

Lemma 7. *Suppose that*

$$c_1(j) = -\sum_{mn=j} \Lambda(m)\chi(n),$$

$$b_1(j) = -\sum_{h\le y}\ \sum_{hmn=j} a(h)\Lambda(m)\chi(n)d(n),$$

and

$$b_2(j) = \sum_{h\le y}\ \sum_{hn=j} a(h)\chi(n)d(n).$$

Then if $y = T^{\eta}$ *with* $\eta < 1/2$,

$$\sum_{k\le y}\frac{\bar{a}(k)}{k}\sum_{j<qKT/2\pi} c_1(j)e\left(\frac{-j}{qk}\right) = \sum_{k\le y}\frac{\bar{a}(k)}{k}\operatorname*{res}_{s=1}\left(\frac{R(s,\chi,-1/qk)}{s}\left(\frac{qkT}{2\pi}\right)^s\right) + O_\varepsilon\left(y^{1/2}T^{3/4+\varepsilon} + TL^{-1}\right), \tag{3.6}$$

$$\sum_{k\le y}\frac{\bar{a}(k)}{k}\sum_{j\le qKT/2\pi} b_1(j)e\left(\frac{-j}{qk}\right) = \sum_{h,k\le y}\frac{a(h)\bar{a}(k)}{k}\operatorname*{res.}_{s=1}\left(\frac{Q(s,\chi,\frac{-H}{qK})}{s}\left(\frac{qKT}{2\pi}\right)^s\right) + O_\varepsilon\left(y^{1/2}T^{3/4+\varepsilon} + TL^{-1}\right), \tag{3.7}$$

and

$$\sum_{k\le y}\frac{\bar{a}(k)}{k}\sum_{j\le qkT/2\pi} b_2(j)e\left(\frac{-j}{qk}\right) = \sum_{h,k\le y}\frac{a(h)\bar{a}(k)}{k}\operatorname*{res.}_{s=1}\left(\frac{D(s,\chi,\frac{-H}{qK})}{s}\left(\frac{qK\tau}{2\pi H}\right)^s\right) + O\left(y^{1/2}T^{3/4+\varepsilon} + TL^{-1}\right), \tag{3.8}$$

where $\tau \ll T$ *in* (3.8) *and* R, Q *and* D *are as in* Lemmas 3, 4 *and* 6.

Proof. All three formulae are proved by the method used to estimate the sum M_2 in Conrey, Ghosh and Gonek [1;Sec.5]. Since the method is rather complicated and lengthy, we shall only indicate the idea of the proof of (3.7) here; the interested reader is referred to sections 5-7 of the afore mentioned paper for details. Had we assumed GRH, the lemma could be established with considerably less work; the reader may wish to consult Lemma 6 in [3] for the proof of a similar result.

First we set

$$B(s,\frac{-j}{qk}) = \sum_{j=1}^{\infty} b_1(j)e(\frac{-j}{qk})j^{-s} \qquad (\sigma > 1).$$

Then the sum on the left in (3.7) is

$$\sum_{k\leq y} \frac{\bar{a}(k)}{k} \left(\frac{1}{2\pi i} \int_{(c)} B(s,\frac{-j}{qk})\left(\frac{jkT}{2\pi}\right)^{s} \frac{ds}{s}\right), \tag{3.9}$$

where c depends on T and $c > 1$. Now by the definitions of $b_1(j)$ and $Q(s,x,.)$, we see that

$$B(s,\frac{-j}{qk}) = \sum_{h\leq y} a(h)Q(s,\chi,\frac{-H}{qK}), \tag{3.10}$$

where $H = h/(h,k)$ and $K = k/(h,k)$. From this and Lemma 4 it follows that $B(s,-j/(qk))$ is a meromorphic function whose only pole in $\sigma \geq 1$ is at $s=1$. Inserting (3.10) into (3.9), we see that this pole should give rise to the main term

$$\sum_{h,k\leq y}\sum \frac{a(h)\bar{a}(k)}{k} \operatorname*{res.}_{s=1} \left\{(Q(s,\chi,\frac{-H}{qK}) \frac{\left(\frac{qKT}{2\pi H}\right)^{s}}{s}\right\}. \tag{3.11}$$

To prove that this is the case we need to replace the exponential (additive character) in $B(s,\frac{-j}{qk})$ by a character sum. We may then proceed as in the proofs of the Bombieri-Vinogradov theorem given by Vaughan [15] and Gallagher[6].

By (5.12) in [1] we find that

$$e(\frac{-j}{qk}) = \sum_{q'|qk} \sideset{}{^*}\sum_{\psi \bmod q'} \tau(\bar{\psi}) \sum_{d|(qk,j)} \psi(\frac{j}{d})\,\delta(q',qk,d,\psi),$$

where

$$\delta(q',qk,d,\psi) = \frac{\psi\left(\frac{d}{(d,qk/q')}\right)\ \psi\left(\frac{-k}{(d,qk/q')q'}\right)\ \mu\left(\frac{qk(d,qk/q')}{q'}\right)}{\phi\left(\frac{qk}{(d,qk/q')}\right)}.$$

Clearly, we may suppose that $(k,q) = 1$ (otherwise $a(k) = 0$ in (3.7)). Hence, the divisors q' and d split as $q' = q_1q_2$ and $d = d_1d_2$ with $q_1|q$, $q_2|k$, $d_1|q$, $d_2|k$, and $(q_1,q_2) = (d_1,d_2) = 1$. Also, since ψ mod q_1q_2 is primitive, there is a unique pair of primitive characters ψ_1 mod q_1, ψ_2 mod q_2 such that $\psi = \psi_1\psi_2$. From this and the coprimality of q_1 and q_2 it is easy to show that

$$\tau(\bar{\psi}) = \bar{\psi}_1(q_2)\bar{\psi}_2(q_1)\tau(\bar{\psi}_1)\tau(\bar{\psi}_2).$$

Using these factorizations for q', d, ψ and $\tau(\bar{\psi})$, we may now write

$$e\left(\frac{-j}{qk}\right) = \sum_{q_1|q} \sum_{d_1|q} \sideset{}{^*}\sum_{\psi_1 \bmod q_1} \tau(\bar{\psi}_1) \sum_{q_2|k} \sideset{}{^*}\sum_{\psi_2 \bmod q_2} \tau(\bar{\psi}_2)\, \bar{\psi}_1(q_2)\, \bar{\psi}_2(q_1)$$
$$\sum_{\substack{d|k \\ d_1d_2|j}} \psi_1\psi_2\left(\frac{j}{d_1d_2}\right) \delta(q_1q_2,\, qk,\, d_1d_2, \psi_1\psi_2).$$

Substituting this in the definition of $B(s,\frac{-j}{qk})$ and using the result in (3.9), we find after rearranging the sums that the right-hand side of (3.9) equals

$$\sum_{d_1|q} \sum_{q_1|q} \sideset{}{^*}\sum_{\psi_1 \bmod q_1} \tau(\bar{\psi}_1) \sum_{k\le y} \frac{1}{k} \Big\{ \sum_{q_2\le y/k} \frac{\bar{a}(q_2k)}{q_2} \sideset{}{^*}\sum_{\psi_2 \bmod q_2} \tau(\bar{\psi}_2)$$

$$\sum_{d_2|q_2k} \delta(q_1q_2,\, qq_2k,\, d_1d_2,\, \psi_1\psi_2)\, \bar{\psi}_1(q_2)\, \bar{\psi}_2(q_1)$$

$$\left(\frac{1}{2\pi i}\right) \int_{(c)} \left(\sum_{m=1}^{\infty} \frac{b_1(d_1d_2m)\, \psi_1(m)\, \psi_2(m)}{m^s} \right) \left(\frac{qkT}{2\pi d_1d_2}\right)^s \frac{ds}{s} \Big)\Big\} .$$

The expression inside the brackets is analogous to E_2 in (5.15) of [1] and is treated in precisely the same way. That is, we distinguish between the cases $q_2 \le L^A$ for some $A > 0$, and $L^A < q_2 \le y/k$. The integrand above has a pole at $s=1$ if and only if $q_1 = q_2 = 1$, so the contribution of this term must be identical to (3.11). For $q_2 \le L^A$ we move the contour to the left and use Siegel's theorem as in the proof of the prime number theorem for arithmetic progressions. For the remaining cases we use a Vaughan-type identity and the large sieve. If we assumed GRH, it is this last part that could be dispensed with (and so the analysis is much easier).

We now state some elementary lemmas, the proofs of which we omit.

Lemma 8. *Let* $\sigma_{-1/2}(m) = \sum_{d|m} d^{-1/2}$. *Then*

(i) $\frac{m}{\phi(m)} \ll \sigma_{-1/2}(m)$,

and for $x \geq 1$

(ii) $\sum_{m\leq x} \sigma_{-1/2}(m)^2/m = c\log x + O(1)\ ;\ c = \zeta(2)\zeta(3/2)/\zeta(3)^2.$

Lemma 9. *For a fixed character* χ mod q, m *a positive integer, and* $x \geq 1$, *we have*

(i) $$\sum_{p|m} \frac{(\log p)^j}{p} \ll \begin{cases} \text{logloglog } 30m & \text{if } j=0, \\ (\text{loglog } 3m)^j & \text{if } j=1,2, \end{cases}$$

and

(ii) $$\sum_{\substack{p<x \\ p\nmid m}} \frac{\chi(p)}{p}(\log p)^j \ll \begin{cases} \text{logloglog } 30m & \text{if } j=0 \\ (\text{loglog } 3m)^j & \text{if } j=1,2. \end{cases}$$

Lemma 10. *Let* $G_k(s,\chi)$ *be as in* (3.4) *with* χ mod q *being a fixed character and* k *a positive integer. Then*

$$G_k(1,\chi) = -\chi(-1)\frac{\phi(q)}{q} \sum_{p|k} \bar{\chi}(p)\log p + O(\text{loglog } 3k)$$

and

$$G_k'(1,\chi) = \chi(-1)\frac{\phi(q)}{q} \sum_{p|k} \bar{\chi}(p)\log p \log \frac{k}{p} + O(\log 2k \text{ loglog } 3k)$$

Lemma 11. *For* $x \geq 1$ *and* q *fixed,*

$$\sum_{\substack{m\leq x \\ (m,q)=1}} \frac{\mu^2(m)}{\phi(m)} = \frac{\phi(q)}{q}\log x + O(1)$$

Lemma 12. *Let* y, $a(h)$ *and* $F_h(s,\chi)$ *be as in Sec.2 and let* $g(h)$ *be as* *in* (3.3). *Then*

$$\text{(i)} \quad \sum_{h\le y/m} \frac{a(mh)}{h} \ll \sigma_{-\frac{1}{2}}(m),$$

$$\text{(ii)} \quad \sum_{h\le y/m} \frac{\mu(h)\bar{a}(h)F_h(0,\chi)}{\phi(h)} \ll L^{-1},$$

$$\text{(iii)} \quad \sum_{h\le y/m} \frac{\mu(h)\bar{a}(mh)g(h)}{h\phi(h)} \ll \sigma_{-\frac{1}{2}}(m)L^{-1}.$$

Lemma 13. *Let* y *and* a(h) *be as before. Then*

$$\text{(i)} \quad \sum_{h\le y/m} \frac{a(mh)\bar{\chi}(h)}{h} = -\frac{qm\mu(m)\chi(m)}{\phi(q)\phi(m)\log y} + O\left(\frac{\sigma_{-\frac{1}{2}}(m)}{\log y \log^4 2y/m}\right)$$

and

$$\text{(ii)} \quad \sum_{h\le y/m} \frac{a(mh)\bar{\chi}(h)\log h}{h} = -\frac{qm\mu(m)\chi(m)\log y/m}{\phi(q)\phi(m)\log y} + O\left(\frac{\sigma_{-\frac{1}{2}}(m)\log L}{\log y}\right).$$

Proof. We may base a proof on the formula (see Graham [9])

$$\sum_{\substack{k\le x\\(k,r)=1}} \frac{\mu(k)}{k}\log\frac{x}{k} = \frac{r}{\phi(r)} + O(\sigma_{-\frac{1}{2}}(r)\log^{-4}2x),$$

For (i) we have

$$\sum_{h\le y/m} \frac{a(mh)\bar{\chi}(h)}{h} = \frac{\mu(m)\chi(m)}{\log y} \sum_{\substack{h\le y/m\\(h,mq)=1}} \frac{\mu(h)}{h}\log\frac{y}{mh}$$

$$= \frac{\mu(m)\chi(m)mq}{\phi(mq)\log y} + O\left(\frac{\sigma_{-1/2}(mq)}{\log y \log^4 2y/m}\right).$$

The original sum vanishes if $(m,q) > 1$ so the result follows from the multiplicativity of ϕ and $\sigma_{-\frac{1}{2}}$.

In a similar way (ii) follows on noting that h is squarefree and so $\log h = \sum_{p|h} \log p$.

4. The estimation of N.

Recall from (2.13) that

$$N = N_1 + N_2 - N_3 + O_\varepsilon(yT^{1/2+\varepsilon}), \qquad (4.1)$$

where the N_i are given by (2.14) - 2.16).

We first consider N_1. Using the functional equation (2.22) in (2.14), we have

$$N_1 = \frac{1}{2\pi i}\int_{a+i}^{a+iT} \frac{\zeta'}{\zeta}(s)L(s,\chi)A(1-s,\bar{\chi})X(1-s,\chi)\, ds.$$

Setting

$$c_1(j) = -\sum_{mn=j} \Lambda(m)\chi(n)$$

and using Lemma 2, we obtain

$$N_1 = \frac{\chi(-1)}{\tau(\chi)} \sum_{k\le y} \frac{\bar{a}(k)}{k} \sum_{j\le qkT/2\pi} c_1(j)e\left(\frac{-j}{qk}\right) + O_\varepsilon(yT^{1/2-\varepsilon}).$$

Now by (3.6) we find that

$$N_1 = \frac{\chi(-1)}{\tau(\chi)} \sum_{k\le y} \frac{\bar{a}(k)}{k} \operatorname*{res.}_{s=1}\left(\frac{R(s,\chi,\frac{-1}{qk})}{s} \left(\frac{qkT}{2\pi}\right)^s\right)$$

$$+ O_\varepsilon(y^{1/2}T^{3/4+\varepsilon}) + O(TL^{-1}).$$

Here the sum may be taken over squarefree k coprime to q (other-wise $a(k) = 0$), hence the residue may be computed by means of Lemma 6. The result is , after simplification,

$$N_1 = \frac{T}{2\pi}\left(\frac{L'}{L}(1,\bar{\chi}) - \frac{\chi(-1)}{\tau(\chi)}\frac{\mu(q)\,q}{\phi(q)}L(1,\chi)\sum_{k\le y}\frac{\mu(k)\,\bar{a}(k)\,f_k(0,\chi)}{\phi(k)}\right.$$

$$\left. - \sum_{p\le y}\frac{\bar{a}(p)\log p}{p-\bar{\chi}(p)}\right) + O(y^{1/2}T^{3/4+\varepsilon}) + O(TL^{-1}).$$

By Lemma 12(ii) the sum over k is bounded by L^{-1}, while the sum over p equals

$$\frac{-1}{\log y}\sum_{p\le y}\frac{\bar{\chi}(p)\log p\log y/p}{p} + O\Big(\sum_{p\le y}\frac{\log p}{p^2}\Big).$$

The error term is clearly bounded and, by Lemma 9(ii), with m=1, so is the first term. Hence

$$N_1 = O_\varepsilon(y^{1/2}T^{3/4+\varepsilon}) + O(T). \qquad (4.2)$$

Next, by (2.15), we have

$$N_1 = \frac{1}{2\pi i}\int_{a+i}^{a+iT} \frac{\zeta'}{\zeta}(s)L(s,\chi)A(s,\chi)\,ds$$

$$= \sum_{n=2}^{\infty} c_2(n)\, n^{-a}\left(\frac{1}{2\pi}\int_1^T n^{-it}\,dt\right) \ll L^3 \qquad (4.3)$$

since

$$c_2(n) = -\sum_{hjk=n} \Lambda(h)\chi(j)a(k) \ll d_3(n)\log n.$$

Finally we come to N_3. Taking the logarithmic derivative of (2.12) it is easily shown that

$$\frac{X'}{X}(1-s) = -\log\frac{t}{2\pi} + O\left(\frac{1}{t}\right) \qquad (4.4)$$

for $t\geq 1$, $0 \leq \sigma \leq 2$, say. Inserting this into (2,16), we obtain

$$N_3 = \frac{-1}{2\pi}\int_1^T L(a+it,\chi)A(a+it,\chi)\log(t/2\pi)\,dt + O(L^3)$$

since

$$L(a+it,\chi)A(a+it,\chi) < \zeta^2(a) < L^2.$$

The main term can be written as

$$-\sum_{n=1}^{\infty} c_3(n)n^{-a}\left(\frac{1}{2\pi}\int_1^T n^{-it}\log(t/2\pi)\,dt\right),$$

with

$$c_3(n) = \sum_{hk=n} \chi(h)a(k) < d(n).$$

The term n=1 contributes $TL/2\pi + O(T)$ to N_3, while the remaining terms contribute an amount of $O(L^3)$, so that

$$N_3 = -\frac{T}{2\pi}L + O(T).$$

Combining this with (4.1)-(4.3) we see that

$$\mathcal{N} = \frac{T}{2\pi} L + O(T) + O_\varepsilon(y^{1/2} T^{3/4+\varepsilon}). \tag{4.5}$$

5. The estimation of $\mathcal{D}_1$

We now turn to the first term $\mathcal{D}_1$ in the denominator $\mathcal{D}$ see (2.07), (2.17), and (2.19)). By the functional equation (2.22) we have

$$\mathcal{D}_1 = \frac{1}{2\pi i} \int_{a+i}^{a+iT} \frac{\zeta'}{\zeta}(s) L^2(s,\chi) A(s,\chi) A(1-s,\bar{\chi}) X(1-s,\chi)\, ds,$$

where $a = 1 + L^{-1}$. We define

$$\sum_{j=1}^{\infty} b_1(j) j^{-s} = \frac{\zeta'}{\zeta}(s) L^2(s,\chi) A(s,\chi) \qquad (\sigma > 1).$$

Then by Lemma 2,

$$\mathcal{D}_1 = \frac{\chi(-1)}{\tau(\chi)} \sum_{k \le y} \frac{\bar{a}(k)}{k} \sum_{j \le qkT/2\pi} b_1(j) e\left(-\frac{j}{qk}\right) + O_\varepsilon(yT^{1/2+\varepsilon}).$$

To evaluate this we use (3.7) of Lemma 7 and find that

$$\mathcal{D}_1 = \frac{\chi(-1)}{\tau(\chi)} \sum_{h,k \le y} \frac{a(h)\, \bar{a}(k)}{k} \operatorname*{res.}_{s=1} \left\{ \frac{1}{s} \left(\frac{qkT}{2\pi}\right)^s Q\left(s,\chi,\frac{-H}{qk}\right) \right\}$$

$$+ O_\varepsilon(y^{1/2} T^{3/4+\varepsilon}) + O(TL^{-1}),$$

where

$$H = \frac{h}{(h,k)} \quad \text{and} \quad K = \frac{k}{(h,k)} .$$

Observe that in the sum above we may suppose that both H and K are square-free and that $(H,q) = (K,q) = 1$. Therefore, Lemma 4 is applicable and we may write the residue as

$$\bar{\chi}(H) \chi(K) \tau(\chi) \operatorname*{res.}_{s=1} \left\{ \left(\frac{T}{2\pi H}\right)^s G_K(s,\chi) \zeta(s)^2 / s \right\}$$

$$+ \frac{\mu(qK)g(K)}{\phi(qK)} L^2(1,\chi) \underset{s=1}{\text{res.}} \{ (\frac{qKT}{2\pi H})^s \frac{\zeta'}{\zeta}(s)\frac{1}{s} \}.$$

If we use the expansion $\zeta(s) = \frac{1}{s-1} + \gamma + \ldots$ near s=1 to evaluate these residues and insert the result into $\mathcal{D}_1$, we obtain

$$\mathcal{D}_1 = \chi(-1)\frac{T}{2\pi} \sum_{h,k\leq y}\sum \frac{a(\bar{h})a(\bar{k})\chi(h)\chi(k)(h,k)}{hk}(G_K(1,\chi)\log \frac{Te^{2\gamma-1}}{2\pi H} + G_K'(1,\chi))$$

$$- \frac{\chi(-1)}{\tau(\chi)} \frac{\mu(q)\phi(q)}{q} L^2(1,\chi) \frac{T}{2\pi} \sum_{h,k\leq y}\sum \frac{a(h)\bar{a}(k)}{hk} \cdot \frac{\mu(\frac{k}{(h,k)})\, g(\frac{k}{(h,k)})}{\phi(\frac{k}{(h,k)})}$$

$$+ O_\varepsilon(y^{1/2} T^{3/4+\varepsilon}) + O(TL^{-1}).$$

We next apply the Möbius inversion formula in the form

$$f((h,k)) = \sum_{\substack{m|h\\ m|k}} \sum_{n|m} \mu(n)f(\frac{m}{n}) .$$

On applying this to $\mathcal{D}_1$ and simplifying, we find that

$$\mathcal{D}_1 = \chi(-1) \frac{T}{2\pi} \sum_{m\leq y} \frac{1}{m} \sum_{n|m} \frac{\mu(n)}{n} \sum_{h,k\leq y/m}\sum \frac{a(mh)\bar{a}(mk)\bar{\chi}(h)\chi(k)}{hk}$$

$$\cdot(G_{kn}(1,\chi) \log \frac{Te^{2\gamma-1}}{2\pi hn} + G_{kn}'(1,\chi))$$

$$- \frac{\chi(-1)\mu(q)\phi(q)}{\tau(\chi)q} L^2(1,\chi)\frac{T}{2\pi} \sum_{m\leq y} \frac{1}{m} \sum_{h|m} \mu(n) \sum_{h,k<\frac{y}{m}}\sum \frac{a(mh)\bar{a}(mk)\mu(nk)g(nk)}{hk\phi(nk)}$$

$$+ O_\varepsilon(y^{1/2} T^{3/4+\varepsilon}) + O(TL^{-1}),$$

or, say

$$\mathcal{D}_1 = \mathcal{D}_{11} - D_{12} + O_\varepsilon(y^{1/2} T^{3/4+\varepsilon}) + O(TL^{-1}). \qquad (5.1)$$

We first treat $\mathcal{D}_{11}$. By Lemma 10, the expression in brackets is

$$G_k(1,\chi) \log \frac{Te^{2\gamma-1}}{2\pi hn} + G_k'(1,\chi) + O(L \text{ loglog } 3kn)$$

$$= -\chi(-1)\,\frac{\phi(q)}{q}\Big(\log\frac{T}{h}\sum_{p|k}\bar{\chi}(p)\log p - \sum_{p|k}\bar{\chi}(p)\log p\,\log\frac{k}{p}\Big)$$

$$+ O(L\,\text{loglog}\,3kn) + O(\log 2k\,\log 2n).$$

Since neither n nor k is greater than y, the error terms here are $O(L\ \log\ 2nL)$. Hence, using the identity $\sum_{n|m}\frac{\mu(n)}{n} = \frac{\phi(m)}{m}$, we have

$$\mathcal{D}_{11} = -\frac{\phi(q)}{q}\,\frac{T}{2\pi}\sum_{m\leq y}\frac{\phi(m)}{m^2}\sum_{h,k<y/m}\sum\frac{a(mh)\bar{a}(mk)\bar{\chi}(h)\chi(k)}{hk}$$

$$\cdot\sum_{p|k}\bar{\chi}(p)\log p\,\log\frac{Tp}{hk}$$

$$+ O\Big(TL\sum_{m\leq y}\frac{1}{m}\sum_{n|m}\frac{\log 2nL}{n}\Big|\sum_{h\leq y/m}\frac{a(mh)\bar{\chi}(h)}{h}\Big|^2\Big).$$

Notice that in each sum over m we may assume that m is square-free. With this in mind, we see from Lemma 13(i) and Lemma 8 that the O-term is

$$\ll \frac{TL}{\log^2 y}\sum_{m\leq y}\frac{\mu^2(n)\sigma^2_{-1/2}(m)}{m}\sum_{n|m}\frac{\log 2nL}{n} \ll T\log L\,.$$

We may therefore rewrite $\mathcal{D}_{11}$ equals

$$\frac{\phi(q)T}{2q\pi}\sum_{m\leq y}\frac{\phi(m)}{m^2}\sum_{\substack{p\leq y/m\\(p,q)=1}}\frac{\log p}{p}\Big(L\sum_{h\leq y/m}\frac{\bar{a}(mh)\bar{\chi}(h)}{h}\sum_{\ell\leq y/mp}\frac{\bar{a}(mp\ell)\chi(\ell)}{\ell}$$

$$-\sum_{k\leq y/m}\frac{a(mh)\bar{\chi}(h)\log h}{h}\sum_{\ell\leq y/mp}\frac{\bar{a}(mp\ell)\chi(\ell)}{\ell}$$

$$-\sum_{h\leq y/m}\frac{a(mh)\bar{\chi}(h)}{h}\sum_{\ell\leq y/mp}\frac{\bar{a}(mp\ell)\chi(\ell)\log\ell}{\ell}\Big) + O(T\log L).$$

Using Lemma 13 to estimate the sums over h and ℓ and noting that we may suppose that $(p,m) = (q,m) = 1$, we find that the expression in parenthesis is

$$- \frac{\mu^2(m)m^2q^2}{\phi^2(m)\phi^2(q)\log^2 y} \frac{p\bar{\chi}(p)}{\phi(p)} (L + \log y/m + \log y/mp)$$

$$+ O\left(\frac{\sigma_{-1/2}(m)^2}{\log y \log^4 2y/mp}\right) \quad + \quad O\left(\frac{\sigma_{-1/2}(m)^2 \log L}{\log^2 y}\right).$$

The first O-term contributes

$$\ll \frac{T}{\log y} \sum_{m \le y} \frac{\sigma_{-1/2}(m)^2}{m} \int_1^{y/m} \log^{-4} 2y/mu \;\; du/u$$

by the prime number theorem. The integral is easily seen to be O(1). So by Lemma 8(ii) the contribution is O(T).
The second error term is

$$\ll \frac{T \log L}{\log y} \sum_{m \le y} \frac{\sigma_{-1/2}(m)^2}{m} \ll T \log L.$$

Finally, by Lemmas 9 and 11, the main term contributes an amount

$$\ll \frac{T}{\log^2 y} \sum_{m \le y} \frac{\mu^2(m)}{\phi(m)} (L \text{ logloglog } 30m + \text{loglog } 3m) \ll T \text{ loglog } L .$$

Thus,

$$\mathcal{V}_{11} \ll T \log L. \tag{5.2}$$

We now turn to $\mathcal{V}_{12}$. We have

$$\mathcal{V}_{12} = \frac{\chi(-1)}{\tau(\chi)} \frac{\mu(q)\phi(q)}{q} L^2(1,\chi) \frac{T}{2\pi} \sum_{m \le y} \frac{1}{m} \sum_{n|m} \frac{\mu^2(n)g(n)}{\phi(n)}$$

$$\cdot \sum_{h \le y/m} \frac{a(mh)}{h} \sum_{k \le y/m} \frac{\bar{a}(mk)\mu(k)g(k)}{k\phi(k)} ,$$

since we may obviously assume that (n,k)=1. By Lemma 12(i) and (iii) and Lemma 8(ii), this is, on interchanging orders of summation

$$\ll \frac{T}{L} \sum_{m \le y} \frac{\mu^2(m)\sigma^2_{-1/2}(m)}{m} \sum_{n|m} \frac{\mu^2(n)|g(n)|}{\phi(n)}$$

$$\ll T \sum_{n \le y} \frac{\mu^2(n)\sigma^2_{-1/2}(n)|g(n)|}{n\phi(n)} .$$

Now for square-free n,

$$|g(n)| = \prod_{p|n} \left|1 - 2\chi(p) + \frac{\chi^2(p)}{p}\right| \leq 4^{\omega(n)} = d(n)^2 .$$

and

$$\sigma^2_{-1/2}(n) = \prod_{p|n} (1+ p^{-1/2})^2 \leq 2^{2\omega(n)} = d(n)^2 ,$$

where $\omega(n)$ is the number of prime-factors of n. Hence,

$$\mathcal{D}_{12} \ll T \sum_{n\leq y} \frac{\mu^2(n)d(n)^4}{n\phi(n)} \ll T.$$

It follows from this, (5.1) and (5.2) that

$$\mathcal{D}_1 = O_\varepsilon(y^{1/2} T^{3/4+\varepsilon}) + O(T \log L). \tag{5.3}$$

6. The estimation of $\mathcal{D}_2$.

We shall see in this section that the main term in $\mathcal{D}$ is from $\mathcal{D}_2$. Recall from (2.18) that

$$\mathcal{D}_2 = \frac{1}{2\pi i} \int_{a+i}^{a+iT} \frac{X'}{X}(1-s)L(s,\chi)L(1-s,\bar{\chi})A(s,\chi)A(1-s,\bar{\chi})\, ds,$$

Moving the line of integration to $\sigma = 1/2$ and using (2.8), (2.9) and (4.4), we obtain

$$\mathcal{D}_2 = -\frac{1}{2\pi} \int_1^T |L(1/2+ it,\chi)A(1/2+ it,\chi)|^2 \log \frac{t}{2\pi}\, dt \tag{6.1}$$

$$+ O\left(\int_1^T |L(1/2+ it,\chi)A(1/2+ it,\chi)|^2 \frac{dt}{t}\right) + O_\varepsilon(yT^{1/2+\varepsilon}).$$

The mean-values are evaluated using the techniques indicated in Sec.5 to give us

$$\mathcal{D}_2 = -\frac{TL}{2\pi}\left(1+ \frac{L}{\log y}\right) + O_\varepsilon(y^{1/2} T^{3/4+\varepsilon}) + O(T \log L). \tag{6.2}$$

7. Completion of the proof.

By (4.5) we see that

$$\mathcal{N} = \frac{TL}{2\pi} + O_\varepsilon(y^{1/2}T^{3/4+\varepsilon}) + O(T). \tag{7.1}$$

Also, from (5.3) we have

$$\mathcal{D}_1 = O_\varepsilon(y^{1/2}T^{3/4+\varepsilon}) + O(T \log L),$$

and from (6.2) that

$$\mathcal{D}_2 = -\frac{TL}{2\pi}\left(1+\frac{L}{\log y}\right) + O_\varepsilon(y^{1/2}T^{3/4+\varepsilon}) + O(T \log L).$$

Thus, by (2.19) it follows that

$$\mathcal{D} = \frac{TL}{2\pi}\left(1+\frac{L}{\log y}\right) + O_\varepsilon(y^{1/2}T^{3/4+\varepsilon}) + O(T \log L). \tag{7.2}$$

We now take $y = T^{1/2-2\varepsilon}$ in (7.1) and (7.2) and find that

$$\mathcal{N} = \frac{TL}{2\pi} + O_\varepsilon(T) \qquad \text{and} \qquad D = (3 + O(\varepsilon))\frac{TL}{2\pi}. \tag{7.3}$$

This establishes (2.4) and (2.5) and therefore Theorem 2, provided that T is in the sequence Π defined in Sec.2 (preceding (2.6)). To remove this restriction first note that every positive T is within O(1) of some element of Π and that increasing T by O(1) in (2.2) introduces at most O(L) new terms into the sum. However, by (2.9) and (2.10) each of these terms is

$$\ll_\varepsilon y^{1/2+\varepsilon/2}\, T^{1/4+\varepsilon/2} = T^{1/2-\varepsilon/4-\varepsilon^2}$$

if $y = T^{1/2-2\varepsilon}$. Thus (7.3) is valid for all large T. Similarly, increasing T by O(1) introduces at most O(L) new terms into the sum for $\mathcal{D}$ in (2.3). Each of these is

$$\ll_\varepsilon T^{1-\varepsilon/2-2\varepsilon^2},$$

so (7.3) is also valid for all large T. This completes the proof of

Theorem 2.

References

1. J.B. Conrey, A. Ghosh, and S.M. Gonek, Simple zeros of the Riemann zeta-function, submitted.

2. J.B. Conrey, A. Ghosh, and S.M. Gonek, Simple zeros of the zeta-function of a quadratic number field I, *Invent. Math.* **86** (1986), 563-576

3. J.B. Conrey, A. Ghosh, and S.M. Gonek, Large gaps between zeros of the zeta-function, to appear *Mathematika*.

4. H. Davenport, Multiplicative Number Theory, Graduate Texts in Mathematics, v.74, Springer Verlag, New York, 1980.

5. A. Fujii, On the zeros of Dirichlet L-functions (V), *Acta Arith.* **28** (1976), 395-403.

6. P.X. Gallagher, Bombieri's mean value theorem, *Mathematika* **15** (1968), 1-6.

7. S.M. Gonek, The zeros of Hurwitz's zeta-function on $\sigma = 1/2$, Analytic Number Theory(Phil. Pa. 1980) 129-140, Springer Verlag Lecture Notes 899, 1981.

8. S.M. Gonek, Mean values of the Riemann zeta-function and its derivatives, *Invent. Math.* **75** (1984), 123-141.

9. S.W. Graham, An asymptotic estimate related to Selberg's sieve, *Jour. Num. Thy.* **10**(1978), No.1 ,83-94.

10. H.L. Montgomery, The pair correlation of zeros of the zeta-function, Proc. Symp. Pure Math., 24(1973), 181-193.

11. H.L. Montgomery, Distribution of the zeros of the Riemann zeta-function , Proc.Int.Cong.Math., Vancouver 1974 , 379-381.

12. H.L. Montgomery and R.C. Vaughan, The exceptional set in Goldbach's problem, *Acta Arith.* **XXVII** (1975), 353-370.

13. E.C. Titchmarsh, The Theory of the Riemann Zeta-Function, Oxford, Clarendon Press, 1951.

14. R.C. Vaughan, Mean value theorems in prime number theory, *J. London Math. Soc.*(2) **10** (1975), 153-162.

J.B. Conrey and A. Ghosh
Oklahoma State University
Stillwater, OK 74078-0613
U.S.A.

S.M. Gonek
University of Rochester
Rochester, NY 14627
U.S.A.

DIFFERENTIAL DIFFERENCE EQUATIONS ASSOCIATED WITH SIEVES

***H. Diamond, H. Halberstam and H.-E. Richert**

1. Our aim in this note is to analyse the differential difference equations underlying sieves of dimension $\kappa > 1$. A heuristic version of such an analysis together with some valuable numerical information was given by Iwaniec, van de Lune and te Riele [5] (see also te Riele [7]) and what we seek to do here, in effect, is to justify the conclusions of [5]. It has been shown elsewhere (in [2]) how to construct sieves of dimension $\kappa > 1$ on the basis of such information. In this connection we acknowledge also our indebtedness to the important thesis of Rawsthorne [6].

Let $\sigma(u) = \sigma_\kappa(u)$ be the continuous solution of the Ankeny-Onishi differential-difference equation (cf [1], or Chapter 7 of [3])

$$u^{-\kappa}\sigma(u) = C^{-1} \, (0<u\leq 2), \quad (u^{-\kappa}\sigma(u))' = -\kappa u^{-\kappa-1}\sigma(u-2) \, (u>2) \tag{1.1}$$

where $C = (2e^\gamma)^\kappa\Gamma(\kappa + 1)$ and γ denotes Euler's constant. We shall indicate how to prove that there exist numbers $\alpha_\kappa > 1$, $\beta_\kappa > 1$ and continuous functions F_κ, f_κ that satisfy the simultaneous differential-difference equations with retarded argument

$$F_\kappa(u) = 1/\sigma_\kappa(u) \, (0<u\leq\alpha_\kappa), \quad (u^\kappa F_\kappa(u))' = \kappa u^{\kappa-1} f_\kappa(u-1) \, (u>\alpha_\kappa), \tag{1.2}$$

$$f_\kappa(u) = 0 \quad (0<u\leq\beta_\kappa), \quad (u^\kappa f_\kappa(u))' = \kappa u^{\kappa-1} F_\kappa(u-1) \, (u>\beta_\kappa), \tag{1.3}$$

as well as the additional conditions

---

*All three authors acknowledge with gratitude support from the National Science Foundation.

$$F_\kappa(u) = 1 + O(e^{-u}),\ f_\kappa(u) = 1 + O(e^{-u}) \text{ as } u \to \infty \tag{1.4}$$

and

$$0 \leq f_\kappa(u) < 1 < F_\kappa(u),\quad u > 0\ . \tag{1.5}$$

Once we have such a pair of numbers α_κ, β_κ and a pair of functions F_κ, f_κ, we can derive with relative ease (by the method sketched in [2]) the following:

Theorem. *Let* A *be a finite integer sequence whose elements are not necessarily positive or distinct, let* P *be a set of primes and* $z \geq 2$ *a real number. Write*

$$P(z) = \prod_{\substack{p<z \\ p \in P}} p \quad \text{and} \quad A_d = \{a \in A : a \equiv 0 \bmod d\}\ .$$

Suppose there exist an approximation X *to the cardinality* $|A|$ of A, *and a non-negative multiplicative function* $\omega(d)$ *on the squarefree integers (with* $0 \leq \omega(p) < p$ *if* $p \in P$ *and* $\omega(p) = 0$ *if* $p \notin P$), *so that*

$$R_d := |A_d| - \frac{\omega(d)}{d} X\ ,\ \mu(d) \neq 0 \text{ and } p|d \Rightarrow p \in P\ ,$$

are in the nature of remainders. Define

$$S(A, P, z) = |\{a \in A : (a, P(z)) = 1\}|$$

and

$$V(z) = \prod_{p<z} \left(1 - \frac{\omega(p)}{p}\right)\ .$$

If there exists a constant $A \geq 2$ *such that*

$$V(w_1)/V(w) \leq \left(\frac{\log w}{\log w_1}\right)^\kappa \left(1 + \frac{A}{\log w_1}\right) \quad \textit{whenever} \quad 2 \leq w_1 \leq w,$$

then, for any number $y \geq z$,

$$S(A, P, z) \leq XV(z)\left\{F_\kappa\left(\frac{\log y}{\log z}\right) + O\left(\frac{\log \log y}{(\log y)^\nu}\right)\right\} + \sum_{\substack{m|P(z) \\ m<z}} c^+(m) R_m \tag{1.6}$$

and

$$S(A,P,z) \geqslant XV(z)\{f_\kappa(\frac{\log y}{\log z}) + O(\frac{\log\log y}{(\log y)^\nu})\} - \sum_{\substack{m|P(z)\\ m<y}} c^-(m)R_m , \qquad (1.7)$$

where $\nu = 1/(2\kappa + 2)$ *and, in the remainder sums,* $c^\pm(m) \ll 4^{\Omega(m)}$, *with* $\Omega(m)$ *denoting the number of prime factors of* m.

Inequality (1.6) coincides for $\log y/\log z \leqslant \alpha_\kappa$ with the upper bound from the Ankeny-Onishi theory [1], and (1.7) is, of course, non-trivial only if $\log y/\log z > \beta_\kappa$. Our theorem is the natural refinement of [1]: we begin with the Ankeny-Onishi upper sieve up to α_κ, and from there on proceed to improve on [1] by a combinatorial device that has the same effect as infinitely many iterations of Buchstab's identities. The Rosser-Iwaniec theory [4] uses no `start-up' sieve; but what is an advantage when $\kappa \leqslant 1$ turns out to be a defect when $\kappa > 1$. Nevertheless, while the theorem is superior for $\kappa > 1$ to both [1] and [4], it should be said that the gains relative to [1], especially for larger κ, are only modest.

Our theorem may be used to contruct a weighted sieve, as is shown in the first part of Chapter 10 of [3]. The theorem itself, with $\kappa = 2$, may be applied to show (on the basis of [8]) that the maximal number N(n) of pairwise orthogonal Latin squares of order n satisfies, for all sufficiently large n, the inequality

$$N(n) > N^{\frac{1}{14.8}}$$

The details of the work described below will appear elsewhere in due course.

2. From now on we shall use α, β, F and f without the suffix κ when it is clear that we are working with a particular $\kappa > 1$. It is easy to check that if F, f are solutions of (1.2), (1.3) and (1.4) such that

$$Q(u) := F(u) - f(u) > 0 \text{ for } u > 0, \qquad (2.1)$$

then F(u) decreases and f(u) increases, each towards 1, as $u \to \infty$, so that we may replace (1.5) by (2.1). Introduce also

$$P(u) := F(u) + f(u),\ u > 0,$$

so that (1.4) is equivalent to

$$P(u) = 2 + O(e^{-u}) \quad \text{and} \quad Q(u) = O(e^{-u}) \text{ as } u \to \infty ; \tag{2.2}$$

and (1.2), (1.3) together imply that

$$uP'(u) = -\kappa P(u) + \kappa P(u-1) \tag{2.3}$$

and $\left.\right\}\ u > \max(\alpha, \beta).$

$$uQ'(u) = -\kappa Q(u) - \kappa Q(u-1) \tag{2.4}$$

From here on we proceed as far as we can by the method of `adjoint´ equations due to Iwaniec [4], and take full advantage of the analytic tools he fashioned here. Thus the adjoint equations of (2.3) and (2.4) are

$$(up(u))' = \kappa p(u) - \kappa p(u + 1) \tag{2.5}$$

and

$$(uq(u))' = \kappa q(u) + \kappa q(u + 1), \tag{2.6}$$

and these have solutions, regular in the half-plane Re $u > 0$, normalized to satisfy

$$p(u) \sim u^{-1},\ q(u) \sim u^{2\kappa-1} \text{ for } u \text{ real and } u \to \infty . \tag{2.7}$$

The adjoint functions p and q derive importance from the fact that the `inner products´

$$\big(P\cdot p\big)(u) := up(u)P(u) + \kappa \int_{u-1}^{u} p(x + 1)P(x)\,dx \tag{2.8}$$

and

$$\big(\big(Q\cdot q\big)\big)(u) := uq(u)Q(u) - \kappa \int_{u-1}^{u} q(x + 1)Q(x)\,dx \tag{2.9}$$

are constant from $\max(\alpha,\beta)$ onward. Indeed, by (2.2) and (2.7) we have

$$\big(P\cdot p\big)(u) = 2 \text{ and } \big(\big(Q\cdot q\big)\big)(u) = 0 \text{ if } u > \max(\alpha,\beta) \; ; \qquad (2.10)$$

and, conversely, (2.10) and (2.7) together imply (2.2). The functions p and q are representable as Laplace transforms, having rather complicated expressions (see section 5 of [4]), and will not be given here. When $2\kappa \in \mathbf{N}$, q is in fact a polynomial of degree $2\kappa - 1$. For each $\kappa > 1$, q(u) possesses finitely many positive zeros and the largest of these, to be denoted by $\rho = \rho_\kappa$, plays a central role in all subsequent calculations. (One might expect this from the Rosser-Iwaniec theory for $\kappa \leqslant 1$, where $\beta_\kappa = \rho_\kappa + 1)$. For any one κ, ρ_κ has to be computed numerically, but it can be shown that*

$$\begin{cases} 2\kappa - 1 < \rho_\kappa \leqslant \kappa + \sqrt{\kappa(\kappa-1)}, & 1 < \kappa \leqslant 1.5 \; , \\ \kappa + \sqrt{\kappa(\kappa-1)} < \rho_\kappa \leqslant \kappa + 1 + \sqrt{\kappa(\kappa - 3/2)}, & 1.5 < \kappa < 2 \; , \\ 2.843\kappa - 2 < \rho_\kappa < D\kappa \; , & 2 \leqslant \kappa \; , \end{cases} \qquad (2.11)$$

where $D = 3.59112\ldots$ is the solution of $D(\log D - 1) = 1$. In fact, Iwaniec [5] has proved that $\rho_\kappa \sim D\kappa$ $(\kappa \to \infty)$, and we conjecture that ρ_κ/κ is strictly increasing towards D as $\kappa \to \infty$. Some of the inequalities (2.11) can be sharpened if necessary.

Our method is complicated but, from a techincal point of view, rather simple. For the most part we rely heavily on interplay between the differential difference equations satisfied by the various functions in play and use of convexity and Taylor's theorem. One might say that most arguments come down to verification of inequalities linking p(u) and σ(u) at values of u having the form $\rho_\kappa + a$ and $b\kappa + c$. Complications are to be expected since p(u) and σ(u) are strangers to one another, and neither knows about ρ! In particular, we have to study the properties of σ(u) more deeply than was done in [1] and to find out more about ρ than is to be found in [1]; nevertheless we make extensive use of both

---

(*) The value of ρ_k has been thoroughly investigated by Dr. F.Grupp in his Habilitationsschrift, Ulm, January 1986 . We are grateful to him for making some of his results available to us earlier.

these pioneering studies and also of [6].

Our procedure is first to show that $\alpha \leqslant \beta$ cannot occur (for $\kappa > 1$; for $\kappa = 1$ one obtains $a = \beta = 2$). We do so in two steps: we show first that

$$\alpha \leqslant \beta - 1 \text{ is impossible.}$$

Otherwise, necessarily,

$$\frac{\rho p(\rho)}{\sigma(\rho)} \geqslant 2,$$

whereas, on the contrary, we are able to show that

$$\frac{\rho p(\rho)}{\sigma(\rho)} < 2. \tag{2.12}$$

Here is a typical instance of the kind of inequality mentioned earlier. We know that $up(u) \uparrow 1$ as $u \to \infty$, and that $\sigma(u) \uparrow 1$ as $u \to \infty$, so that certainly $up(u)/\sigma(u) < 2$ if u is large enough. Indeed, since $\rho p(\rho) < 1$ trivially, it would suffice to show that $\sigma(\rho) > \frac{1}{2}$, or, for $\kappa \geqslant 2$ at least, that $\sigma(2.843\kappa - 2) > 1/2$. The tables suggest that $\sigma(2\kappa - \frac{2}{10}) > \frac{1}{2}$ for all $\kappa \geqslant 1$. We can prove the first of these inequalities for κ sufficiently large; and so far we have proved only that $\sigma(2\kappa) > 0.4$ for $\kappa \geqslant 8$. Fortunately the behaviour of $(a\kappa + b)p(a\kappa + b)$, $a, b \geqslant 0$, comes to our aid: this quantity *decreases* as a function of κ and is almost constant (close to $\frac{a}{a+1}$) for long ranges of values of κ. Hence we may get away with weaker information about σ. Roughly speaking, we have *least* difficulty with κ beyond 5 or 6, just where numerical computation gets rapidly out of hand; and find the small κ's *hardest* to deal with theoretically, although here numerical computation of the highest precision is available.

Continuing with our story, we next *rule out case*

$$\beta - 1 < \alpha < \beta,$$

This is harder, but here we end up with a necessary condition for this case to exist that is violated if we can find a u_0 between ρ and β such that

$$\left(1 + \frac{1}{u_0 - 1}\right)^{\kappa} \frac{(u_0 - 1)p(u_0 - 1)}{\sigma(u_0)} < 2 . \qquad (2.13)$$

We can prove in this case that $\beta > \rho + \frac{1}{2}$ --and slightly better inequalities even--so we have candidates for the role of u_0 and, once again, we are back to the kind of scenario I've described in connection with (13). (A useful observation here is that the expression on the left of (14) is strictly decreasing as a function of u_0).

Subject to verification of (2.12) and (2.13) we have established now that for each $\kappa > 1$

$$\alpha_\kappa > \beta_\kappa .$$

We distinguish next two cases:

I. $\beta < \alpha < \beta + 1$,

II. $\beta + 1 \leq \alpha$.

3. Case I. From the inner product relations we obtain

$$(\mathrm{I}) \quad \begin{cases} \dfrac{\alpha p(\alpha)}{\sigma(\alpha)} + \kappa \displaystyle\int_{\beta-1}^{\alpha} \frac{p(x+1)}{\sigma(x)}\, dx = 2 \\ \dfrac{\alpha q(\alpha)}{\sigma(\alpha)} - \kappa \displaystyle\int_{\beta-1}^{\alpha} \frac{q(x+1)}{\sigma(x)}\, dx = 0 \end{cases}$$

and from these we are able to deduce that, necessarily,

$$\max(2,\rho) < \beta < \rho + 1 . \qquad (3.1)$$

Inequalities (3.1) tell us, when combined with earlier information about ρ, that α and β are small when κ is small; and numerical evidence tells us that Case I actually corresponds (cf [5]) to the range

$$1 < \kappa < 1.8344323 = \kappa_0$$

with the right-hand limit corresponding to $\alpha = \beta + 1 = 4.8819016$.

We are able to show in this case that *the equations* (I) *have a unique solution pair* α, β *--with* β *satisfying* (3.1). Moreover, it is surprisingly simple to deduce from the fact that $\beta > \rho$, that $Q(u) > 0$ when $u > 0$.

4. Case II $\alpha \geq \beta + 1$.

Here the inner product relations

$$\text{(II)} \quad \begin{cases} \dfrac{\alpha p(\alpha)}{\sigma(\alpha)} + \kappa \displaystyle\int_{\alpha-2}^{\alpha} \dfrac{p(x+1)}{\sigma(x)}\,dx + (\alpha-1)p(\alpha-1)f(\alpha-1) = 2 \\[2ex] \dfrac{\alpha q(\alpha)}{\sigma(\alpha)} - \kappa \displaystyle\int_{\alpha-2}^{\alpha} \dfrac{q(x+1)}{\sigma(x)}\,dx - (\alpha-1)q(\alpha-1)f(\alpha-1) = 0 \end{cases}$$

lead us to the equation

$$\frac{\alpha}{\sigma(\alpha)} \{p(\alpha)q(\alpha-1) + q(\alpha)p(\alpha-1)\} + \kappa q(\alpha-1) \int_{\alpha-2}^{\alpha} \frac{p(x+1)}{\sigma(x)}\,dx$$

$$- \kappa p(\alpha-1) \int_{\alpha-2}^{\alpha} \frac{q(x+1)}{\sigma(x)}\,dx - 2q(\alpha-1) = 0 \tag{4.1}$$

for α, and if we can show that this equation has a solution $\alpha > \rho + 1$ we can show, surprisingly easily again, that $Q(u) > 0$ for $u > 0$. Since β is then uniquely determined from

$$(\alpha-1)^{\kappa} f(\alpha-1) = \kappa \int_{\beta}^{\alpha-1} \frac{x^{\kappa-1}}{\sigma(x-1)}\,dx ,$$

with $f(\alpha-1)$ given by

$$(\alpha-1)q(\alpha-1)f(\alpha-1) = \frac{\alpha q(\alpha)}{\sigma(\alpha)} - \kappa \int_{\alpha-2}^{\alpha} \frac{q(x+1)}{\sigma(x)}\,dx ,$$

we are finished.

To show that (19) has a solution greater than $\rho + 1$ is rather a complicated business: the function of α on the left of (4.1) is negative at $\alpha = \rho + 1$ --this part fairly straightforward--but then to show that at some stage between $\rho + 1$ and 4κ this function becomes positive requires good information about the finer distribution of the values assumed by $\sigma(u)$ and its derivatives.

5. Final discussion.

The critical numbers α_κ, β_κ -- *the sifting limits* in sieve language--have to be computed. From present evidence it appears to be true that

$$\rho_\kappa + 1 < \alpha_\kappa < \rho_\kappa + 2$$

and possibly that

$$\alpha_\kappa \downarrow \rho_\kappa + 1 \quad \text{as } \kappa \to \infty .$$

On the other hand, Iwaniec conjectures that

$$\frac{\beta_\kappa}{\kappa} \to 2.44518586\ldots \text{ as } \kappa \to \infty ,$$

and so converges to the same limit as the Ankeny-Onishi sifting limit. In other words, for large k our sieve is not *significantly better* than the one-step A.-O. sieve.

As Iwaniec remarks, this demonstrates the power of the Selberg upper sieve. The question remains: is there a `start-up´ sieve better than Selberg´s when $\underline{\kappa > 1}$?

References.

1. N. C. Ankeny and H. Onishi, The general sieve, *Acta Arith.* **10**(1964/65), 31-62.

2. H. Diamond and H. Halberstam, The Combinatorial Sieve, to appear in the Proceedings of the Math. Science Conference on Number Theory 1983, Springer Lecture Notes 1985.

3. H. Halberstam and H. -E. Richert, Sieve Methods, Academic Press, 1974.

4. H. Iwaniec, Rosser´s Sieve, *Acta Arith.* **36** (1980), 171-202.

5. H. Iwaniec, J. van de Lune and H. J. J. te Riele, The limits of Buchstab´s iteration sieve, *Indag. Math. Proc.* A **83**(4), (1980).

6. D. Rawsthorne, Improvements in the small sieve estimate of Selberg by iteration, Ph.D. thesis, University of Illinois, 1980.

7. H. J. J. te Riele, Numerical solution of two coupled non linear equations related to the limits of Buchstab's iteration sieve, *Afdeling Numerieke Wiskunde*, **86.** Math. Centrum, Amsterdam, 1980, 15 pp.

8. R. N. Wilson, Concerning the number of mutually orthogonal latin squares, *Discrete Math.* **9** (1974), 181-198.

H.G. Diamond and H. Halberstam
University of Illinois
1409 West Green Street
Urbana, Illinois 61801, U.S.A.

H.-E. Richert
Universitä Ulm (MNH)
Abt. für Mathematik III
7900 Ulm (Donau)
Oberer Eselsberg
West Germany

PRIMES IN ARITHMETIC PROGRESSIONS AND RELATED TOPICS

John Friedlander

0. Introduction.

This paper (talk) has a dual purpose. The first is to report without proof some of the results of recent collaborative work on a number of multiplicative topics. These topics are connected by a thread which we shall follow in the reverse order so that in fact the work in each section was to a greater or lesser extent motivated by the work in the subsequent sections.

The second purpose is to publicize Iwaniec's recent (version of the) proof of Burgess' estimate [4] for character sums. Although this proof uses essentially the same ingredients as the earlier ones, it seems to this author to be much simpler. I am grateful to my friend Henryk Iwaniec for allowing me to include his proof here, and for his comments on the first draft of this paper.

I should like to dedicate this paper to Keith who spent his third birthday without his father who was giving this talk at exactly that time.

1. Primes in Arithmetic Progressions.

The results in this section represent work done jointly by the author with E. Bombieri and H. Iwaniec (to appear in [3] to which we shall refer as B-F-I) as well as related recent results of Fouvry [7,8]. These are concerned with estimates of the type

$$\sum_{q \leqslant Q} \max_{y \leqslant X} \max_{(a,q)=1} |\psi(y;q,a) - y/\phi(q)| \ll_A X(\log X)^{-A}$$

for arbitrary $A > 0$. The famous Bombieri-Vinogradov theorem [2,19] gives the above for $Q = X^{1/2 - \varepsilon}$ while the conjecture of Elliott-

Halberstam predicts that it even holds for $Q = X^{1-\varepsilon}$.

In attempting to prove results with exponent beyond $1/2$, we are first of all led to drop the expression $\max_y$. This is not a serious restriction since it is known (see for example [12, lemma 1]) that the resultant weakening of the inequality is only apparent. A second concession we make is to drop the expression $\max_a$. This is a more serious deficiency which is necessitated by the methods at our disposal, but nevertheless is not a hindrance for most applications.

Problem. We want to show that, for arbitrary weights γ_q not too large (say bounded by a power of $d(q)$), and for some fixed $\delta > 0$, $Q = X^{1/2+\delta}$, we have

$$\sum_{q \leqslant Q} \gamma_q \big(\psi(X;q,a) - X/\phi(q)\big) \ll_A X \log^{-A} X. \tag{*}$$

The requirement of (*) for arbitary γ_q includes the case of absolute values (take $\gamma_q = \operatorname{sgn}\big(\psi(X;q,a) - X/\phi(q)\big)$ and, by Cauchy's inequality, is no more difficult than this special case. Although, in this generality, the above goal has not yet been reached, there have been a number of successes in proving (*) for certain special classes of γ_q. The first such results are due to Fouvry and Iwaniec [9] and then to Fouvry [6].

Results for even the simplest of weights γ_q have interesting applications. Thus B-F-I and Fouvry [7] independently proved

Theorem. *If* γ_q *is identically one then* (*) *holds with any* $Q \leqslant X^{1-\varepsilon}$.

Corollary. (Titchmarsh divisor problem). *For* $a \neq 0$, $A > 0$, *we have*

$$\sum_{|a| < n \leqslant X} \Lambda(n)d(n+a) = c_1(a)X\log X + c_2(a)X + O(X\log^{-A} X).$$

Previous proofs of this asymptotic formula were not strong enough to give the second main term, giving only an error of order X loglog X .

Definition. We say that the weights λ are well-factorable of level Q if for every decomposition $Q = Q_1Q_2$, $1 \leq Q_1$, $1 \leq Q_2$, there exists a decomposition $\lambda = \lambda_1 * \lambda_2$ (Dirichlet convolution) with λ_j having support on $[1,Q_j]$ and $|\lambda_j| \leq 1$.

Improving previous results of [9,6] B-F-I shows

Theorem. (*) *holds for any weights* $\{\gamma_q\}$ *well-factorable of level* $Q = X^{4/7 - \varepsilon}$.

The importance of the well-factorable weights is due to their appearance in the Iwaniec error term [16] in the linear sieve. This now gives

Corollary. *For* $X > X_o(\varepsilon)$ *the number of pairs of twin primes up to* X *is no more than* $(7/2 + \varepsilon)$ *times the expected number.*

The basic problem described above is attacked by the dispersion method. A combinatorial identity (such as that due to Heath-Brown [13]) is used to replace sums over primes by bilinear forms. These are estimated by a variety of methods appealing mainly to the work of Deshouillers-Iwaniec [5]. In this estimation the degree of flexibility of the weights γ_q becomes significant. For the extremely flexible special weights above the situation is rather favorable. In fact, the bulk of B-F-I is devoted to the extension of (*) to classes of weights far less flexible.

We conclude this section by mentioning some spectacular recent work of Fouvry [8] which makes heavy use of B-F-I (and requires much else besides).

Theorem. (Fouvry). *There exists* $\delta > 0$ *such that for a positive proportion of* $p \leq X$ *the greatest prime factor of* $p-1$ *exceeds* $p^{2/3 + \delta}$. (*In* [8] *Fouvry gives* $2/3 + \delta = 0.6687$.)

The above problem has been studied extensively and the improvement here although quantitatively small was pursued with strong motivation. In fact Fouvry is able to prove that the same result holds even when one restricts p to the arithmetic progression $p \equiv 2 \pmod 3$. Combining this with a generalization, due to Adleman and Heath-Brown, of the Sophie Germain criterion one gets

Corollary. (Adleman, Fouvry, Heath-Brown). *For infinitely many primes* p *the first case of Fermat's last theorem is true; that is*

$$x^p + y^p + z^p = 0 \quad \textit{implies} \quad p \mid xyz.$$

2. Divisor Problems.

The work in this section was done jointly with Iwaniec. We were concerned with the problem of proving the expected asymptotic formulae

$$\sum_{\substack{n \leqslant X \\ n \equiv a(q)}} d_r(n) \sim \frac{X}{\phi(q)} P_r(\log X)$$

(where P_r is a certain polynomial of degree $r-1$) uniformly for $q < X^{\theta_r - \varepsilon}$, for all $\varepsilon > 0$, with the object of making θ_r as large as we could.

With the exception of θ_2 and θ_4 we were able to improve the known results as shown below.

r	old θ_r	due to	new θ_r
2	2/3	Hooley Linnik Selberg	no change
3	1/2	Linnik	1/2 + 1/230
4	1/2	Linnik	no change

5			9/20
6	$\frac{8}{3r+4}$	Lavrik	5/12
$\geqslant 7$			$\frac{8}{3r}$

Here the proof for $r \geqslant 5$ (which will appear in [11]) depends on a result of Iwaniec [17] which in turn rests on the Burgess estimate for character sums and the Halasz-Montgomery method as refined in [14].

The proof for $r = 3$ is completely different and of greater novelty. The fact that $\frac{1}{2} + \frac{1}{230} > \frac{1}{2}$ provided some of the motivation for the work in Sec.1 (although it eventually disappeared from the proof), and in turn the work on θ_3 resulted in consequence of ...

3. Kloosterman Sums.

The results of this section also represent work done jointly with Iwaniec. The details of proof, including the application to θ_3, are given [10].

We let q be prime (although results of the same "essential" strength hold for composite q). The simplest variant of our results here is the following estimate.

Let $1 \leqslant A < N$, $AN < q$, $n\bar{n} \equiv 1 \pmod q$. Then

$$\sum_{1 \leqslant a \leqslant A} \Big| \sum_{\substack{M < n \leqslant M+N \\ (n,q)=1}} e\Big(a\frac{\bar{n}}{q}\Big) \Big| \ll_\varepsilon q^{\frac{1}{8}+\varepsilon} (AN)^{\frac{3}{4}} + q^{\frac{1}{4}+\varepsilon} A^{\frac{5}{4}} N^{\frac{1}{4}}.$$

As an illustration of the strength of this result let us take $N = q^{1/2}$. Here an application of Weil's estimate to the inner sum does not improve the trivial estimate whereas a simple computation shows that the above estimate is non-trivial for $q^{\varepsilon} < A < q^{1/2 - \varepsilon}$.

The above result is proved by modifying the ideas used in the Burgess estimate for character sums and (as does that estimate) appeals to Weil´s "Riemann hypothesis for curves".

For application to θ_3 it was necessary to develop a non-trivial estimate for the sum

$$\sum_{1\leq h\leq H} \sum_{\substack{1\leq m\leq M \\ (m,q)=1}} \sum_{\substack{1\leq n\leq N \\ (n,q)=1}} e\left(\frac{h\overline{mn}}{q}\right) .$$

Given the presence of the extra variable it is not surprising that the proof here was based on a modification of the Burgess ideas which then appealed to Deligne´s "Riemann hypothesis for varieties". For the two varieties considered here the question of the applicability of the Deligne theory was far from straightforward, following in the one case from a result of Hooley [15] and in the other from a result of Birch and Bombieri [1].

4. Character Sums.

We now proceed to Iwaniec´s elegant proof of the Burgess estimate. As already mentioned this proof utilizes essentially the same tools and in particular draws its strength from the same main lemma.

Lemma. (Burgess). *For* χ *a non-principal character modulo the prime* q *and* k *a positive integer we have*

$$\sum_{y(\mathrm{mod}\, q)} \Big|\sum_{1\leq b\leq B} \chi(y+b)\Big|^{2k} \ll_k B^{2k} q^{1/2} + B^k q.$$

For simplicity we restrict to prime modulus q. We seek to estimate the sum

$$S = \sum_{N<n\leq N+H} \chi(n).$$

Employing an idea used by I.M. Vinogradov and by A.A. Karatsuba we translate the interval by a product

$$S = \sum_{N<n\leq N+H} \chi(n+ab) + T(a,b)$$

where a,b are integers and

$$T(a,b) = \sum_{N<n\leqslant N+ab} \chi(n) - \sum_{N+H<n\leqslant N+H+ab} \chi(n).$$

If (a,q)=1 we have

$$S = \chi(a) \sum_{N<n\leqslant N+H} \chi(\bar{a}n+b) + T(a,b).$$

Here T(a,b) consists of two sums of length ab. We think of ab as being less than H and we shall attempt to prove some result by induction. We sum the last identity over a,b with $1 \leqslant a \leqslant A$, $1 \leqslant b \leqslant B$, (a,q)=1. The number J of such pairs is $\gg$ AB and

$$J|S| \leqslant \sum_a \sum_n \Big| \sum_b \chi(\bar{a}n+b)\Big| + \sum_{a,b} |T(a,b)|.$$

In the first sum we make a single "longer" variable $y = \bar{a}n$ getting

$$\sum_a \sum_n \Big| \sum_b \Big| = \sum_{y(\mathrm{mod}\, q)} \nu(y) \Big| \sum_{1\leqslant b\leqslant B} \chi(y+b)\Big|$$

where

$$\nu(y) = \#\ (a,n) : \begin{cases} 1\leqslant a\leqslant A, & (a,q) = 1 \\ N<n\leqslant N+H, & \bar{a}n\equiv y(q) \end{cases} .$$

By Hölder's inequality

$$\sum_a \sum_n \Big| \sum_b \Big| \leqslant \Big\{ \sum_{y(q)} \nu(y)^2\Big\}^{1-\frac{1}{2k}} \Big\{ \sum_{y(q)} \Big| \sum_{1\leqslant b\leqslant B} \chi(y+b)\Big|^{2k}\Big\}^{\frac{1}{2k}}.$$

The latter factor may be estimated at once by the main lemma. To estimate the former we note that

$$\sum_{y(q)} \nu(y)^2 = \#\{(a_1,a_2,n_1,n_2) : \bar{a}_1 n_1 \equiv \bar{a}_2 n_2 \bmod q\}.$$

Now, $\bar{a}_1 n_1 \equiv \bar{a}_2 n_2$ implies $a_2 n_1 \equiv a_1 n_2$ implies $a_1(n_2-n_1) \equiv (a_2-a_1)n_1$.

There are no more than 2AH choices of the pair (a_2-a_1, n_1).

Each such choice determines $(a_2-a_1)n_1$ and hence it determines $a_1(n_2-n_1)$ modulo q. We assume that $AH < q/2$, so then $a_1(n_2-n_1)$ is determined. (It is easily checked that with the choice of A we shall later make, in case $AH \geqslant q/2$ the result follows from the Polya-Vinogradov inequality.) We thus have

$$\sum_{y(q)} \nu(y)^2 \ll AHq^{\varepsilon}.$$

(A more careful estimate would allow q^{ε} to be replaced by log q.) Combining our estimates we have, for some positive $c_k = c(k,\varepsilon)$,

$$|S| \leqslant c_k A^{-\frac{1}{2k}} H^{1-\frac{1}{2k}} \{q^{\frac{1}{4k}} + B^{-1/2} q^{\frac{1}{2k}}\} q^{\varepsilon} \qquad (**)$$

$$+ J^{-1} \sum_{a,b} |T(a,b)| .$$

Assume for the moment that we can ignore (by induction) the last sum and fix attention on the rest. Since A and B occur with negative exponents we should like them large; for the induction we are constrained to $AB < H$, say $AB = H/2$. A little thought shows that $B = q^{1/2k}$ is optimal and this determines A as well. Substituting these values we see that we can do no better than obtain an inequality

$$|S| \leqslant \lambda_k H^{1-\frac{1}{k}} q^{\frac{k+1}{4k^2}+\varepsilon} .$$

and using (**) and induction we do just that.

The induction is begun by noting that, for $H \leqslant 2q^{1/4}$, the result is trivial. To deduce it for H, we assume it up to H/2, choosing A and B as above. Substitution in (**) shows that we require

$$\lambda_k 2^{\frac{1}{k}-1} + c_k 2^{\frac{1}{2k}+1} \leqslant \lambda_k .$$

Provided that $k \geqslant 2$ and λ_k is sufficiently large in terms of c_k, this is clearly possible.

References.

[1] B. J. Birch and E. Bombieri, On some exponential sums, appendix to [10], *Annals of Math.*, **121** (1985), 345-350.

[2] E. Bombieri, On the large sieve, *Mathematika* **12** (1965), 201-225.

[3] E. Bombieri, J. B. Friedlander and H. Iwaniec, Primes in arithmetic progressions to large moduli, to appear in *Acta Math.*

[4] D. A. Burgess, On character sums and L-series II, *Proc. London Math. Soc.* (3) **13** (1963), 524-536.

[5] J.-M. Deshouillers and H. Iwaniec, Kloosterman sums and Fourier coefficients of cusp forms, *Invent. Math.* **70** (1982), 219-288.

[6] E. Fouvry, Autour du théorème de Bombieri-Vinogradov, *Acta Math.* **152** (1984), 219-244.

[7] E. Fouvry, Sur le problème des diviseurs de Titchmarsh, preprint (1984).

[8] E. Fouvry, Théorème de Brun-Titchmarsh, application au théorème de Fermat, *Invent. Math.* **79** (1985), 383-407.

[9] E. Fouvry and H. Iwaniec, Primes in arithmetic progressions, *Acta Arith.* **42** (1983), 197-218.

[10] J. B. Friedlander and H. Iwaniec, Incomplete Kloosterman sums and a divisor problem, *Annals of Math.*, **121** (1985), 319-350.

[11] J. B. Friedlander and H. Iwaniec, The divisor problem for arithmetic progressions, *Acta Arith.*, **XLV** (1985), 273-277.

[12] D. R. Heath-Brown, Primes in 'almost all' short intervals, *J. London Math. Soc.* (2) **26** (1982), 385-396.

[13] D. R. Heath-Brown, Prime numbers in short intervals and a generalized Vaughan identity, *Can. J. Math.* **34** (1982), 1365-1377.

[14] D. R. Heath-Brown and H. Iwaniec, On the difference between consecutive primes, *Invent. Math.* **55** (1979), 49-69.

[15] C. Hooley, On exponential sums and certain of their applications, *Journees Arith. 1980*, Armitage, J. V. ed., Cambridge (1982), pp. 92-122.

[16] H. Iwaniec, A new form of the error term in the linear sieve, *Acta Arith.* **37** (1980), 307-320.

[17] H. Iwaniec, On the Brun-Titchmarsh theorem, *J. Math. Soc. Japan* **34** (1982), 95-123.

[18] A. F. Lavrik, A functional equation for Dirichlet L-series and the problem of divisors in arithmetic progressions, *Izv. Akad. Nauk SSSR Ser. Mat.* **30** (1966), 433-448 (= *Transl.* A.M.S. (2) **82** (1969), 47-65).

[19] A. I. Vinogradov, On the density hypothesis for Dirichlet L-functions, *Izv. Akad. Nauk SSSR Ser. Mat.* **29** (1965), 903-934; correction ibid. **30** (1966), 719-720.

J. Friedlander
Scarborough College
University of Toronto
Scarborough, M1C 1A4 Canada

APPLICATIONS OF GUINAND´S FORMULA

P. X. Gallagher

The explicit formula of Weil [21] connects quite general sums over primes with corresponding sums over the critical zeros of the Riemann zeta function (or more general L-functions). In the earlier version of Guinand [8], there is on the Riemann hypothesis[1)] a kind of Fourier duality between the differentials of the remainder terms in the prime number theorem (suitable renormalized) and in the formula counting critical zeros of the Riemann zeta function.

According to Weil [22], analytic number theory, which deals with inequalities and asymptotic formulas, is not number theory but analysis. Nowhere is this more true than in our first topic, which is the relation between and bounds for these two remainder terms. It is convenient to begin in a general context, consisting of a function $Z = Z(s)$ meromorphic on the s-plane ($s = \sigma + it$), and satisfying for some positive integer k the conditions

(n_k) In $\sigma > 0$, Z has only finitely many zeros; in each vertical strip, Z has only finitely many poles and is of order $< k$.

and

(p) In some right half plane $\sigma > \sigma_1$, the logarithmic derivative of Z is given by an absolutely convergent Dirichlet series,

$$Z'/Z(s) = \sum_{\nu} c(\nu)e^{-\nu s}$$

---

1) In compensation for the extra hypothesis, while in Weil´s formula the function which is summed over zeros must be holomorhpic in a strip containing the critical strip, Guinand can sum certain functions with compact support on the critical line.

Research supported in part by NSF Grant DMS 82-02633

<u>with arbitrary complex coefficients</u>, ν <u>running over a sequence of positive numbers bounded away from</u> 0.

We denote by $\rho = \beta + i\gamma$ a typical zero or pole of Z, by $m(\rho)$ the order of Z at ρ ($= \pm$ the multiplicity of the zero or pole at ρ), and put $N(0) = 0$ and

$$N(T) = \pm \left(\sum_{\beta=0} m(\rho) + 2\sum_{\beta>0} m(\rho)\right) \quad \text{for } T \gtrless 0$$

where the sum is over the ρ with γ between 0 and T, the terms with $\gamma = 0$ or iT weighted by a factor $1/2$. It follows from the argument principle that

$$N(T) = M(T) + S(T) - S(0) \tag{1}$$

with

$$M(T) = -\frac{1}{\pi}\int_0^T \operatorname{Re} Z'/Z(it)\, dt$$

and

$$S(T) = -\frac{1}{\pi}\int_0^\infty \operatorname{Im} Z'/Z(\sigma + iT)\, d\sigma .$$

In fact, since $Z'/Z(s) \to 0$ exponentially as $\sigma \to \infty$, N(T) is finite and

$$N(T) = \frac{1}{2\pi i}\int_C Z'/Z(s)\, ds,$$

where C goes in straight lines from $\infty + iT$ to iT to 0 to ∞ and a Cauchy principal value is taken at each ρ on C. On taking the real part (1) follows. The integrands in the formula for M and S are undefined at the ρ but these singularities may be removed, and then the integrands are real-analytic. It follows that M is real-analytic.

For real U, let

$$P(U) = \sum_{\nu \leqslant U}{}' \; c(\nu)$$

where the dash indicates that only half of the possible term with $\nu = U$ is taken, and let

$$Q(U) = \sum_{\beta>0} \left(\frac{e^{\rho U}-1}{\rho} + \frac{e^{\rho^* U}-1}{\rho^*} - 2\,\frac{e^{i\gamma U}-1}{i\gamma} \right) m(\rho) ,$$

where $s^* = -\sigma + it$ for $s = \sigma + it$. We define R by

$$P(U) = Q(U) + R(U) \qquad (U \geq 0) \tag{2}$$

and $R(-U) = -\overline{R}(U)$ for $U < 0$.

For our applications it will suffice to have a Guinand formula with weight functions $f = f(u)$ defined on $\mathbf{R}$ and satisfying the condition

(W_k) $f,\dots,f^{(k-2)}$ <u>are continuous</u>, $f^{(k-1)}$ <u>and</u> $f^{(k)}$ <u>piecewise continuous</u>[2] <u>and</u> $f,\dots,f^{(k)}$ <u>are</u> $\ll e^{-\sigma_2|u|}$ <u>for some</u> $\sigma_2 > \sigma_1$.

For such f, the function

$$\tilde{f}(s) = \int_{-\infty}^{\infty} e^{su} f(u)du$$

is holomorphic in $|\sigma| < \sigma_2$ and is $O(|t|^{-k})$ in each closed substrip. In particular, the Fourier transform $g(t) = \tilde{f}(it)$ is real analytic.

Theorem 1. *If* Z *satisfies* (n_k) *and* (p) *and* f *satisfies* (W_k) *and has no discontinuities at the* $\pm\nu$, *then*

$$\int_{-\infty}^{\infty} f(u)\, dR(u) = \int_{-\infty}^{\infty} g(t)\, dS(t), \tag{3}$$

both integrals existing as symmetric limits.

The right side of (3) is

$$g(0)\big(S(0^+) - S(0^-)\big) + \int_{-\infty}^{\infty} g(t)dS_0(t),$$

where $S_0(0) = 0$ and $S_0(t) = S(t) - S(0\pm)$ for $t \gtrless 0$. Thus with

[2] A function is <u>piecewise continuous</u> if it has only finitely many discontinuities at each of which its value is the average of left and right limits.

$R_0(u) = R(u) - \big(S(0+) + S(0-)\big)u$, we have (under the same hypotheses as above)

$$\int_{-\infty}^{\infty} f(u)dR_0(u) = \int_{-\infty}^{\infty} g(t)dS_0(t). \qquad (3_0)$$

If also $f(0) = 0$, then

$$\int_{-\infty}^{\infty} f(u)dR_0 = \int_{-\infty}^{\infty} g(t)dS_{00}(t) \qquad (3_{00})$$

where $S_{00}(t) = S_0(t) + M'(0)t$. At their origins, R_0 and S_0 vanish to first order, while R and S_{00} vanish to second order. We have

$$N_0(t) = M(t) + S_0(t) \qquad (1_0)$$

and

$$P(u) = Q_0(u) + R_0(u) \qquad (2_0)$$

where explicitly N_0 differs from N by the omission of the real non-negative zeros and poles.

The proof of (3) given in Section 1 follows Weil [21] with a simplification arising from the hypothesis of only finitely many zeros and poles in $\sigma > 0$. Another real part argument replaces the use of the functional equation in [21] and thus allows us to defer the definition of the gamma factor to Section 2.

Strengthening conditions (n_k) and (p), we now assume

(N_k) Z is of order $< k + 1$. Z has only zeros on $\sigma = 0$; in $\sigma \neq 0$, Z has only real zeros and poles[3]: only zeros or only poles on $\sigma > 0$; on $\sigma < 0$, with only finitely many exceptions, only zeros if $k \equiv 1$ or 2 (mod 4) and only poles if $k \equiv 0$ or 3 (mod 4);

and either

---

[3] For k=1, this could be weakened to read: in $\sigma < 0$, the zeros and poles of Z have bounded imaginary part.

(P) <u>the coefficients</u> $c(\nu)$ <u>in</u> (p) <u>are real and all positive or all negative according as</u> Z <u>has only zeros or only poles on</u> $\sigma > 0$,

or

(P#) <u>there is a function</u> $Z^{\#}$ <u>satisfying</u> (N_k) <u>and</u> (P) <u>for which</u> $|c(\nu)| \ll |c^{\#}(\nu)|$.

In Section 2 we show by the usual gamma factor arguments that (N_k) and (p) imply (n_k), and also derive a trivial bound for S and some qualitative properities of M´. In Section 3, we use these facts, together with (3_{00}) and (3), to give rather parallel proofs of "dual" bounds for R_0 and S, each in terms of the two functions M´ and Q´:

Theorem 2. *If* Z *satisfies* (N_k) *and* (P), *then for* $T \geq 2$ *and* $U \geq 2$,

$$R_0(U) < \frac{|Q'(u)| + 1}{T} + \int_1^T \frac{|M'(t)| + t^{k-1}}{t}\,dt, \tag{5}$$

and

$$S(T) \ll \frac{|M'(T)| + T^{k-1}}{U} + \int_1^U \frac{|Q'(u)| + 1}{u}\,du. \tag{6}$$

If Z *satisfies* (N_k) *and* $(P^{\#})$, *then the same bounds hold with* $|Q'|$ *and* $|M|$ *replaced by* $|Q'| + |Q^{\#\prime}|$ *and* $|M'| + |M^{\#\prime}|$.

In the proof, first (5) is derived, using Theorem 1, from the trivial bound on S mentioned above. Then the analogous trivial special case of (5) is used with Theorem 1, in an analogous way, to get (6).

If there are no positive β, so that $Q' = 0$, then the optimal choices for T and U in (5) and (6) are $T \cong 1$ and $U \cong |M'(T)| + T^{k-1}$, giving

$$R_0(U) \ll 1 \tag{5_0}$$

$$S(T) \ll \log\,(|M'(T)| + 2T^{k-1}). \tag{6_0}$$

If there are positive β and b is the largest of these, so that $Q'(u) < e^{bu}$, then suitable choices of T and U in (5) and (6) give e.g.

$$R_0(U) \ll \begin{cases} U^2 & \text{if } M'(T) \simeq \log T \text{ and } k = 1; \\ e^{(1 - 1/k)bU}, & \text{if } M'(T) \ll T^{k-1}. \end{cases} \qquad (5_b)$$

$$S(T) \ll (|M'(T)| + 2T^{k-1})/\log(|M'(T)| + 2T^{k-1}). \qquad (6_b)$$

(For $k > 1$, the logs in (6_0) and (6_b) are $\cong \log T$.)

The simplest example of a function Z satisfying the hypothesis (N_1) and (P) is given by $Z(s) = 1 - e^{-s}$. Here Theorem 1 gives the Poisson sum formula. In this example, the sawtooth functions R_0 and S are bounded, but do not tend to zero. Here $Q' = 0$ and M' is constant, so (5_0) and (6_0) are best possible in this case.

A second example with $k = 1$ is given by $Z(s) = \zeta(s + 1/2)$ where ζ is, on a Riemann hypothesis, the Riemann zeta function, or the zeta function of an algebraic number field, or an ordinary primitive Hecke L-function, or a Hecke L-function with grossencharacter. Artin L-functions, as quotients of products of Hecke L-functions, are then indirectly covered, directly on Artin's conjecture. In these examples,

$$P(u) = -\sum_{N\mathfrak{a} \leqslant e^u}' \Lambda(\mathfrak{a})\chi(\mathfrak{a})(N\mathfrak{a})^{-1/2}$$

and

$$Q(u) = -4 \sinh \frac{u}{2} + 2u \quad \text{or} \quad 0$$

according as χ is principal or nonprincipal. Here Theorem 1 is Guinand's formula (for weights satisfying (W_1)). In these cases $M'(t) \cong \log t$ for $t \to \infty$, so Theorem 2 gives the standard R.H. estimates $R(u) \ll u^2$ of von Koch [11] and $S(t) \ll (\log t)/\log\log t$ of Littlewood [12]; for the latter, there is also a proof due to Selberg [18], using his approximate formula for S.

A third example, for $k = 2$, is given by $Z(s) = Z_\Gamma(s + 1/2)$ where Z_Γ is the Selberg zeta function attached to a compact Riemann surface of genus $\geqslant 2$. Here Theorem 1 is a version of the Selberg trace formula[4]. In this case $Q'(t) \sim 2e^{\eta/2}$ and $M'(t) \cong t$ for

[4] This does not give a new proof of the trace formula, which logically preceeds the definition of Z_Γ.

$t \to \infty$, so (5_b) and (6_b) give $R_o(U) \ll e^{U/4}$ and $S(T) < T/\log T$. The bound on R_o is due to Randol [15], improving by a factor of $U^{1/2}$ an earlier estimate of Huber [10]. Randol [16] and Hejhal [9] have given proofs, analogous to those of Littlewood and Selberg mentioned in the previous example, for the bound on S.

More generally, for $k \geq 2$, we may take $Z(s) = Z_\Gamma(s+\rho_o,\chi)$ where Z_Γ is the Selberg zeta function attached to a compact space form of a k-dimensional symmetric space of rank 1; here ρ_o and χ are as in Gangolli [4]. For compact hyperbolic space forms, there are only finitely many negative zeros or poles if k is odd, while there are $\cong t^k$ of them on $[-t,0]$ for large t if k is even. By (14) and (15) below, this gives $M'(t) \ll t^{k-1}$ in both cases. The corresponding bound for $R_0(U)$ in (5_b) is due to Randol [17] (whose proof suggested our proof of (5)); the bound $S(T) < T^{k-1}/\log T$ from (6_b) was proved, in even greater generality, by Berard [1].

Guinand derived (3) (in the case $Z(s) = \zeta(s + 1/2)$ on R.H.) for a wider class of weights starting with a special case of (3) which he reformulated to show that $R_0(u)/u$ and $S_0(t)/t$ are connected by a unitary operator closely related to the Fourier transform. In Section 3 we use a similar operator to give a correspondence between second moments in the distribution of zeros and primes in corresponding short intervals:

Theorem 3. *If, besides satisfying* (N_1) *and* (P), Z *has order* $< \frac{3}{2}$, *then for positive* $\varepsilon \to 0$

$$\int_{-\infty}^{\infty} \left|\frac{R_0\big((1+\varepsilon)u\big)-R_0(u)}{u}\right|^2 du \sim \frac{1}{2\pi}\int_{-\infty}^{\infty}\left(\frac{S_0\big((1+\varepsilon)t\big)-S_0(t)}{t}\right)^2 dt. \tag{7}$$

Supposing in addition that $M'(t) \sim c \log|t|$ as $|t|\to\infty$, we show by a method of Mueller [14] that the integral on the right is $\gtrsim c\varepsilon \log^2 \frac{1}{\varepsilon}$, with asymptotic equality if and only if

$$\sum_{0<\gamma\leq T} \frac{m(i\gamma)}{\gamma} \sum_{\gamma\leq\gamma'\leq(1+\varepsilon)\gamma} m(i\gamma') \sim \sum_{0<\gamma\leq T}\frac{m(i\gamma)}{\gamma} \tag{8}$$

$$\left(\sim \frac{1}{2} c \log^2 T\right)$$

for $T \to \infty$, $\varepsilon T \to 0$. In the case $Z(s) = \zeta(s + 1/2)$ on R.H., it follows from Fujii's second moment estimates for zeros in short intervals [2] that the left side of (8), and therefore also (7), is $\ll \varepsilon \log^2 \frac{1}{\varepsilon}$. In this case, condition (8) is a consequence of Mueller's "essential simplicity" condition for which she finds an arithmetic equivalent, and which in turn is a consequence of Montgomery's pair correlation conjecture [13]. For other second moment correspondences, with various normalizations, see Goldston and Montgomery [6] and the papers cited there and in [3]. An unachieved goal of analytic number theory is to find some pair of equivalent second moment asymptotic evaluations which can be proved in the case $Z(s) = \zeta(s + 1/2)$ on R.H. making use of the arithmetic nature of the coefficients. This is of course part of the more general goal of explicating duality in this part of analytic number theory: translating completely a definition of prime number into an understanding of the critical zeros of the Riemann zeta function.

1. Guinand's formula

It suffices to prove Theorem 1 for real g, i.e. for $f(-u) = \overline{f}(u)$. Since $R(-u) = -\overline{R}(u)$, this gives

$$\int_{-\infty}^{\infty} f(u)dR(u) = 2\mathrm{Re} \int_{0}^{\infty} f(u)dR(u). \tag{9}$$

Following Weil [21], we next write, for $\sigma_1 < \sigma < \sigma_2$,

$$\int_{0}^{\infty} f(u)dP(u) = \lim_{T\to\infty} \frac{1}{2\pi i} \int_{\sigma - iT}^{\sigma + iT} \tilde{f}(s)Z'/Z(s)ds. \tag{10}$$

This follows by Fourier inversion, at $u = 0$, from the formula

$$\tilde{f}(s)Z'/Z(s) = \int_{-\infty}^{\infty} e^{su} \sum_{\nu} c(\nu)f(u+\nu)\, du,$$

which is gotten from the product of the absolutely convergent integral for $\tilde{f}(s)$ by the absolutely convergent series for $Z'/Z(s)$ by making changes of variable and reversing the order of summation and integration. To justify the Fourier inversion, it suffices to observe that both

$$e^{\sigma u}\sum_{\nu} c(\nu)f(u+\nu)$$

and its derivative belong to $L^1(\mathbf{R})$ and are piecewise-continuous near $u = 0$. In fact, the series converges uniformly, as does the series with f replaced by f′ since

$$e^{\sigma u}\sum_{\nu} |c(\nu)|e^{-\sigma_2|u+\nu|} \leqslant \sum_{\nu} |c(\nu)|e^{-\sigma\nu} .$$

Next, the order condition on Z in vertical strips together with the fact that $Z'/Z(s) \to 0$ exponentially as $\sigma \to \infty$ implies that for $|T| \to \infty$

$$N(T + 1) - N(T) + |S(T)| + \int_T^{T+1} \int_0^{\infty} |Z'/Z(\sigma + i\, t)|d\sigma dt = o(T^k). \tag{11}$$

In fact, these bounds follow easily from the following standard partial fraction approximation:

Lemma 1. (Jensen, Landau): *If* $h = h(z)$ *is analytic and* $|h(z)/h(0)| \leqslant B$ *in* $|z| \leqslant r$, *then* h *has* $\ll_{\lambda} \log B$ *zeros* z_i *in* $|z| < \lambda r$, *for each* $\lambda < 1$, *and*

$$\frac{h'}{h}(z) = \sum \frac{1}{z - z_i} + O_{\lambda,\mu}\left(\frac{\log B}{r}\right)$$

uniformly in $|z| \leqslant \mu r$, *for each* $\mu < \lambda$.

The bound $\tilde{f}(s) \ll |t|^{-k}$ together with (11) justifies moving the line of integration in (10) to $\sigma = 0$, giving

$$\int_0^{\infty} f(u)dP(u) = \lim_{T\to\infty} \{\sum_{\substack{\beta\geqslant 0\\ |\gamma|\leqslant T}} \tilde{f}(\rho)m(\rho) + \frac{1}{2\pi} \int_{-T}^{T} g(t)Z'/Z(it)dt\}, \tag{12}$$

where each term with $\beta = 0$ is weighted with the factor ½ and at each pole of Z'/Z on $\sigma = 0$ a principal value is taken. By the reflection principle, $\overline{\tilde{f}(\rho)} = \tilde{f}(\rho^*)$, so

$$2\mathrm{Re} \int_0^{\infty} f(u)dP(u) = \sum_{\beta>0} \left(\tilde{f}(\rho) + \tilde{f}(\rho^*) - 2g(\gamma)\right)m(\rho) \tag{13}$$

$$+ \lim_{T\to\infty} \{\sum_{\substack{\beta\geq 0 \\ |\gamma|\leq T}} g(\gamma)m(\rho) - \int_{-T}^{T} g(t)dM(t)\},$$

the dash indicating that in the last sum terms with $\beta > 0$ are multiplied by two. The first sum is

$$\int_{-\infty}^{\infty} f(u)dQ(u) = 2 \text{ Re} \int_{0}^{\infty} f(u)dQ(u),$$

since $Q(-u) = -Q(u)$ via $\rho^* = -\bar{\rho}$. Using $dN - dM = dS$ and combining (13) with (9) gives

$$\int_{-\infty}^{\infty} f(u)dR(u) = \int_{-\infty}^{\infty} g(t)dS(t),$$

both integrals existing as symmetric limits.

2. Consequences of (N_k) and (p) for M′ and S.

For each nonzero function Z meromorphic of order $< k + 1$, there is a gamma factor, i.e. a function G meromorphic of order $< k + 1$ with all zeros and poles in $\sigma < 0$ for which $X = GZ$ is real on $\sigma = 0$. With the normalizations

$$X'/X(s) = \frac{m(0)}{s} + O(s^k), \qquad X(s) \sim (is)^{m(0)}$$

for $s\to 0$, G is uniquely determined. In fact X is $(is)^{m(0)}$ times the standard genus k Weierstrass product over the $\rho \neq 0$ with $\beta \geq 0$ and the ρ^* corresponding to ρ with $\beta > 0$. On $\sigma = 0$, the factors with $\beta = 0$ are real, and the factors corresponding to ρ and ρ^* for $\beta > 0$ are conjugate. Explicitly,[5)]

$$G(s) = e^{p(s)} \prod_{\beta<0} E_k(s/\rho)^{-m(\rho)} \prod_{\beta>0} E_k(s/\rho^*)^{m(\rho)},$$

where p is a polynomial of degree k (essentially the negative of the

---

5) In certain cases of the Selberg zeta function of order 2, Vigneras [20] has written G explicitly as a finite product of Barnes double gamma functions.

analogous polynomial in the corresponding expression for Z), and $E_k(z)$ is the standard genus k Weierstrass factor. From $E'_k/E_k(z) = z^k/(z-1)$ it follows that

$$G'/G(s) = p'(s) - \sum_{\beta<0} \frac{(s/\rho)^k}{(s-\rho)} m(\rho) + \sum_{\beta>0} \frac{(s/\rho^*)^k}{(s-\rho^*)} m(\rho).$$

At this point we invoke hypothesis (N_k) and get

$$M'(t) = L(t) + O(|t|^{k-1}), \tag{14}$$

where

$$L(t) = \begin{cases} \dfrac{i^k}{\pi} \displaystyle\sum_{\beta<0}' \dfrac{t^k}{(\beta^2 + t^2)\beta^{k-1}} m(\beta) & (k \text{ even}); \\ \\ \dfrac{i^{k+1}}{\pi} \displaystyle\sum_{\beta<0}' \dfrac{t^{k+1}}{(\beta^2 + t^2)\beta^k} m(\beta) & (k \text{ odd}). \end{cases} \tag{15}$$

In (15) the dash indicates that the sum in either over real negative zeros or real negative poles. The contribution of any exceptional poles or zeros, terms with $\beta > 0$, and p' has been put in the O-term in (14).

The function L is even and non-negative in all cases. Since in each sum on the right in (15) all terms have the same sign, it follows that for $t > 0$

$L(t)/t^{k-2}$ increases and $L(t)/t^k$ decreases (k even);

$L(t)/t^{k-1}$ increases and $L(t)/t^{k+1}$ decreases (k odd).

In particular, in both cases L <u>changes slowly</u>, i.e. changes only by a bounded factor when t changes by a bounded factor. Since G has $\ll r^{k+1-\delta}$ zeros and poles in $|s| \leqslant r$, (15) also gives

$$L(t) \ll |t|^{k-\delta} \tag{16}$$

in both cases.

We next show that in each vertical strip $|\sigma| \leq \sigma_2$,

$$\operatorname{Re} G'/G(s) \ll L(t) + |t|^{k-1} \qquad (t \text{ large}). \tag{17}$$

Since p' has degree k-1, it suffices to observe that

$$\operatorname{Re} \frac{s^k}{(s - \beta)\beta^k} = \frac{(\sigma - \beta)\operatorname{Re}(s^k) + t \operatorname{Im}(s^k)}{((\sigma - \beta)^2 + t^2)\beta^k},$$

which for $|\sigma| \leq \sigma_2$ and $\beta \leq -2\sigma_2$ is

$$\ll \frac{t^k}{(\beta^2 + t^2)\beta^{k-1}} \quad \text{or} \quad \frac{t^{k+1} + t^{k-1}|\beta|}{(\beta^2 + t^2)\beta^k}$$

according as k is even or odd. The exceptional terms and other terms with $\beta > -2\sigma_2$ contribute $\ll |t|^{k-1}$ in both cases.

Next we conclude from (17), using also (p), that

$$|Z(s)| \leq \exp\left(O(L(t) + |t|^{k-1})\right) \qquad (t \text{ large}). \tag{18}$$

We may suppose that $\sigma_2 > \sigma_1$, so by (p) Z is bounded on $\sigma = \sigma_2$. By the reflection principle, Z satisfies $GZ(s^*) = \overline{GZ}(s)$ (functional equation !), so

$$Z(-\sigma_2 + it) \ll \exp\left(\int_{-\sigma_2}^{\sigma_2} \operatorname{Re} G'/G(\sigma + it)d\sigma \right).$$

It follows from (17) that (18) holds also for $\sigma = -\sigma_2$. For $|\sigma| < \sigma_2$, put $F_s(w) = Z(w) \exp(-(w-s)^{4k})$. On the horizontal sides of the rectangle (in the $w = u + iv$ plane) bounded by $u = \pm \sigma_2$ and $v - t = \pm \frac{1}{2} t$ this function is bounded for large t since Z has order $< k + 1$ and $\operatorname{Re}(w-s)^{4k} \cong t^{4k}$. On the vertical sides it is bounded by the right side of (18) by what we have shown and the fact that L changes slowly. Since $Z(s) = F_s(s)$, the maximum principle now gives (18) for $|\sigma| < \sigma_2$.

In particular, (18) and (16) show that Z has order $< k$ in vertical strips. Thus (N_k) and (p) imply (n_k).

Using (18), the proof of (11) now gives (for large T)

$$N(T+1) - N(T) + |S(T)| + \int_T^{T+1} \int_0^\infty |Z'/Z(\sigma + it)| \, d\sigma dt \tag{19}$$

$$\ll L(T) + |T|^{k-1}.$$

It follows that

$$\int_T^{T+1} |dS_{00}(t)| \ll L(T) + |T|^{k-1} \qquad \text{(for large T).} \qquad (20)$$

In fact, with $M_{00}(t) = M(t) - M'(0)t$, the integral is

$$\int_T^{T+1} |dN_0(t)| + \int_T^{T+1} |dM_{00}(t)| \ll |\Delta_T^{T+1}N_0| + |\Delta_T^{T+1}M| + |T|^{k-1}$$

$$\ll |\Delta_T^{T+1}N| + |\Delta_T^{T+1}S| + |T|^{k-1};$$

here we have used the monotonicity of N_0 and the positivity of L.

3. Bounds for R_0 and S.

Beginning with R_0 in case (P), we have for $U > 1$, supposing all $c(\nu) \geqslant 0$,

$$P(U) - P(1) = \int_{-\infty}^{\infty} \phi_{1,U}(u)dP(u) \lesseqgtr \int_{-\infty}^{\infty} f^{\pm}(u)dP(u)$$

for any compactly supported continuous majorant/minorant of the characteristic function $\phi_{1,U}$ of $[1,U]$. On subtracting

$$Q_0(U) - Q_0(1) = \int_{-\infty}^{\infty} \phi_{1,U}(u)dQ_0(u).$$

From this and using (3_{00}), we get, provided $f^{\pm}$ is sufficiently differentiable and $f^{\pm}(0) = 0$,

$$R_0(U) - R_0(1) \lesseqgtr \int_{-\infty}^{\infty} (f^{\pm}(u) - \phi_{1,U}(u))dQ_0(u) + \int_{-\infty}^{\infty} g^{\pm}(t)dS_{00}(t) \qquad (21)$$

where $g^{\pm}$ is the Fourier transform of f. (If all $c(\nu) \leqslant 0$, the inequalities are reversed).

For $U > 2$ and $T > 2$ we take for $f^{\pm}$ in (21) the characteristic function of the interval $[1 \mp T^{-1}, U \pm T^{-1}]$ convolved with $Tf(Tu)$, where f is any nonnegative C^{∞} function supported in $(-1,1)$ and satisfying $\int_{-\infty}^{\infty} f(\alpha)d\alpha = 1$. The first integral in (21) is then in modulus at most

$$\int_{1-T^{-1}}^{1+T^{-1}} |Q_0'(u)|du + \int_{U-T^{-1}}^{U+T^{-1}} |Q_0'(u)|du \ll \frac{|Q'(U)| + 1}{T}.$$

The Fourier transform of $\phi_{1,U}$ is $\ll |t|^{-1}$, and the Fourier transform of f satisfies $g(\beta) \ll (1 + |\beta|)^{-(k+2)}$, from which

$$g^{\pm}(t) \ll |t|^{-1}(1 + |t/T|)^{-(k+2)}.$$

Thus the second integral in (21) is

$$\ll \int_{-T}^{T} \frac{|dS_{00}(t)|}{|t|} + T^{k+2} \int_{|t|>T} \frac{|dS_{00}(t)|}{|t|^{k+3}}.$$

Because of the double zero of S_{oo} at $t=0$, the interval $[-1,1]$ contributes $O(1)$ to the first integral. Using (20), we thus get the bound

$$\ll \int_{1}^{T} \frac{L(t) + t^{k-1}}{t} dt + T^{k+2} \int_{T}^{\infty} \frac{L(t) + t^{k-1}}{t^{k+3}} dt .$$

The t^{k-1} terms here contribute $\ll T^{k-1} + \log T$; since $L(t)/t^{k+1}$ is decreasing and L changes slowly, we have

$$T^{k-2} \int_{T}^{\infty} \frac{L(t)}{t^{k+3}} dt \leqslant L(T) \ll \int_{1}^{T} \frac{L(t)}{t} dt.$$

Using $L(t) \ll |M'(t)| + t^{k-1} + 1$, this gives

$$R_0(U) \ll \frac{|Q'(U)| + 1}{T} + \int_{1}^{T} \frac{|M'(t)| + t^{k-1}}{t} dt. \tag{22}$$

For example, the choice $T = 2$ gives the trivial bound

$$R_0(U) \ll |Q'(U)| + 1 \tag{23}$$

which will play the same role in getting a refined bound for S as (19) did in getting the bound (22) for R_0. First, it follows from (23) that

$$\int_{U}^{U+1} |dR(u)| \ll |Q'(U)| + 1 \qquad \text{(for large U)} \tag{24}$$

since the integral here is

$$\int_U^{U+1} |dP(u)| + \int_U^{U+1} |dQ(u)| = |\Delta_U^{U+1} P| + |\Delta_U^{U+1} Q|$$

$$\ll |\Delta_U^{U+1} Q| + |\Delta_U^{U+1} R_0| + 1 ;$$

here we have made use of the monotonicity of P and Q, which follow from the assumptions (P) and (N).

Now we bound S. For each $T > 0$,

$$N(T) \lesseqgtr \int_{-\infty}^{\infty} g^{\pm}(t)dN(t),$$

where this time $g^{\pm}$ is a majorant/minorant of the characteristic function ψ_T of $[0,T]$ and is the Fourier transform of a function $f^{\pm}$ in (W_k). On subtracting

$$M(T) = \int_{-\infty}^{\infty} \psi_T(t)dM(t)$$

from this and using Theorem 1 in reverse, we get

$$S(T) - S(0) \lesseqgtr \int_{-\infty}^{\infty} \big(g^{\pm}(t) - \psi_T(t)\big)dM(t) + \int_{-\infty}^{\infty} f^{\pm}(u)dR(u). \qquad (25)$$

For the functions $g^{\pm}$ we use the following construction, which was suggested by Goldston's use [5] of Selberg's kernels[6)] [7], [19] for a related purpose in the case of the Riemann zeta function on R.H.:

Lemma 2. *Let* k *be a positive integer. For each* $L > 0$ *there are real functions* $f_L^{\pm}$ *supported on* (-1,1) *with* k *continuous derivatives, whose Fourier transforms* $g_L^{\pm}$ *satisfy* $g_L^{\pm} \gtrless \phi_L$ *where* ϕ_L *is the characteristic function of* $[0,L]$, *and for which* $f_L^{\pm}(\alpha) \ll |\alpha|^{-1}$ *and*

---

6) For k=1 we could use Selberg's kernels.

$$g_L^+(\beta) - g_L^-(\beta) \ll (1 + |\beta|)^{-(k+1)} + (1 + |\beta - L|)^{-(k+1)} ,$$

the implied constants depending only on k.

Proof. For each positive integer k, we put

$$\delta_k(\beta) = d_k\left(\frac{\sin \beta/k}{\beta}\right)^k ,$$

with d_k chosen so that

$$\int_{-\infty}^{\infty} \delta_k(\beta)d\beta = 1.$$

Thus δ_k is even and is the Fourier transform of a function in C^{k-1} supported in $(-1,1)$. For $\beta > 0$, we have

$$\int_{\beta}^{\infty} \delta_{k+2}(b)db \ll (1 + |\beta|)^{-(k+1)}$$

Since

$$\begin{aligned}(\delta_{k+2} * \phi_L)(\beta) &= \int_{\beta -L}^{\beta} \delta_{k+2}(b)db, \\ &= 1 - \left(\int_{\beta}^{\infty} + \int_{-\infty}^{\beta-L}\right) \delta_{k+2}(b)db,\end{aligned}$$

it follows that

$$\delta_{k+2} * \phi_L(\beta) \ll (1 + |\beta|)^{-(k+1)} + (1 + |\beta-L|)^{-(k+1)} .$$

For odd k, we have

$$(1 + |\beta|)^{-(k+1)} \cong \int_{\beta-1}^{\beta} \delta_{k+1}(b)db = (\delta_{k+1} * \phi_1)(\beta) .$$

It follows that for sufficiently large c_k, the functions

$$g_L^{\pm}(\beta) = (\delta_{k+2} * \phi_L) \pm c_k\big((\delta_{k+1} * \phi_1)(\beta) + (\delta_{k+1} * \phi_1)(\beta-L)\big)$$

have all the required properties. This completes the proof, since we may suppose k odd.

In (25), we take $f^{\pm}(u) = U^{-1}f_{TU}^{\pm}(uU^{-1})$. Thus $f^{\pm} \in C^k(-U,U)$, and $f^{\pm}(u) \ll |u|^{-1}$. Using (24), it follow that the second integral in (25) is

$$\ll \int_{-U}^{U} \frac{|dR(u)|}{|u|} \ll \int_{1}^{U} \frac{|Q'(u)| + 1}{u}\, du,$$

the interval $[-1,1]$ contributing $\ll 1$.

From $g^{\pm}(t) = g_{TU}^{\pm}(tU)$, it follows that $g^{\pm}(t) \gtrless \psi_T(t)$, and

$$g^{+}(t) - g^{-}(t) \ll (1 + |t|U)^{-(k+1)} + (1 + |t-T|U)^{-(k+1)}.$$

Since $M'(t) \ll L(t) + |t|^{k-1} + 1$, the first integral in (25) is

$$\ll \int_{-\infty}^{\infty} \frac{L(t) + |t|^{k-1} + 1}{(1 + |t|U)^{k+1}}\, dt + \int_{-\infty}^{\infty} \frac{L(t) + |t|^{k-1} + 1}{(1 + |t-T|U)^{k+1}}\, dt.$$

The $|t|^{k-1} + 1$ parts here contribute $\ll T^{k-1}/U$. Since L changes slowly,

$$\int_{T/2}^{2T} \frac{L(t)\, dt}{(1 + |t-T|U)^{k+1}} \ll \frac{L(T)}{U}.$$

The rest of the L part of the second integral above is bounded by the L part of the first integral, which is

$$\ll \int_{0}^{\infty} \frac{L(t)\, dt}{(1 + tU)^{k+1}} \ll \int_{0}^{\infty} \frac{t^{k-\delta} dt}{(1 + tU)^{k+1}} \ll \frac{1}{U}.$$

This gives

$$S(T) \ll \frac{|M'(T)| + T^{k+1}}{U} + \int_{1}^{U} \frac{|Q'(u)| + 1}{u}\, du. \tag{26}$$

In case $(P^{\#})$ in the argument bounding R_0, the same choice of $f^{\pm}$ gives

$$P(U) - P(1) = \int_{-\infty}^{\infty} f^{\pm}(u)dP(u) + O\Big(\int_{1-T^{-1}}^{1+T^{-1}} |dP^{\#}(u)| + \int_{U-T^{-1}}^{U+T^{-1}} |dP^{\#}(u)|\Big).$$

Since

$$\int_{\cdot}^{\cdot} |dP^{\#}(u)| = |\Delta P^{\#}| \leqslant |\Delta Q^{\#}| + |\Delta R^{\#}|,$$

we get on using the bound corresponding to (22) for $R^{\#}$ that (22) also holds for R, with $|Q'|$ replaced by $|Q'| + |Q^{\#\prime}|$ and $|M'|$ by $|M'| + |M^{\#\prime}|$. In particular, we get the correspondingly modified (23). To get the correspondingly modified (24), we use

$$\int_U^{U+1} |dP(u)| \ll |\Delta_U^{U+1} P^{\#}|$$

in the displayed line below (24). From this point, the argument proceeds as with case (P) to get the correspondingly modified (26).

4. A second moment correspondence

For k = 1, we may take for f in (3_0) the characteristic function of the interval [0,u] (u > 0) renormalized to take the value ½ at 0 and u. This gives the special formula

$$R_0(u) = \lim_{T\to\infty} \int_{-T}^{T} \frac{e^{itu} - 1}{it} \, dS_0(t),$$

which is valid for all $u \neq \pm \nu$ by our definition of R(u) for $u \leqslant 0$. Following Guinand [8], this may be reformulated as

$$\frac{R_0(u)}{u} = -\lim_{T\to\infty} \int_{-T}^{T} \frac{S_0(t)}{t} \, h(tu)dt, \tag{27}$$

with

$$h(\theta) = \theta \frac{d}{d\theta} \frac{e^{i\theta} - 1}{i\theta} = e^{i\theta} - \frac{e^{i\theta} - 1}{i\theta} . \tag{28}$$

In fact, after integrating by parts and dividing by u, we get

$$\frac{R_0(u)}{u} = - \lim_{T\to\infty} \int_{-T}^{T} S_0(t) \frac{d}{dt} \frac{e^{itu} - 1}{itu} \, dt,$$

which is (27), the $\pm$ T terms vanishing in the limit since $S_0(t) = o(t)$ for large $|t|$.

Lemma 3. *The kernel* h *defined in* (28) *gives a linear isometry*

$$H: \ \phi(t) \to \lim_{T\to\infty} \frac{1}{\sqrt{2\pi}} \int_{-T}^{T} \phi(t)h(tu)dt$$

from $L^2(\mathbf{R})$ *to* $L^2(\mathbf{R})$. *(In fact, the operator* H *is also invertible,* *with* $H^{-1} = H^*$, *i.e. unitary, but we don't need this*).

Proof. It suffices to show that for $U \to \infty$,

$$\int_{-U}^{U} |H\phi(u)|^2 du \to \int_{-\infty}^{\infty} |\phi(t)|^2 dt \tag{29}$$

for $\phi \in C_0^\infty(\mathbf{R})$, since H then extends by continuity from this dense subset to a linear isometry on all of $L^2(\mathbf{R})$. From (28) we get the identity

$$h(t_1u)\bar{h}(t_2u) = e^{i(t_1-t_2)u} - \frac{\partial}{\partial u}\frac{(e^{it_1u}-1)(e^{-it_2u}-1)}{t_1t_2u}$$

from which it follows that

$$\int_{-U}^{U} |\phi(u)|^2 du = \int_{-U}^{U} |F\phi(u)|^2 du - \left(|G\phi(U)|^2 + |G\phi(-U)|^2\right)/U,$$

where F is the Fourier transform and $G = H - F$. For $\phi \in C_0^\infty(\mathbf{R})$, $G\phi$ is bounded, so (29) follow from the corresponding (Plancherel) formula for F.

Supposing now that, besides satisfying (N_1) and (P) Z has order $< 3/2$, we have $L(t) \ll t^{1/2-\delta}$ so by (19) $S(t) < |t|^{1/2-\delta}$ for large $|t|$. It follows that $S_0(t)/t \in L^2(\mathbf{R})$ from which (27) and the lemma give $R_0(u)/u \in L^2(\mathbf{R})$.

For each $\lambda > 0$, (27) gives

$$\frac{R_0(\lambda u)}{\lambda u} = -\lim_{T\to\infty}\int_{-T}^{T}\frac{S_0(t/\lambda)}{t}h(tu)dt.$$

Combining this with (27) and the lemma gives

$$\int_{-\infty}^{\infty}\left(\frac{\lambda R_0(u) - R_0(\lambda u)}{\lambda u}\right)^2 du = \frac{1}{2\pi}\int_{-\infty}^{\infty}\left(\frac{S_0(t) - S_0(t/\lambda)}{t}\right)^2 dt. \tag{30}$$

In this equation, R_0 and S_0 could be replaced by R and S.

Next we will work towards an asymptotic evaluation of the integral on the right as $\lambda \to 1+$. For convenience we replace t by λt and put $\lambda = 1 + \varepsilon$.

Supposing that $M'(t) \to \infty$ as $t \to \infty$, Theorem 2 gives $S(t) \ll M'(t)/\log M'(t)$, from which for $\lambda \cong 1$

$$\int_T^\infty \left(\frac{S(\lambda t) - S(t)}{t}\right)^2 dt \ll \int_T^\infty \left(\frac{M'(t)}{t \log M'(t)}\right)^2 dt. \tag{31}$$

Next, we have

$$\int_0^T \left(\frac{M(\lambda t) - M(t)}{t}\right)^2 dt \leqslant \varepsilon^2 \int_0^{\lambda T} (M'(t))^2 dt. \tag{32}$$

In fact, the integral on the left is

$$\int_0^T \left(\frac{1}{t} \int_t^{\lambda t} M'(\tau)d\tau\right)^2 dt \leqslant \varepsilon \int_0^{\lambda T} \int_t^{\lambda t} (M'(\tau))^2 d\tau \frac{dt}{t}$$
$$\leqslant \varepsilon \int_0^{\lambda T} (M'(\tau))^2 \int_{\tau/\lambda}^{\tau} \frac{dt}{t} d\tau ,$$

from which (32) follows.

Finally, we have

$$J(T) \leqslant \int_0^T \left(\frac{N(\lambda t) - N(t)}{t}\right)^2 dt \leqslant J(\lambda T), \tag{33}$$

with

$$J(T) = \sum_{0<\gamma,\gamma'\leqslant T}{}' m(i\gamma)m(i\gamma') \int_{\max(\gamma,\gamma')/\lambda}^{\min(\gamma,\gamma')} dt/t^2 ,$$

the dash indicating that the sum is over all pairs γ, γ' for which $\max(\gamma,\gamma') \leqslant \lambda \min(\gamma,\gamma')$. Thus

$$J(T) = \varepsilon \sum_{0<\gamma\leqslant T} \frac{m(i\gamma)^2}{\gamma} + 2 \sum_{0<\gamma\leqslant T} m(i\gamma) \sum_{\gamma<\gamma'\leqslant\lambda\gamma} m(i\gamma')\left(\frac{\lambda}{\gamma'} - \frac{1}{\gamma}\right).$$

On the hypothesis $M'(t) \to \infty$, we have

$$\sum_{0<\gamma\leqslant T} \frac{m(i\gamma)}{\gamma} \sim \int_1^T \frac{M'(t)}{t} dt \qquad (T \to \infty). \tag{34}$$

In fact, the two sides differ by

$$\int_1^T \frac{dS(t)}{t} + O(1) = \frac{S(T)}{T} + \int_1^T \frac{S(t)}{t^2} dt + O(1),$$

and the bound $S(t) \ll M'(t)/\log M'(t)$ shows that this is of smaller order than the right side of (34). It follows that

$$J(T) \gtrsim \varepsilon \int_1^T \frac{M'(t)}{t} dt, \qquad (T \to \infty,\ \varepsilon T \to 0) \tag{35}$$

with asymptotic equality if and only if (for $T \to \infty$, $\varepsilon T \to 0$)

$$\text{(A)} \qquad \sum_{0<\gamma\leq T} \frac{m(i\gamma)}{\gamma} \sum_{\gamma\leq\gamma'\leq(1+\varepsilon)\gamma} m(i\gamma') \sim \sum_{0<\gamma\leq T} \frac{m(i\gamma)}{\gamma} .$$

Combined with (31), (32), (33), this gives

$$\int_0^\infty \left(\frac{S(\lambda t) - S(t)}{t}\right)^2 dt \gtrsim \varepsilon \int_1^T \frac{M'(t)}{t}\, dt \qquad (T \to \infty,\ \varepsilon T \to 0) \qquad (36)$$

with asymptotic equality if and only if (A) holds, provided also

$$\int_T^\infty \left(\frac{M'(t)}{t \log M'(t)}\right)^2 dt = o\left(\varepsilon \int_1^T \frac{M'(t)}{t}\, dt\right)$$

and

$$\varepsilon \int_0^T \left(M'(T)\right)^2 dt = o\left(\int_1^T \frac{M'(t)}{t}\, dt\right).$$

For $M'(t) \sim c \log t$, both of these provisos are satisfied if only $\varepsilon T \to 0$ sufficiently slowly, which we may suppose.

References

[1] Berard, P.H., On the wave equation on a compact Riemannian manifold without conjugate points. *Math. Z.* **155** (1977), 249-276.

[2] Fujii. A., On the zeros of Dirichlet L-function I. *T.A.M.S.* **196** (1977), 249-276.

[3] Gallagher, P.X., Pair correlation of the zeros of the zeta function. *J. Reine Agnew Math* **362** (1985), 72-86.

[4] Gangolli, R., Zeta functions of Selberg's type for compact space forms of symmetric spaces of rank 1. *Illinois J. Math* **21** (1977), 1-41.

[5] Goldston, D.A., Lecture at the 1984 Stillwater Conference on Analytic Number Theory and Diophantine Problems.

[6] Goldston, D.A. and Montgomery, H.L., Pair correlation of zeros and primes in short intervals, Proc. 1984 Stillwater Conference on Analytic Number Theory and Diophantine Problems, Birkhauser Verlag (this volume).

[7] Graham, S.W. and Vaaler, J.D., A class of extremal functions for the Fourier transform, *T.A.M.S.* **265** (1981), 283-302.

[8] Guinand, A.P., A summation formula in the theory of prime numbers, *Proc. London Math. Soc.* (2) **50** (1984), 107-119.

[9] Hejhal, D.A., The Selberg trace formula for PSL(2,**R**), Vol. 1, *Lecture Notes in Mathematics*, **584** (1976).

[10] Huber, H., Zur analytischen theorie hyperbolisher Raumformen und Bewegungsgruppen II, *Math. Ann.* **142** (1961), 385-398 and **143** (1961), 463-464.

[11] von Koch, H., Sur la distribution des nombres premiers, *Acta Math*, **24** (1901), 159-182.

[12] Littlewood, J.E., On the zeros of the Riemann zeta function, *Proc. London Math. Soc.* (2) **24** (1924), 295-318.

[13] Montgomery, H.L., The pair correlation of zeros of the zeta function. Analytic Number Theory (Proc. Sympos. Pure Math. **24** St. Louis Univ., St. Louis, MO. 1972 A.M.S. Providence, R.I. 1973.

[14] Mueller, J.H., Arithmetic equivalent of essential simplicity of zeta zeros, *T.A.M.S.*, **275** (1983), 175-183.

[15] Randol, B., On the asymptotic distribution of closed geodesics on compact Riemann surfaces, *T.A.M.S.*, **233** (1977), 241-247.

[16] Randol, B., The Riemann hypothesis for Selberg´s zeta function and the asymptotic behavior of eigenvalues of the Laplace operator. *T.A.M.S.* **236** (1978), 209-223.

[17] Randol. B., The Selberg trace formula, in Eigenvalues in Riemannian geometry, by Isaac Chavel (to appear).

[18] Selberg, A., On the remainder term for N(T), *Avhandlinger Norske Vid. Akad. Oslo* (1944) No. 1.

[19] Vaaler, J.D., Some extremal functions in Fourier analysis, *B.A.M.S.*, **12** (1985), 183-212.

[20] Vigneras, M. F., L´equation functionelle de la fonction zeta de Selberg de la group modulaire PSL(2,**Z**), *Asterisque*, **61** (1979), 235-249.

[21] Weil, A., Sur les "formules explicites" de la Theorie des nombres premiers, *Comm. Sem. Math. Univ. Lund* [Medd. Lunds Univ. Mat. Sem.] Tome Supplementaire 1952, 252-265.

[22] Weil, A., Two lectures on number theory, past and present, *Enseignment Mat.* (2) **20** (1974), 87-110.

P.X.Gallagher
Department of Mathematics
Columbia University
New York, N.Y. 10027

ANALYTIC NUMBER THEORY ON GL(r,R)

Dorian Goldfeld*

with an appendix by **Solomon Friedberg**

1. Introduction.

There has been much progress in recent years on some classical questions in analytic number theory. This has been due in large part to the fusion of harmonic analysis on GL(2,**R**) with the techniques of analytic number theory, a method inspired by A. Selberg [17]. A lot of impetus has been gained by the trace formula of Kuznetsov [11], [12], which relates Kloosterman sums with eigenfunctions of the Laplacian on GL(2,**R**) modulo a discrete subgroup. We cite some of the most striking applications.

Letting

$$S(m,n;c) = \sum_{\substack{a=1\\ (a,c)=1\\ a\bar{a} \equiv 1 \bmod c}}^{c} e^{2\pi i\left(\frac{am+\bar{a}n}{c}\right)}$$

denote the classical Kloosterman sum, Kuznetsov [12] has shown that

$$\sum_{c \leq X} \frac{S(m,n;c)}{c} = O\left(X^{1/6}(\log X)^{1/3}\right)$$

Where the O-constant depends at most on m and n. This is the first result of its kind showing a cancellation between Kloosterman sums. A simpler proof of this, with a higher power of (log x), is given in Goldfeld-Sarnak [5]. It is based on the study of the zeta function

* The author gratefully acknowledges the generous support of the Vaughn Foundation

$$\sum_{c=1}^{\infty} \frac{S(m,n;c)}{c^{2s}}$$

as initiated by Selberg [19].

If P_n denotes the nth prime, then Iwaniec and Pintz [8] have proved

$$P_{n+1} - P_n = O\left(P_n^{1/2 + 1/21 + \varepsilon} \right) .$$

Also, Fouvry [4] has shown that there exist infinitely many primes p such that p-1 has a prime factor greater than $p^{2/3}$. This, together with some unpublished results of L.M. Adleman and R. Heath-Brown (extensions of Sophie Germaine's criterion) enable one to show that

$$x^p + y^p = z^p \qquad (p \nmid xyz)$$

is impossible for positive integers x,y,z for infinitely many primes p.

The excellent survey article of Iwaniec [7] lists many more applications of harmonic analysis on GL(2,**R**) to analytic number theory. In view of these advances, it is natural , therefore, to ask if the fusion of harmonic analysis on GL(2,**R**) ($r \geq 2$) with analytic number theory will yield further results and improvements. We believe that this is the case.

The object of these lectures is to provide a brief and elementary introduction to harmonic analysis on GL(r,**R**) with $r \geq 2$. Stress has been laid on those aspects of the theory which are particularly useful to analytic number theory; namely, Fourier expansions, L-functions, Eisenstein and Poincaré series, and arithmetic sums such as Kloosterman sums. We have followed the elegant classical exposition of Jacquet [9] that was further developed by Bump [1] (for the special case of GL(3,**R**)), which we believe is particularly suited to the types of explicit calculations that arise in analytic number theory.

The author would like to thank D. Bump and S. Friedberg for many helpful discussions.

2. Iwasawa decomposition.

The Iwasawa decomposition for $GL(2,\mathbf{R})$ states that every $g \in GL(2,\mathbf{R})$ can be written in the form

$$g = \begin{pmatrix} y & x \\ 0 & 1 \end{pmatrix} \begin{pmatrix} \alpha & \beta \\ \gamma & \delta \end{pmatrix} \begin{pmatrix} d & \\ & d \end{pmatrix} \tag{2.1}$$

where $y > 0$, x, $d \in \mathbf{R}$, and

$$\begin{pmatrix} \alpha & \beta \\ \gamma & \delta \end{pmatrix} \in O(2)$$

where

$$O(r) = \{g \in GL(r,\mathbf{R}) \ ; \ g\,{}^t g = I \} \tag{2.2}$$

is the orthogonal group. Setting

$$Z_r = \{ \begin{pmatrix} d & & 0 \\ & \ddots & \\ 0 & & d \end{pmatrix} \quad GL(r,\mathbf{R}) \} \tag{2.3}$$

to be the group of scalar matrices, we can then identify the upper half plane

$$h = \{ x+iy; \ x \in \mathbf{R}, \ y > 0 \}$$

as the group of 2 by 2 matrices of type

$$\{ \begin{pmatrix} y & x \\ 0 & 1 \end{pmatrix} \ ; \ y > 0, \ x \in \mathbf{R} \},$$

or by the isomorphism

$$h \cong GL(2,\mathbf{R})/O(2)\ Z_2 \ .$$

We seek to generalize the decomposition (2.1) to the group $GL(r,\mathbf{R})$ for $r \geqslant 2$. To this end, we define the generalized upper half space H^r to be the set of all matrices

$$H^r = \left\{ \begin{pmatrix} 1 & x_{1,2} & \cdots & x_{1,r} \\ & 1 & & \vdots \\ & & \ddots & x_{r-1,r} \\ & & & 1 \end{pmatrix} \begin{pmatrix} y_1 y_2 \cdots y_{r-1} & & & & \\ & y_1 y_2 \cdots y_{r-2} & & & \\ & & \ddots & & \\ & & & y_1 & \\ & & & & 1 \end{pmatrix} \right\} \tag{2.4}$$

where $x_{i,j} \in \mathbf{R}$ for $i < j \leqslant r$ and $y_i \geqslant 0$ for $1 \leqslant i \leqslant r-1$.

Proposition 2.1 (Iwasawa decomposition)

$$GL(r,\mathbf{R}) \cong H^r \cdot O(r) \cdot Z_r \ .$$

Proof. Let $g \in GL(r,\mathbf{R})$. Then $g^t g$ is a positive definite symmetric non-singular matrix. It is not difficult to show that there exist u and ℓ in $GL(r,\mathbf{R})$ where u is upper triangular with ones on the diagonal, ℓ is lower triangular with ones on the diagonal, such that

$$u g^t g = \ell d \tag{2.5}$$

$$d = \begin{pmatrix} d_1 & & \\ & \ddots & \\ & & d_r \end{pmatrix}$$

Hence $u^{-1} \ell d = g^t g = d^t \ell ({}^t u)^{-1}$ or

$$\underbrace{\ell d^t u}_{\text{lower } \Delta} = \underbrace{u d^t \ell}_{\text{upper } \Delta} = d.$$

Consequently $\ell d = d({}^t u)^{-1}$. Substituting into (2.5) gives $u g^t g^t u = d$ $= a^{-1}({}^t u)^{-1}$ for

$$a = \begin{pmatrix} \sqrt{d}_1 & & \\ & \ddots & \\ & & \sqrt{d}_r \end{pmatrix} .$$

Hence $a u g^t g^t u^t a = I$ so that $aug \in O(r)$. Consequently,

$$g = (au)^{-1}(aug) \quad H^r \cdot O(r) \cdot Z_r.$$

Q.E.D.

To illustrate the Iwasawa decomposition, we consider an arbitrary matrix

$$g = \begin{pmatrix} A & B & C \\ \alpha & \beta & \gamma \\ a & b & c \end{pmatrix} \in GL(3,\mathbf{R}) \ .$$

Then

$$g \equiv \begin{pmatrix} y_1y_2 & y_1x_3 & x_3 \\ & y_1 & x_1 \\ & & 1 \end{pmatrix} \quad (\mathrm{mod}\ O(3).Z_3)$$

where

$$x_1 = \frac{a\alpha + b\beta + c\gamma}{a^2 + b^2 + c^2} = \frac{\Sigma\, a\alpha}{\Sigma\, a^2}$$

$$x_2 = \frac{\Sigma\, A\alpha\ \Sigma\, a^2 - \Sigma\, Aa\ \Sigma\, \alpha a}{\Sigma\, \alpha^2\ \Sigma\, a^2 - (\Sigma\, a\alpha)^2}$$

$$y_1 = \frac{[\Sigma\, \alpha^2\ \Sigma\, a^2 - (\Sigma\, a\alpha)^2]^{1/2}}{\Sigma\, a^2}$$

$$y_2 = \frac{(\Sigma\, a^2)^{1/2}\ S^{1/2}}{\Sigma\, \alpha^2\ \Sigma\, a^2 - (\Sigma\, a\alpha)^2}$$

with

$$S = \Sigma\, A^2\ \Sigma\, \alpha^2\ \Sigma\, a^2 - \Sigma\, A^2(\Sigma\, \alpha a)^2 - \Sigma\, a^2\ (\Sigma\, A\alpha)^2 - \Sigma\, \alpha^2(\Sigma\, Aa)^2 + 2\ \Sigma\, Aa\ \Sigma\, \alpha a\ \Sigma\, A\alpha\ .$$

3. Automorphic forms.

Let I_r denote the identity matrix in $GL(r,\mathbf{R})$. For a positive integer M, we let

$$\Gamma_r(M) = \{\gamma \in SL(r,\mathbf{Z})\ ;\ \gamma \equiv I_r(\mathrm{mod}\ M)\ \}$$

denote the principal congruence subgroup (mod M) of $SL(r,\mathbf{R})$. This will be a discrete subgroup of $GL(r,\mathbf{R})$, and it acts on the generalized upper half space H^r by left multiplication. That is, for $\gamma \in \Gamma_r(M)$, $\tau \in H^r$, we let $\gamma\tau = \tau^*$ where $\tau^* \in H^r$ is uniquely chosen so that $\gamma\tau \equiv \tau^*$ (mod $O(r)Z_r$).

Let $v_1,\ldots,v_{r-1}$ be complex numbers. For $\tau \in H^r$ given by

$$\tau = \begin{pmatrix} 1 & x_{1,2} & \cdots & x_{1,r} \\ & 1 & & \vdots \\ & & \ddots\ 1 & x_{r-1,r} \\ & & & 1 \end{pmatrix} \begin{pmatrix} y_1y_2 \cdots y_{r-1} & & & \\ & y_1 \cdots y_{r-2} & & \\ & & \ddots\ y_1 & \\ & & & 1 \end{pmatrix} \tag{3.1}$$

let us define

$$I_{v_1,\dots,v_{r-1}}(r) = \prod_{i=1}^{r-1} \prod_{j=1}^{r-1} y_i^{c_{ij}v_j}$$

where

$$c_{ij} = \begin{cases} (r-i)j & 1 \leqslant j \leqslant i \\ (r-j)i & i \leqslant j \leqslant r-1 \end{cases} .$$

If $\mathcal{D}$ denotes the algebra of G_r-invariant differential operators on H^r, then $I_{v_1,\dots,v_{r-1}}$ is an eigenfunction of $\mathcal{D}$, and hence determines a character $\lambda_{v_1,\dots,v_{r-1}}$ on $\mathcal{D}$ by the formula

$$DI_{v_1,\dots,v_{r-1}} = \lambda_{v_1,\dots,v_{r-1}}(D)\, I_{v_1,\dots,v_{r-1}} \qquad (D \in \mathcal{D}).$$

For example, when $r = 2$, $\mathcal{D}$ is generated by the Laplacian

$$\Delta = -y^2\left(\frac{\partial^2}{\partial x^2} + \frac{\partial^2}{\partial y^2} \right) \text{ and } \lambda_v(\Delta) = v(1-v) \quad .$$

When $r = 3$, $\mathcal{D}$ is generated by two elements Δ_1, Δ_2 (see [1], pp. 33-34) where

$$\Delta_1 = y_1^2 \frac{\partial^2}{\partial y_1^2} + y_2^2 \frac{\partial^2}{\partial y_2^2} - y_1 y_2 \frac{\partial^2}{\partial y_1 \partial y_2} + y_1^2 (x_2^2 + y_2^2) \frac{\partial^2}{\partial x_3^2}$$
$$+ y_1^2 \frac{\partial^2}{\partial x_1^2} + y_2^2 \frac{\partial^2}{\partial x_2^2} + y_1^2 x_2 \frac{\partial^2}{\partial x_1 \partial x_3} ,$$

$$\Delta_2 = -y_1^2 y_2 \frac{\partial^3}{\partial y_1^2\, \partial y_2} + y_1 y_2^2 \frac{\partial^3}{\partial y_1 \partial y_2^2} - y_1^3 y_2^2 \frac{\partial^3}{\partial x_3^2 \partial y_1} + y_1 y_2^2 \frac{\partial^3}{\partial x_2^2 \partial y_1}$$
$$- 2y_1^2 y_2 x_2 \frac{\partial^3}{\partial x_1 \partial x_3 \partial y_2} + (-x_2^2 + y_2^2)\, y_1^2 y_2 \frac{\partial^3}{\partial x_3^2 \partial y_2} - y_1^2 y_2 \frac{\partial^3}{\partial x_1^2 \partial y_2}$$
$$+ 2y_1^2 y_2^2 \frac{\partial^3}{\partial x_1 \partial x_2 \partial x_3} + 2y_1^2 y_2 x_2 \frac{\partial^3}{\partial x_2 \partial x_3^2} + y_1^2 \frac{\partial^2}{\partial y_1^2} - y_2^2 \frac{\partial^2}{\partial y_2^2}$$
$$+ 2y_1^2 x_2 \frac{\partial^2}{\partial x_1 \partial x_3} + (x_2^2 + y_2^2)\, y_1^2 \frac{\partial^2}{\partial x_3^2} + y_1^2 \frac{\partial^2}{\partial x_1^2} - y_2^2 \frac{\partial^2}{\partial x_2^2} .$$

Here Δ_1 is the Laplacian and Δ_2 is a third order operator. We have

$$\lambda_{v_1,v_2}(\Delta_1) = 3(v_1^2+v_1v_2+v_2^2-v_1-v_2)$$

$$\lambda_{v_1,v_2}(\Delta_2) = -2v_1^3-3v_1^2v_2+3v_1v_2^2+2v_2^3+3v_1^2-3v_2^2-v_1+v_2 .$$

We now define the notion of an automorphic form for the principal conguence subgroup $\Gamma_r(M)$.

Definition 3.1 *Fix complex numbers* $v_1,\ldots,v_{r-1}$. *A smooth function* $\phi : H^r \to \mathbf{C}$ *is called an automorphic form of type* $(v_1,\ldots,v_{r-1})$ *for* $\Gamma_r(M)$ *if*

(i) $\phi(\gamma\tau) = \phi(\tau)$ *for* $\gamma \in \Gamma_r(M)$, $\tau \in H^r$.

(ii) $D\phi = \lambda_{v_1,\ldots,v_{r-1}}(D)\phi$ *for* $D \in \mathcal{D}$.

(iii) $\phi(\rho\tau)$ *has polynomial growth in* $y_1,\ldots,y_{r-1}$ *on the region* $\{\tau \mid y_i \geq 1 \ (i = 1,2,\ldots,r-1)\}$, *for every* ρ in $\Gamma_r(M)\backslash SL(r,\mathbf{R})$. *Further,* ϕ *is called a cusp form if it satisfies the additional condition*

(iv) $$\int_{\Gamma_r(M)\cap U\backslash U} \phi(\rho u\tau)du = 0$$

for every $\rho \in \Gamma_r(M)\backslash SL(r,\mathbf{R})$ *and every group* U *of the form*

$$U = \begin{pmatrix} I_{r_1} & & & * \\ & I_{r_2} & & \\ & & \ddots & \\ 0 & & & I_{r_s} \end{pmatrix} \in GL(r,\mathbf{R}) .$$

Generalized Ramanujan conjecture: *If* ϕ *is a cusp form of type* $(v_1,\ldots,v_{r-1})$ *for* $\Gamma_r(M)$ *, then*

$$\mathrm{Re}(v_1) = \ldots = \mathrm{Re}(v_{r-1}) = \frac{1}{r} .$$

This conjecture was first explicitly stated by Selberg [19] for the case $r = 2$. Using Weil's [22] estimates for Kloosterman sums, Selberg [19] obtained

$$\frac{1}{4} < \mathrm{Re}(v) < \frac{3}{4}$$

for the case $r = 2$. It is known [1] that

$$\frac{1}{6} \leq \mathrm{Re}(v_1),\ \mathrm{Re}(v_2) \leq \frac{1}{2}$$

for $r = 3$. By developing a $GL(2,\mathbf{R})$ generalization of the "large sieve", Deshouillers and Iwaniec [3] have shown that the generalized Ramanujan conjecture is true on the average (over M) for the case $r = 2$. Very little has been done when $r \geq 3$.

4. Fourier expansions of automorphic forms.

Let ϕ be an automorphic form for $\Gamma_r(M)$. In view of the non-commutativity of the situation, it is remarkable that ϕ has a Fourier expansion. These expansions were first found independently by I.Piatetski-Shapiro [15] and J. Shalika [20]. We follow, however, the more classical approach given in [9] for the special case of an automorphic form ϕ for $GL(r,\mathbf{Z})$. A proof of these expansions for the principal congruence subgroup $\Gamma_r(M)$ is given in the appendix by S. Friedberg.

Let $N_r \subset SL(r,\mathbf{R})$ denote the group of upper triangular matrices of type

$$\begin{pmatrix} 1 & x_{1,2} & \cdot & \cdot & \cdot & x_{1,r} \\ & 1 & & & & \vdots \\ & & \cdot & & & \vdots \\ & & & \cdot & & \vdots \\ & & & & 1 & x_{r-1,r} \\ & & & & & 1 \end{pmatrix} \tag{4.1}$$

For integers $n_1,\ldots,n_{r-1}$, let θ denote a character of N_r defined by

$$\theta(x) = \exp\big(\, 2\pi i(n_1 x_{1,2} + \ldots + n_{r-1} x_{r-1,r})\big). \tag{4.2}$$

Proposition 4.1 *Let* ϕ *be an automorphic cusp form for* $\Gamma_r = GL(r,\mathbf{Z})$. *Then*

$$\phi(\tau) = \sum_{n_1=1}^{\infty} \cdots \sum_{n_{r-1}=1}^{\infty} \sum_{\gamma \in N_{r-1} \cap \Gamma_{r-1} \backslash \Gamma_{r-1}} \phi_{n_1,\ldots,n_{r-1}}\left(\begin{pmatrix} \gamma & 0 \\ 0 & 1 \end{pmatrix}\tau\right)$$

where

$$\phi_{n_1,\ldots,n_{r-1}}(\tau) = \int_{N_r \cap \Gamma_r \backslash N_r} \phi(u\tau)\bar{\theta}(u)du \, .$$

with θ *given by* (4.2) .

Now, if ϕ is of type $(v_1,\ldots,v_{r-1})$, then it follows that $\phi_{n_1,\ldots,n_{r-1}}$ is also of the same type. Consequently, $\phi_{n_1,\ldots,n_{r-1}}$ must satisfy the two properties

$$D\phi_{n_1,\ldots,n_{r-1}} = \lambda_{v_1,\ldots,v_{r-1}}(D)\, \phi_{n_1,\ldots,n_{r-1}} \, , \quad D \in \mathcal{D} \qquad (4.3a)$$

$$\phi_{n_1,\ldots,n_{r-1}}(x\tau) = \theta(x)\, \phi_{n_1,\ldots,n_{r-1}}(\tau) \qquad (4.3b)$$

for every x of type (4.1). Furthermore, in view of Definition (3.1) (iii), we see that $\phi_{n_1,\ldots,n_{r-1}}(\tau)$ must have polynomial growth in $y_1,\ldots,y_{r-1}$ on the region $\{\tau \mid y_i \geqslant 1 \ (i = 1,2,\ldots,r-1)\}$. The multipilicity one theorem of Shalika [20] states that up to a constant multiple, there is a unique function $W_{n_1,\ldots,n_{r-1}}(\tau)$ satisfying conditions (4.3a), (4.3b) and having polynomial growth at the cusp $y_1,\ldots,y_{r-1} \to \infty$. Moreover,

$$W_{n_1,\ldots,n_{r-1}}(\tau) = c((n))W_{1,\ldots,1}((n)\tau)$$

where

$$(n) = \begin{pmatrix} n_1 n_2 \cdots n_{r-1} & & & & \\ & n_1 \cdots n_{r-2} & & & \\ & & \ddots & & \\ & & & n_{r-1} & \\ & & & & 1 \end{pmatrix} \qquad (4.4)$$

and c((n)) is a constant depending on (n) and $v_1,\ldots,v_{r-1}$. We set

$$W(\tau) = W_{1,\ldots,1}(\tau). \tag{4.5}$$

The function $W(\tau)$ is called a Whittaker function. This is due to the fact that in the case $r = 2$, $W(\tau)$ satisfies the classical Whittaker equation. We have now shown

Proposition 4.2 *Let* ϕ *be an automorphic cusp form of type* $(v_1,\ldots v_{r-1})$ *for* $GL(r,\mathbf{Z})$. *Then there exist constants* $a_{n_1,\ldots,n_{r-1}}$ *such that*

$$\phi(\tau) = \sum_{n_1=1}^{\infty} \cdots \sum_{n_r=1}^{\infty} \sum_{\gamma \in B_{r-1}} a_{n_1,\ldots,n_{r-1}} \prod_{i=1}^{r-1} |n_i|^{i(r-i)/2} \times$$

$$\times W\left((n)\begin{pmatrix} \gamma & 0 \\ 0 & 1 \end{pmatrix}\tau\right) \tag{4.6}$$

where $B_{r-1} = N_{r-1} \cap \Gamma_{r-1} \backslash \Gamma_{r-1}$ *and* (n) *is given by* (4.4).

As an example, we take $r = 2$. It is known that the unique Whittaker function is given by

$$W(\tau) = 2\sqrt{y}\, K_{v-1/2}(2\pi y)\, e^{2\pi i x}$$

where

$$\tau = \begin{pmatrix} y & x \\ 0 & 1 \end{pmatrix}$$

and

$$K_s(y) = 1/2 \int_0^{\infty} e^{-\frac{1}{2}\left(t + \frac{1}{t}\right)} t^{s-1}\, dt .$$

Proposition (4.2) says that any cusp form ϕ of type v for $SL(2,\mathbf{Z})$ has the Fourier expansion

$$\phi(\tau) = \sum_{n \neq 0} a_n |n|^{-1/2} W\left(\begin{pmatrix} n & 0 \\ 0 & 1 \end{pmatrix}\tau\right) .$$

Let

$$w_r = \begin{pmatrix} 0 & & & -1 \\ & & 1 & \\ & -1 & & 0 \\ \cdot^{\cdot^{\cdot}} & & & \end{pmatrix}.$$

For $\tau \in H^r$, this induces an involution

$$\tau \to w_r({}^t\tau)^{-1}$$

which has the effect of interchanging the y_i $(i = 1,\ldots,r-1)$ and the $x_{i,i+1}$ $(i = 1,\ldots,r-1)$ if τ is given by (3.1). Hence, if we denote

$$\tilde{\phi}(\tau) = (w_r({}^t\tau)^{-1})$$

where ϕ is a cusp form of type $(v_1,\ldots,v_{r-1})$ then it is easily seen that $\tilde{\phi}$ is a cusp form of type $(v_{r-1},v_{r-2},\ldots,v_1)$. Moreover

$$\tilde{a}_{n_1,\ldots,n_{r-1}} = a_{n_{r-1},\ldots,n_1}$$

where $\tilde{a}_{n_1,\ldots,n_{r-1}}$ denotes the Fourier coefficient in the expansion (4.6) for $\tilde{\phi}$.

Now, associated to ϕ , we have an L-series

$$L_\phi(s) = \sum_{n=1}^{\infty} a_n n^{-s}$$

where

$$a_n = a_{n,1,\ldots,1}$$

As shown in [15], [16], [20]. this has a functional equation $s \to 1 - s$, $\phi \to \tilde{\phi}$ when multiplied by suitable gamma factors.

Generalized Ramanujan conjecture: *If* ϕ *is a cusp form of type* $(v_1,\ldots,v_{r-1})$ *for* $GL(r,\mathbf{Z})$, *then for every* $\varepsilon > 0$

$$a_n = O(n^\varepsilon)$$

where the O*-constant is independent of* n.

If in addition ϕ is an eigenform for the Hecke algebra, then the generalized Ramanujan conjecture combined with the multiplicative properties of a_n actually imply

$$|a_p| \leqslant r|a_1|$$

for every prime p.

While the functional equation for $L_\phi(s)$ is somewhat tricky to prove, we can associate with ϕ a different object

$$Z_\phi(s) = \int \cdots \int_{Y \geqslant 0} \phi((Y))\, Y^{s-1}\, dy_1 \ldots dy_{r-1} \tag{4.7}$$

where

$$Y = y_1 \cdots y_{r-1}$$

and

$$(Y) = \begin{pmatrix} y_1 \cdots y_{r-1} & & & & \\ & y_1 \cdots y_{r-2} & & & \\ & & \ddots & & \\ & & & y_1 & \\ & & & & 1 \end{pmatrix} .$$

Splitting the integral on the dexter side of (4.7) into two integrals defined by the regions

$$0 \leqslant Y \leqslant 1 , \qquad 1 \leqslant Y,$$

and using the identity

$$\phi((Y)) = \phi(w_r(Y)) = \tilde{\phi}\left(\frac{1}{(Y)}\right)$$

where

$$\frac{1}{(Y)} = \begin{pmatrix} (y_1 \cdots y_{r-1})^{-1} & & & \\ & \ddots & & \\ & & y_1^{-1} & \\ & & & 1 \end{pmatrix} ,$$

we see that

$$Z_\phi(s) = \int \cdots \int_{Y \geqslant 1} [\phi((Y))Y^{s-1} + \tilde{\phi}((Y))Y^{-s-1}]\, dy_1 \ldots dy_{r-1}$$

from which it follows that

$$Z_\phi(s) = Z_{\tilde{\phi}}(-s) .$$

5. Eisenstein and Poincare series.

Let $\tau \in H^r$ be given by (3.1). Recall that for fixed complex numbers $v_1, \ldots, v_{r-1}$,

$$I_{v_1,\ldots,v_{r-1}}(\tau) = \prod_{i=1}^{r-1} \prod_{j=1}^{r-1} y_i^{c_{ij} v_j}$$

where

$$c_{ij} = \begin{cases} (r-i)j & 1 \leqslant j \leqslant i \\ (r-j)i & i \leqslant j \leqslant r-1 \, . \end{cases}$$

The minimal parabolic Eisenstein series for $\Gamma = SL(r,\mathbf{Z})$ is defined as

$$E(\tau; v_1, \ldots, v_{r-1}) = \sum_{\gamma \in N_r \cap \Gamma \backslash \Gamma} I_{v_1,\ldots,v_{r-1}}(\gamma\tau) \quad , \qquad (5.1)$$

where N_r is given by (4.1). The series on the dexter side of (5.1) converges absolutely and uniformly on compact subsets of H^r if $\mathrm{Re}(v_i) > 2/r$, $i = 1,2,\ldots,r-1$. General methods for obtaining the meromorphic continuation and functional equations of Eisenstein series were first given by Selberg [17], [18]. More detailed proofs appeared in [14], [10]. Langlands [13] obtained, for the first time, proofs of these results for the case of an arbitrary reductive group.

Now, we consider the Hilbert space $L^2(\Gamma\backslash H^r)$ with inner product given by

$$\langle \phi, \psi \rangle = \int_{\Gamma\backslash H^r} \phi(\tau)\, \overline{\psi(\tau)}\, d^*\tau \qquad (5.2)$$

for any two square-integrable automorphic forms for Γ, and where

$$d^*\tau = \prod_{1<i<j<r-1} dx_{i,j} \prod_{i=1}^{r-1} \frac{dy_i}{y_i^{i(r-i)-1}} \qquad (5.3)$$

is the $GL(r,\mathbf{R})$ invariant measure. It can be shown that $E(\tau, v_1, \ldots, v_{r-1})$ is an automorphic form of type $(v_1, \ldots, v_{r-1})$ which is not square-integrable, but lies in the continuous spectrum of $\mathcal{D}$.

The Fourier expansion

$$\zeta^*(2v)E(\tau,v) = \zeta^*(2v)y^v+3^*(2v-1)y^{1-v}+2\sqrt{y}\sum_{n=1}^{\infty} n^{v-1/2}\sigma_{1-2v}(2\pi ny) \times$$
$$\times K_{v-1/2}(2\pi ny)\cos(2\pi nx)$$

where

$$\zeta^*(v) = \pi^{-\frac{v}{2}}\Gamma(\tfrac{v}{2})\zeta(v)$$

$$\sigma_v(n) = \sum_{d|n} d^v$$

$$K_v(y) = \int_0^{\infty} e^{-y\cosh u}(\cosh vu)\, du \qquad (y > 0)$$

$$\tau = \begin{pmatrix} y & x \\ 0 & 1 \end{pmatrix}$$

for the case $r = 2$ is classical. Fourier expansions of Eisensten series for $GL(3,\mathbf{Z})$ were given in [1], [6], [21], and recently [23] has obtained the Fourier expansion of Eisenstein series for $GL(r,\mathbf{Z})$, $r \geq 2$. The arithmetic part of the Fourier coefficient involves certain general divisor or Ramanujan sums.

We now consider a generalization of Eisenstein series, namely Poincaré series. To this end, it is first of all necessary to define the notion of an E-function.

For fixed integers $n_1,\ldots,n_{r-1}$, let θ denote the character of N_r given by (4.2). An E-function is a smooth function $E : H^r \to \mathbf{C}$ satisfying

$$E(x\tau) = \theta(x)E(\tau) \qquad \text{for } x \in N_r,\ \tau \in H^r \tag{5.4}$$

$$E(\tau) = O(1) \qquad \text{for } \tau \in H^r. \tag{5.5}$$

By abuse of notation, we have not specified the dependence of E on $n_1,\ldots,n_{r-1}$.

Now, let $v_1,\ldots,v_{r-1}$ be complex parameters. Let $n_1,\ldots,n_{r-1}$ be integers. We define the Poincaré series $P_{n_1,\ldots,n_{r-1}}(\tau;v_1,\ldots,v_{r-1})$ by the infinite series

$$P_{n_1,\ldots,n_{r-1}}(\tau;v_1,\ldots,v_{r-1}) = \sum_{\gamma\in N_r\cap\Gamma\backslash\Gamma} I_{v_1,\ldots,v_{r-1}}(\gamma\tau) \times \times E_{n_1,\ldots,n_{r-1}}(\gamma\tau) \tag{5.6}$$

where $E_{n_1,\ldots,n_{r-1}}$ is an E-function satisfying (5.4) and (5.5). Again, by (5.5), the series on the dexter side of (5.6) converges absolutely and uniformly on compact subsets of H^r if $\mathrm{Re}(v_i) > 2/r$ for $i = 1,2,\ldots,r-1$.

In order to obtain the Fourier expansion of the Poincare series (5.6), it is necessary to introduce the Bruhat decomposition. Let W denote the Weyl group of $GL(r,\mathbf{R})$, which is simply the group of $r \times r$ matrices with exactly one 1 in each row and column, and zeros everywhere else (i.e. the regular representation of the symmetric group on r symbols). We also let N_r be the group of upper triangular matrices with ones on the diagonal, and we let $D_r \subset GL(r,\mathbf{R})$ be the group of diagonal matrices. For $w \in W$, let

$$G_w = N_r w D_r N_r$$

so that

$$GL(r,\mathbf{R}) = \bigcup_{w\in W} G_w .$$

Similarly,

$$\Gamma = \bigcup_{w\in W} G_w \cap \Gamma$$

The sets $G_w \cap \Gamma$ are called Bruhat cells. The cell corresponding to

$$w_r = \begin{pmatrix} & & & -1 \\ & & 1 & \\ & -1 & & \\ \cdot^{\cdot^{\cdot}} & & & \end{pmatrix}$$

is called the big cell. We can now break up the Poincare series into pieces corresponding to Bruhat cells, namely

$$P_{n_1,\ldots,n_{r-1}}(\tau;v_1,\ldots,v_{r-1}) = \sum_{w\in W} \sum_{\substack{\gamma\in N_r\cap\Gamma\backslash\Gamma \\ \gamma\in G_w}} I_{v_1,\ldots,v_{r-1}}(\gamma\tau) \times \times E_{n_1,\ldots,n_{r-1}}(\gamma\tau) .$$

The Fourier expansion of $P_{n_1,\ldots,n_{r-1}}(\tau;v_1,\ldots,v_{r-1})$ will also break up into pieces corresponding to Bruhat cells. The Fourier coefficients corresponding to a cell will be infinite sums of $SL(r,\mathbf{Z})$ Kloosterman sums weighted by certain integrals which are higher dimensional generalizations of hypergeometric functions. We now describe the Kloosterman sums associated to the big cell for Γ.

Fix integers $n_1,\ldots,n_{r-1}$ and $m_1,\ldots,m_{r-1}$. For x in N_r given by

$$x = \begin{pmatrix} 1 & x_{1,2} & \cdots\cdots & x_{1,r} \\ & 1 & & \vdots \\ & & \ddots\ 1 & x_{r-1,r} \\ & & & 1 \end{pmatrix}$$

let

$$\theta_{n_1,\ldots,n_r}(x) = e^{2\pi i(n_1x_{1,2}+\ \cdots\ +\ n_rx_{r-1,r})} \tag{5.7}$$

Then for $d\in D_r$, we define the big cell Kloosterman sum

$$S_{w_r}(m_1,\ldots,m_{r-1};n_1,\ldots,n_{r-1};d) = \sum_{\substack{\gamma\in N_r\backslash\Gamma/N_r \\ \gamma=b_1w_rdb_2 \\ b_1,b_2\in N_r}} \theta_{m_1,\ldots,m_{r-1}}(b_1) \times \times\ \theta_{n_1,\ldots,n_{r-1}}(b_2)$$

Similarly, there will be Kloosterman sums for all the other Bruhat cells. The Kloosterman sums will have multiplicative properties in the d aspect. For d of the form

$$d = \begin{pmatrix} d_1 & & & \\ & d_2 & & \\ & & \ddots & \\ & & & d_r \end{pmatrix}$$

the d_i $(i = 1,\ldots,r)$ will be rational numbers. If we assume each d_i is a positive or negative integer power of a fixed prime p, then the Kloosterman sum $S_w(m_1,\ldots,m_{r-1};n_1,\ldots,n_{r-1};d)$ will be associated to a certain algebraic variety over $\mathbf{F}_p$. The complete determination of these varieties has only been affected for $r = 2,3$ (see [22], [2]).

If the E-function defined by (5.4) and (5.5) has exponential decay in $y_i \to \infty$ $(i = 1,\ldots,r-1)$ then the Poincaré series (5.6) will be square-integrable. It will not be an eigenfunction for $\mathcal{D}$, however. We will now show that the inner product of a cusp form with $P_{n_1,\ldots,n_{r_1}}(\tau;v_1,\ldots,v_{r-1})$ picks off a certain transform of the $(n_1,\ldots,n_{r-1})^{th}$ Fourier coefficient of the cusp form.

Proposition 5.1 *Let* $\phi \in L^2(\Gamma\backslash H^r)$. *Then*

$$< P_{n_1,\ldots,n_{r-1}}(*;v_1,\ldots,v_{r-1}),\phi > = \int_0^\infty \cdots \int_0^\infty$$

$$\phi_{n_1,\ldots,n_{r-1}}(Y)\, I_{v_1,\ldots,v_{r-1}}(Y)\, E_{n_1,\ldots,n_{r-1}}(Y) \prod_{i=1}^{r-1} \frac{dy_i}{y_i^{i(r-i)-1}}$$

where

$$Y = \begin{pmatrix} y_1\cdots y_{r-1} & & & \\ & y_1\cdots y_{r-2} & & \\ & & \ddots & \\ & & & y_1 & \\ & & & & 1 \end{pmatrix}$$

and

$$\phi_{n_1,\ldots,n_{r-1}}(\tau) = \int_{N_r\cap\Gamma\backslash N_r} \overline{\phi(u\tau)}\, \theta_{n_1,\ldots,n_{r-1}}(u)du$$

and $\theta_{n_1,\ldots,n_{r-1}}$ *is given by* (5.7)

Proof: By the Rankin-Selberg unfolding method

$$< P_{n_1,\ldots,n_{r-1}},\phi > = \int_{\Gamma\backslash H^r} P_{n_1,\ldots,n_{r-1}}(\tau;v_1,\ldots,v_{r-1})\bar{\phi}(\tau)d^*\tau$$

$$= \int_{N_r\backslash H^r} \bar{\phi}(\tau)\, I_{v_1,\ldots,v_{r-1}}(\tau)\, E_{n_1,\ldots,n_{r-1}}(\tau)d^*\tau\ .$$

To complete the proof, we note that

$$\int_{N_r\backslash H^r} = \int_{y_1=0}^{\infty} \cdots \int_{y_{r-1}=0}^{\infty} \cdot \int_{N_r\cap\Gamma\backslash N_r}$$

and that

$$E_{n_1,\ldots,n_{r-1}}(\tau) = \theta_{n_1,\ldots,n_{r-1}}(x)\, E_{n_1,\ldots,n_{r-1}}(Y)$$

for $\tau = xY$ with x given by (4.1).

Let ϕ be a cusp form of type $(\lambda_1, \lambda_2, \ldots, \lambda_{r-1})$. Then proposition (5.1) shows that the inner product

$$\langle P_{n_1,n_2,\ldots,n_{r-1}}(*;\ v_1,v_2,\ldots,v_{r-1}),\phi\rangle$$

has a meromorphic continuation in $v_1, v_2, \ldots, v_{r-1}$ with polar divisors depending on $\lambda_1, \lambda_2, \ldots, \lambda_{r-1}$. In [2], we show how this can be applied to the generalized Ramanujan conjecture.

Appendix

The Fourier expansion on a congruence subgroup of SL(r,Z)

Solomon Friedberg

Let M be a positive integer. We give here the Fourier expansion of a function $\phi : H^r \to \mathbf{C}$ invariant under the congruence subgroup

$$\Gamma_r(M) = \left\{\gamma \in SL(r,\mathbf{Z}) \mid \gamma \equiv \begin{pmatrix} 1 & & 0 \\ & \ddots & \\ 0 & & 1 \end{pmatrix} \pmod{M}\right\} .$$

This Fourier expansion was first developed in an adelic setting by Piatetski-Shapiro [15] and Shalika [20] and this is simply a translation of one of their results into a non-adelic framework.

Let $\mathcal{R}$ be a fixed set of coset representatives for

$$N_{r-1} \cap SL(r-1,\mathbf{Z})\backslash SL(r-1,\mathbf{Z}) .$$

For each γ in $SL(r-1,\mathbf{Z})$ denote by ρ_γ an element of $SL(r-1,\mathbf{Z})$ such that $\rho_\gamma\gamma$ is in $\Gamma_{r-1}(M)$ (the coset of ρ_γ in $\Gamma_{r-1}(M)\backslash SL(r-1,\mathbf{Z})$ is thus uniquely determined by γ). Given integers $n_1,\ldots,n_{r-1}$, let $\theta_{\frac{n_1}{M},\ldots,\frac{n_{r-1}}{M}}$ denote the character of N_r given by

$$\theta_{\frac{n_1}{M},\ldots,\frac{n_{r-1}}{M}}(x) = \exp(2\pi i(n_1x_{1,2}+\ldots+n_{r-1}x_{r-1,r})/M) ,$$

and choose the Haar measure du on N_r such that the measure of $(N_r \cap \Gamma_r(M)\backslash N_r) = 1$.

Theorem (A.1) *Suppose* ϕ: $H^r \to \mathbf{C}$ *is* $\Gamma_r(M)$ *invariant. Then*

$$\phi(\tau) = \sum_{n_1=-\infty}^{\infty} \cdots \sum_{n_{r-2}=-\infty}^{\infty} (\phi^{I_{r-1}})_{\frac{n_1}{M},\ldots,\frac{n_{r-2}}{M},\frac{0}{M}}(\tau)$$

$$+ \sum_{\gamma \in R} \sum_{n_1=-\infty}^{\infty} \cdots \sum_{n_{r-2}=-\infty}^{\infty} \sum_{n_{r-1}=1}^{\infty} (\phi^{\rho}\gamma)_{\frac{n_1}{M},\ldots,\frac{n_{r-1}}{M}} \left(\begin{pmatrix} \gamma & 0 \\ 0 & 1 \end{pmatrix}\tau\right)$$

where

$$(\phi^{\rho}\gamma)_{\frac{n_1}{M},\ldots,\frac{n_{r-1}}{M}}(\tau) = \int_{N_r \cap \Gamma_r(M)\backslash N_r} \phi\left(\begin{pmatrix} \rho_\gamma & 0 \\ 0 & 1 \end{pmatrix} u\tau\right) \bar{\theta}_{\frac{n_1}{M},\ldots,\frac{n_{r-1}}{M}}(u)\, du$$

Proof: Denote by $u(a_1,\ldots,a_{r-1})$ the element

$$u(a_1,\ldots,a_{r-1}) = \begin{pmatrix} 1 & & 0 & a_1 \\ & \ddots & & \vdots \\ & & & a_{r-1} \\ & & & 1 \end{pmatrix}$$

of N_r. First, since ϕ is invariant under the subgroup

$$\{u(n_1,\ldots,n_{r-1}) \mid n_1 \equiv \cdots \equiv n_{r-1} \equiv 0 \pmod M\}$$

of $\Gamma_r(M)$, we may write

$$\phi(\tau) = \sum \cdots \sum_{n_1,\ldots,n_{r-1} \in \mathbf{Z}} \tilde{\phi}_{\frac{n_1}{M},\ldots,\frac{n_{r-1}}{M}}$$

with

$$\tilde{\phi}_{\frac{n_1}{M},\ldots,\frac{n_{r-1}}{M}}(\tau) = \int_{(\mathbf{R}/M\mathbf{Z})^{r-1}} \phi(u(a_1,\ldots,a_{r-1})\tau) \times$$
$$\times\, e\big(-(n_1a_1+\ldots+n_{r-1}a_{r-1})/M\big)da_1\ldots da_{r-1}$$

and

$$e(x) = \exp(2\pi i x) .$$

Next, suppose $\gamma \in SL(r-1,\mathbf{Z})$ has bottom row $(\gamma_1\ldots\gamma_{r-1})$, and m is an integer. Note that such a γ is determined modulo the $SL(r-1,\mathbf{Z})$ *maximal* parabolic

$$P_{r-1} = \left(\begin{array}{c|c} * & * \\ \hline 0 & 1 \end{array}\right) \begin{array}{c} r-2 \\ 1 \end{array}$$

Then we claim that

$$\tilde{\phi}_{\frac{m\gamma_1}{M},\ldots,\frac{m\gamma_{r-1}}{M}}(\tau) = ({\phi}^{\rho_\gamma})_{0,\ldots,0,\frac{m}{M}}\left(\begin{pmatrix} \gamma & 0 \\ 0 & 1 \end{pmatrix}\tau\right) \tag{A.1}$$

with ${\phi}^{\rho_\gamma}(\tau) = \phi(\begin{pmatrix} \rho_\gamma & 0 \\ 0 & 1 \end{pmatrix}\tau)$. To see this, observe that since $\begin{pmatrix} \rho_\gamma\gamma & 0 \\ 0 & 1 \end{pmatrix}$ is in $\Gamma_r(M)$, the left hand side equals

$$\int_{(\mathbf{R}/M\mathbf{Z})^{r-1}} \phi(\begin{pmatrix} \rho_\gamma\gamma & 0 \\ 0 & 1 \end{pmatrix} u(a_1,\ldots,a_{r-1})\tau)e\big(-m(\gamma_1 a_1+\ldots+ \gamma_{r-1}a_{r-1})/M\big) \times \\ \times\, da_1 \ldots da_{r-1}$$

$$= \int_{(\mathbf{R}/M\mathbf{Z})^{r-1}} \phi(\begin{pmatrix} \rho_\gamma & 0 \\ 0 & 1 \end{pmatrix} u(a_1',\ldots,a_{r-1}')\begin{pmatrix} \gamma & 0 \\ 0 & 1 \end{pmatrix}\tau) \times \\ \times\, e\big(-m(\gamma_1 a_1+\ldots\gamma_{r-1}a_{r-1})/M\big)\, da_1\ldots da_{r-1}$$

with

$$\begin{pmatrix} a_1' \\ \vdots \\ a_{r-1}' \end{pmatrix} = \gamma \begin{pmatrix} a_1 \\ \vdots \\ a_{r-1} \end{pmatrix} .$$

Further

$$\gamma_1 a_1+\ldots+ \gamma_{r-1}a_{r-1} = (0,\ldots,0,1)\gamma^t(a_1,\ldots,a_{r-1}) = a_{r-1}' .$$

Thus changing variables gives (A.1).

Now, iterating these steps, replacing u successively by

$$u_i(a_1,\ldots,a_{r-1}) = \begin{pmatrix} 1 & & & & a_1 & & \\ & \cdot & 0 & & & & \\ & & \cdot & & \vdots & & 0 \\ & & & \cdot & a_{r-i} & & \\ & & & & 1 & & \\ & & & & & \cdot & \\ & & & & & & 1 \end{pmatrix}$$

for i = 1,2,...,r-1 completes the proof (for example, when i=2 the γ

to be used run over

$$\left(\begin{array}{c|c} P_{r-2} & * \\ \hline 0 & 1 \end{array}\right) \backslash P_{r-1} \,).$$

Bibliography

[1] D. Bump, Automorphic Forms on GL(3,**R**), Lecture Notes in Math.1983, Springer, (1984).

[2] D. Bump, S. Friedberg, D. Goldfeld, Poincaré series and Kloosterman sums for SL(3,**Z**), to appear in *Acta Arithmetica*.

[3] J. M. Deshouillers, H. Iwaniec, Kloosterman sums and Fourier coefficients of cusp forms, *Invent. Math.*, **70** (1982), 219-288.

[4] E. Fouvry, Brun-Titchmarsh theorem on average, to appear.

[5] D. Goldfeld, P. Sarnak, Sums of Kloosterman sums, *Invent. Math.*, **71** (1983), 243-250.

[6] K. Imai, A. Terras, The Fourier expansions of Eisenstein series for GL(3,**Z**), *Trans. A.M.S.* **273** (1982), #2, 679-694.

[7] H. Iwaniec, Non-holomorphic modular forms and their applications, Modular Forms (R. Rankin, Ed.), Ellis Horwood, West Sussex, (1984), 197-156.

[8] H. Iwaniec, J. Pintz, Primes in short intervals, Mathematics Institute of the Hungarian Academy of Sciences, preprint no. 37, (1983).

[9] H. Jacquet, Dirichlet series for the group GL(n), Automorphic Forms, Representation Theory and Arithmetic, Springer-Verlag, (1981), 155-164.

[10] T. Kubota, Elementary Theory of Eisenstein series, New York, John Wiley and Sons (1973).

[11] N. V. Kuznetsov, The arithmetic form of Selberg's trace formula and the distribution of the norms of the primitive hyperbolic classes of the modular group (in Russian) Preprint, Khabarovsk (1978).

[12] N. V. Kuznetsov, Petersson's conjecture for cusp forms of weight zero and Linnik's conjecture; sums of Kloosterman sums [in Russian], at. *Sb.* (*N.S.*), **39** (1981), 299-342.

[13] R. Langlands, On the Functional Equations Satisfied by Eisenstein Series, Springer Verlag, Lecture Notes in Math. #544 (1976).

[14] H. Maass, Siegel's Modular Forms and Dirichlet Series, Springer Verlag, Lecture Notes in Math. #216 (1971).

[15] I. I. Piatetski-Shapiro, Euler subgroups, Lie Groups and their Representations, John Wiley and Sons, (1975), 597-620.

[16] I. I. Piatetski-Shapiro, Multiplicity one theorems, Automorphic Forms, Representations, and L-Functions, Proc. Symp. in Pure Math. XXXII, (A. Borel, Ed.), Part II, 209-212.

[17] A. Selberg, Harmonic analysis and discontnuous groups in weakly symmetric Riemannian spaces with applications to Dirichlet's series, *J. Indian Math. Soc.*, **20**, (1956), 47-87.

[18] A. Selberg, Discontinuous groups and harmonic analysis, *Proc. Internat. Congr. Math.*, Stockholm, (1962), 177-189.

[19] A. Selberg, On the estimation of Fourier coefficients of modular forms, Proc. Symp. Pure Math. VII, A.M.S., Providence, R.I., (1965), 1-15.

[20] J. Shalika, The multiplicity one theorem for GL(n), *Annals of Math.* **100**, (1974), 171-193.

[21] L. Takhtadzhyan, I. Vinogradov, Theory of Eisenstein series for the group SL(3,**R**), and its application to a binary problem, *J. Sov. Math.* **18** (1982), #3, 293-324.

[22] A. Weil, On some exponential sums, *Proc. Nat. Acad. Sci. U.S.A.*, **34** (1948), 204-207.

[23] A. Yukie, Ph.D. Thesis, Harvard (1985).

D.Goldfeld
Harvard University
Cambridge, Mass.02138

S.Friedberg
Harvard University
Cambridge, Mass.02138

University of Texas at Austin
Austin, Texas 78712

Columbia University
New York, N.Y. 10027 U.S.A.

PAIR CORRELATION OF ZEROS AND PRIMES IN SHORT INTERVALS

Daniel A. Goldston and Hugh L. Montgomery*

1. Statement of results.

In 1943, A. Selberg [15] deduced from the Riemann Hypothesis (RH) that

$$\int_1^X (\psi((1+\delta)x) - \psi(x) - \delta x)^2 x^{-2}\, dx \ll \delta(\log X)^2 \tag{1}$$

for $X^{-1} \leqslant \delta \leqslant X^{-1/4}$, $X \geqslant 2$. Selberg was concerned with small values of δ, and the constraint $\delta \leqslant X^{-1/4}$ was imposed more for convenience than out of necessity. For larger δ we have the following result.

Theorem 1. *Assume* RH. *Then*

$$\int_1^X (\psi((1+\delta)x) - \psi(x) - \delta x)^2 x^{-2}\, dx \ll \delta(\log X)(\log 2/\delta) \tag{2}$$

for $0 < \delta \leqslant 1$, $X \geqslant 2$.

In this estimate, the error term for the number of primes in the interval $(x, (1+\delta)x]$ is damped by the factor x^{-2}, and the length of the interval, δx, varies with x. Saffari and Vaughan [14] considered the undamped integral, and derived from RH the estimates

$$\int_1^X (\psi((1+\delta)x) - \psi(x) - \delta x)^2\, dx \ll \delta X^2 (\log 2/\delta)^2 \tag{3}$$

for $0 < \delta \leqslant 1$, and

$$\int_1^X (\psi(x+h) - \psi(x) - h)^2\, dx \ll hX(\log 2X/h)^2 \tag{4}$$

*Research supported in part by NSF Grant MCS82-01602.

for $0 < h \leqslant X$. It may be similarly shown that RH gives the estimate

$$\int_1^X (\psi(x) - x)^2\, dx \ll X^2 . \tag{5}$$

Gallagher and Mueller [5] showed that if one assumes not only RH but also the pair correlation conjecture

$$\begin{aligned} \# \{(\gamma,\gamma') : 0 < \gamma \leqslant T,\ 0 < \gamma - \gamma' \leqslant 2\pi a/\log T\} \\ = \Big(\frac{1}{2\pi} \int_0^a 1 - \Big(\frac{\sin \pi u}{\pi u}\Big)^2 du + o(1)\Big)\, T \log T \end{aligned} \tag{6}$$

then it can be deduced that

$$\int_1^X (\psi((1 + \delta)x) - \psi(x) - \delta x)^2\, x^{-2}\, dx \sim \delta(\log 1/\delta)(\log X\sqrt{\delta}) \tag{7}$$

for $X^{-1} \leqslant \delta \leqslant X^{-\varepsilon}$. Here γ denotes the ordinate of a non-trivial zero of the Riemann zeta function. Thus it seems likely that the estimate of Theorem 1 is best possible.

In the course of formulating the conjecture (6), Montgomery [13] also proposed a more precise estimate, namely that

$$F(X,T) \sim \frac{1}{2\pi} T \log T \tag{8}$$

uniformly for $T \leqslant X \leqslant T^A$, for any fixed $A > 1$, where

$$F(X,T) = \sum_{0 < \gamma,\gamma' \leqslant T} X^{i(\gamma-\gamma')} w(\gamma-\gamma') \tag{9}$$

and $w(u) = 4/(4 + u^2)$. We now relate this conjecture to the size of the integral in (3).

Theorem 2. *Assume* RH. *If* $0 < B_1 \leqslant B_2 \leqslant 1$, *then*

$$\int_1^X (\psi((1 + \delta)x) - \psi(x) - \delta x)^2\, dx \sim \frac{1}{2} \delta X^2 \log 1/\delta \tag{10}$$

uniformly for $X^{-B_2} \leqslant \delta \leqslant X^{-B_1}$, *provided that* (8) *holds uniformly for*

$$X^{B_1} (\log X)^{-3} \leqslant T \leqslant X^{B_2} (\log X)^3. \qquad (11)$$

Conversely, if $1 \leqslant A_1 \leqslant A_2 < \infty$, *then* (8) *holds uniformly for* $T^{A_1} \leqslant X \leqslant T^{A_2}$, *provided that* (10) *holds uniformly for*

$$X^{-1/A_1} (\log X)^{-3} \leqslant \delta \leqslant X^{-1/A_2} (\log X)^3 . \qquad (12)$$

Previously Mueller [12] derived (10) from RH and a strong quantitative form of (8). Heath-Brown and Goldston [11] showed that RH and (8) for $T^a \leqslant X \leqslant T^b$, $a < 2 < b$, imply

$$p_{n+1} - p_n = o(p_n^{1/2} (\log p_n)^{1/2}) .$$

This estimate follows easily from Theorem 2 by taking $\delta = \varepsilon X^{-1/2} (\log X)^{1/2}$ in (10). In deriving (10) from (8) we also use the weaker estimate (3). In the case of very small δ, say $\delta \approx (\log X)/X$, we can do better by appealing instead to the bound

$$\int_1^X (\psi((1 + \delta)x) - \psi(x) - \delta x)^2 \, dx \ll \delta X^2 \log X + \delta^2 X^3 \qquad (13)$$

which follows from sieve estimates (see the proof of Lemma 7). In this way we could show that

$$\int_1^X (\pi(x + h) - \pi(x) - h/\log x)^2 \, dx \sim hX/\log X \qquad (14)$$

for $h \approx \log X$, given RH and (8) for $T \leqslant X \leqslant f(T)T \log T$. Here $f(T)$ tends to infinity arbitrarily slowly with T. From this it follows easily that

$$\liminf \, (p_{n+1} - p_n) / \log p_n = 0 .$$

Heath-Brown [10] derived this from a slightly stronger hypothesis.

In assessing the depth of the estimates (8) and (10), we note that (10) is a logarithm sharper than (3), and that (8) is a logarithm sharper than the trivial bound

$$|F(X,T)| \leqslant F(1,T) \sim \frac{1}{2\pi} T(\log T)^2 \quad . \tag{15}$$

(See Lemma 8.) As in (4), we can relate (10) to primes in intervals of constant length. In summary we have the following

Corollary. *Assume* RH. *Then the following assertions are equivalent:*

(a) *For every fixed* $A \geqslant 1$, (8) *holds uniformly for* $T \leqslant X \leqslant T^A$.

(b) *For every fixed* $\varepsilon > 0$, (10) *holds uniformly for* $X^{-1} \leqslant \delta \leqslant X^{-\varepsilon}$.

(c) *For every fixed* $\varepsilon > 0$,

$$\int_0^X (\psi(x+h) - \psi(x) - h)^2 \, dx \sim hX \log X/h \tag{16}$$

holds uniformly for $1 \leqslant h \leqslant X^{1-\varepsilon}$.

It is not hard to show that either (b) or (c) implies RH. Gallagher [4] has shown that a weak quantitative form of the prime k-tuple hypothesis gives (16) when $h \approx \log X$.

The path we take between (8) and (10) involves elementary arguments of Abelian and Tauberian character; these are of two sorts. First, we consider the connection between the assertion

$$\int_{-\infty}^{+\infty} e^{-2|y|} f(Y+y) \, dy = 1 + o(1) \tag{17}$$

as $Y \to +\infty$, and the more general assertion

$$\int_a^b R(y) f(Y+y) \, dy = \int_a^b R(y) \, dy + o(1) \tag{18}$$

as $Y \to +\infty$ where R is any Riemann-integrable function. (These two statements are equivalent if f is bounded and non-negative.) This interplay reflects the choice of the weighting function w(u) in the definition (9) of F(X,T). Second, and more intrinsically, we consider a question of Riemann summability (R_2), namely the connection between the two assertions

$$\int_0^\infty \left(\frac{\sin \kappa u}{u}\right)^2 f(u)du = (\pi/2 + o(1))\kappa \log 1/\kappa \qquad (19)$$

as $\kappa \to 0^+$, and

$$\int_0^U f(u)du = (1 + o(1))\, U \log U \qquad (20)$$

as $U \to +\infty$. Because of the intricacies of the (R_2) method, neither of these assertions implies the other, although they are equivalent for non-negative functions f. The lemmas we formulate below are complicated by the fact that we specify the relation between the parameters κ and U.

2. Lemmas of summability.

Lemma 1. *If*

$$I(Y) = \int_{-\infty}^{+\infty} e^{-2|y|} f(Y + y)dy = 1 + \varepsilon(Y) ,$$

and if $f(y) \geq 0$ *for all* y, *then for any Riemann-integrable function* R(y),

$$\int_a^b R(y)\, f(Y + y)dy = \left(\int_a^b R(y)dy\right)\left(1 + \varepsilon'(y)\right) . \qquad (21)$$

If R *is fixed then* $|\varepsilon'(Y)|$ *is small provided that* $|\varepsilon(y)|$ *is small uniformly for* $Y + a - 1 \leq y \leq Y + b + 1$.

In terms of Wiener's general Tauberian theorem, the truth of this lemma hinges on the fact that the Fourier transform of the kernel $k(y) = e^{-2|y|}$, namely the function

$$\hat{k}(t) = \int_{-\infty}^{+\infty} k(y)\, e(-ty)dy = \frac{1}{\pi^2 t^2 + 1} , \quad \left(e(u) = e^{2\pi i u}\right),$$

never vanishes.

Proof. Let $K_c(y) = \max(0, c - |y|)$. By comparing Fourier transforms, or by direct calculation, we see that

$$K_c(y) = \frac{1}{2} e^{-2|y|} - \frac{1}{4} e^{-2|y-c|} - \frac{1}{4} e^{-2|y+c|}$$

$$+ \int_{-c}^{c} (c - |z|)\, e^{-2|y-z|}\, dz .$$

Hence

$$\int_{-c}^{c} K_c(y)\, f(Y + y)\, dy = \frac{1}{2} I(Y) - \frac{1}{4} I(Y + c) - \frac{1}{4} I(Y - c)$$

$$+ \int_{-c}^{c} (c - |z|) I(Y + z)\, dz$$

$$= c^2 + \varepsilon_1(Y)$$

where $|\varepsilon_1|$ is small if $c > 0$ is fixed and if $|\varepsilon(y)|$ is small for $Y - c \leqslant y \leqslant Y + c$. Since

$$\frac{1}{\eta}\big(K_c(y) - K_{c-\eta}(y)\big) \leqslant \chi_{[-c,c]}(y) \leqslant \frac{1}{\eta}\big(K_{c+\eta}(y) - K_c(y)\big) ,$$

and since $f \geqslant 0$, we deduce that (21) holds in the case of the step function $R(y) = \chi_{[-c,c]}(y)$. Since the general R can be approximated above and below by step functions, we obtain (21).

Lemma 2. *Suppose that* $f(t)$ *is a continuous non-negative function defined for all* $t \geqslant 0$, *with* $f(t) \ll \log^2 (t + 2)$. *If*

$$J(T) = \int_0^T f(t)dt = \big(1 + \varepsilon(T)\big)T \log T ,$$

then

$$\int_0^\infty \Big(\frac{\sin \kappa u}{u}\Big)^2 f(u)du = \big(\pi/2 + \varepsilon'(\kappa)\big)\kappa \log 1/\kappa \qquad (22)$$

where $|\varepsilon'(\kappa)|$ *is small as* $\kappa \to 0^+$ *if* $|\varepsilon(T)|$ *is small uniformly for* $\kappa^{-1} (\log \kappa)^{-2} \leqslant T \leqslant \kappa^{-1}(\log \kappa)^2$.

Proof. We divide the range of integration in (22) into four subintervals: $0 \leqslant u \leqslant \kappa^{-1}(\log \kappa)^{-2} = U_1$, $U_1 \leqslant u \leqslant C\kappa^{-1} = U_2$, $U_2 \leqslant u \leqslant \kappa^{-1}(\log \kappa)^2 = U_3$, and $U_3 \leqslant u < \infty$.
Since $f(t) \ll \log^2(t + 2)$, we see that

$$\int_0^{U_1} < \int_0^{U_1} \kappa^2 \log^2(u+2)\,du < \kappa^2 U_1 \log^2 U_1 \ll \kappa ,$$

and similarly that

$$\int_{U_3}^{\infty} < \int_{U_3}^{\infty} u^{-2} \log^2 u\,du \ll U_3^{-1} \log^2 U_3 < \kappa .$$

By writing $f(u) = \log 1/\kappa + \log \kappa u + (f(u) - \log u)$, we express the integral from U_1 to U_2 as a sum of three integrals. We note that

$$\int_{U_1}^{U_2} \left(\frac{\sin \kappa u}{u}\right)^2 du = \int_0^{\infty} \left(\frac{\sin \kappa u}{u}\right)^2 du + O(\kappa(\log \kappa)^{-2})$$

$$= \frac{\pi}{2}\kappa\left(1 + O(\log \kappa)^{-2}\right),$$

and that

$$\int_{U_1}^{U_2} \left(\frac{\sin \kappa u}{u}\right)^2 \log \kappa u\,du < \int_0^{\infty} \min(\kappa^2, u^{-2}) \log \kappa u\,du < \kappa .$$

Put $r(u) = J(u) - u \log u + u$. Then by integrating by parts we see that

$$\int_{U_1}^{U_2} \left(\frac{\sin \kappa u}{u}\right)^2 (f(u) - \log u)du < \kappa\Big(1 + (\log \frac{1}{\kappa}) \max_{U_1 \leqslant u \leqslant U_2} |\varepsilon(u)|\Big)\log(C+2) .$$

As for the range $U_2 \leqslant u \leqslant U_3$, we see that if $\varepsilon(u) \leqslant 1$ then

$$\int_{U_2}^{U_3} \leqslant \int_{U_2}^{U_3} f(u)u^{-2}\,du \ll U_2^{-1} \log U_2 \ll C^{-1}\kappa \log 1/\kappa .$$

We make this small by taking C large. Then the remaining error terms are small if $\varepsilon(u)$ is small.

Lemma 3. *If* K *is even,* K" *continuous,* $\int_{-\infty}^{+\infty} |K| < \infty$, $K(x) \to 0$ *as* $x \to +\infty$, $K' \to 0$ *as* $x \to +\infty$, *and if* $K''(x) \ll x^{-3}$ *as* $x \to +\infty$, *then*

$$\hat{K}(t) = \int_0^{\infty} K''(x)\left(\frac{\sin \pi t x}{\pi t}\right)^2 dx . \tag{23}$$

Proof. Integrate by parts twice.

Lemma 4. *If f is a non-negative function defined on* $[0, +\infty)$, $f(t) \ll \log^2(t+2)$, *and if*

$$I(\kappa) = \int_0^\infty \left(\frac{\sin \kappa t}{t}\right)^2 f(t)dt = (\pi/2 + \varepsilon(\kappa))\kappa \log 1/\kappa$$

then

$$J(T) = \int_0^T f(t)dt = (1 + \varepsilon')T \log T$$

where $|\varepsilon'|$ *is small if* $|\varepsilon(\kappa)| \leqslant \varepsilon$ *uniformly for*

$$T^{-1}(\log T)^{-1} \leqslant \kappa \leqslant T^{-1}(\log T)^2 .$$

Proof. Let K be a kernel with the properties specified in Lemma 3. Replace t by t/T in (23), multiply by $f(t) - \log t$, and integrate over $0 \leqslant t < \infty$. Then we find that

$$\int_0^\infty \big(f(t) - \log t\big) \hat{K}(t/T)dt = \pi^{-2} T^2 \int_0^\infty K''(x)\, R(\pi x/T)dx$$

where

$$R(\kappa) = I(\kappa) - \int_0^\infty \left(\frac{\sin \kappa t}{t}\right)^2 \log t\, dt$$

$$= I(\kappa) - \frac{1}{2}\pi\kappa \log 1/\kappa + O(\kappa).$$

Since

$$I(\kappa) < \int_0^\infty \min(\kappa^2, t^{-2})\log^2(t+2)dt < \kappa \log^2(2 + 1/\kappa)$$

for all $\kappa > 0$, on taking $x_1 = (\log T)^{-1}$ we see that

$$\int_0^{x_1} K''R \ll \int_0^{x_1} xT^{-1} \log^2 T/x\, dx \ll T^{-1}.$$

On taking $x_2 = \tfrac{1}{4}(\log T)^2$ we find that

$$\int_{x_2}^\infty K''R \ll \int_{x_2}^\infty x^{-3}(x/T) \log^2 T\, dx \ll T^{-1}.$$

Assuming, as we may, that $\varepsilon \geqslant (\log T)^{-1}$, we have $R(\pi x/T) \ll \varepsilon\, xT^{-1}\log T$ for $x_1 \leqslant x \leqslant x_2$. Hence

$$\int_{x_1}^{x_2} K''R \ll \varepsilon\, T^{-1}(\log T) \int_0^\infty \min(1, x^{-3})\, x\, dx \ll \varepsilon\, T^{-1}\log T.$$

For $\eta > 0$ take

$$K(x) = K_\eta(x) = \big(\sin 2\pi x + \sin 2\pi(1+\eta)x\big)\big(2\pi x(1 - 4\eta^2x^2)\big)^{-1} ,$$

so that

$$\hat{K}(t) = \begin{cases} 1 & \text{if } |t| \leqslant 1, \\ \cos^2\big(\pi(|t| - 1)/(2\eta)\big) & \text{if } 1 \leqslant |t| \leqslant 1 + \eta, \\ 0 & \text{if } |t| \geqslant 1 + \eta . \end{cases}$$

Thus

$$\int_0^\infty f(t)\hat{K}_\eta(t/T)dt = (1 + O(\eta))T \log T + O_\eta(T) + O_\eta(\varepsilon\, T \log T) .$$

Since f is non-negative, we see that

$$\int_0^\infty f(t)\hat{K}_\eta\big((1+\eta)t/T\big)dt \leqslant J(T) \leqslant \int_0^\infty f(t)\hat{K}_\eta(t/T)dt ,$$

and we obtain the desired result by taking η small.

In this argument we have made free use of existing treatments of Riemann summability. We note especially Hardy [8, pp. 301, 316, 365] and Hardy and Rogosinski [9, Theorem III].

3. Lemmas of analytic number theory.

As is customary, we write $s = \sigma + it$, and we let $\rho = \beta + i\gamma$ be a typical non-trivial zero of the Riemann zeta function. We first note a simple result of Gallagher [3]:

Lemma 5. *Let* $S(t) = \sum_{\mu \in M} c(\mu)e(\mu t)$ *where* M *is a countable set of real numbers and* $\sum |c(\mu)| < \infty$. *Then*

$$\int_{-T}^{T} |S(t)|^2 \, dt < T^2 \int_{-\infty}^{+\infty} \Big| \sum_{\substack{\mu \in M \\ |\mu-u| \leq (4T)^{-1}}} c(\mu)\Big|^2 \, du.$$

When a main term is desired, we use the following more elaborate estimate.

Lemma 6. *Let* $S(t)$ *be as above. If* $\delta \geq T^{-1}$ *then*

$$\int_0^T |S(t)|^2 \, dt = (T + O(\delta^{-1})) \sum_{\mu \in M} |c(\mu)|^2$$

$$+ \; O\Big(T \sum_{\substack{\mu,\nu \in M \\ 0 < |\mu-\nu| < \delta}} |c(\mu)c(\nu)|\Big) .$$

Proof. Selberg (see Vaaler[17]) has constructed functions $F_-(t)$ and $F_+(t)$ such that $F_-(T) \leq \chi_{[0,T]}(t) \leq F_+(t)$, $\hat{F}_\pm(x) = 0$ for $|x| \geq \delta$, and $\int_{-\infty}^{+\infty} F_\pm(t)dt = T \pm \delta^{-1}$. Hence

$$\int_0^T |S|^2 \leq \int_{-\infty}^{+\infty} |S|^2 \, F_+ = \sum_{\mu,\nu} c(\mu)\overline{c(\nu)} \, \hat{F}_+(\nu - \mu) .$$

The terms $\mu = \nu$ contribute $(T + \delta^{-1}) \sum_\mu |c(\mu)|^2$. Since

$$|\hat{F}_+| \leq \int |F_+| = T + \delta^{-1} \leq 2T ,$$

the terms $\mu \neq \nu$ contribute at most

$$2T \sum_{0 < |\mu-\nu| < \delta} |c(\mu)c(\nu)|.$$

This gives an upper bound, and a corresponding lower bound is derived similarly using F_-.

Lemma 7. *Let* $C(x) > 0$ *be a continuous function such that* $C(x) \approx C(y)$ *whenever* $x \approx y$. If $|c(p)| \leqslant C(p)$ *for all primes* p, *and if* $\delta \geqslant T^{-1}$, *then*

$$\int_0^T |\sum_p c(p)p^{it}|^2\, dt = (T + O(\delta^{-1})) \sum_p |c(p)|^2 + O\Big(\delta T \int_{\delta^{-1}}^{\infty} C(u)^2\, u(\log u)^{-2}\, du\Big)$$

Proof. We appeal to the previous lemma. In the second error term, the primes $p \in (X,2X]$ contribute

$$T\, C(X)^2 \sum_{X<p\leqslant 2X} \sum_{p<p'\leqslant(1+2\delta)p} 1 \ll T\, C(X)^2 \sum_{1\leqslant k\leqslant 4\delta X} \pi_2(2X,k)$$

where $\pi_2(x,k)$ denotes the number of primes $p \leqslant x$ for which $p + k$ is also prime. It is well-known (see Halberstam and Richert [7, p.117]) that

$$\pi_2(x,k) \ll \big(k/\phi(k)\big)x(\log x)^{-2}$$

uniformly for $x \geqslant 2$, $k \neq 0$. Since $\sum_{k \leqslant K} k/\phi(k) \ll K$, it follows that our upper bound is

$$\ll T\, C(X)^2\, \delta X^2\, (\log X)^{-2} \ll \delta T \int_X^{2X} C(u)^2\, u(\log u)^{-2}\, du\,.$$

We put $X = \delta^{-1} 2^r$ and sum over $r \geqslant 0$ to obtain the desired result.

We now present the main known properties of $F(X,T)$.

Lemma 8. *Assume* RH, *and let* $F(X,T)$ *be as in* (9). *Then* $F(X,T) \geqslant 0$, $F(X,T) = F(1/X,T)$, *and*

$$F(X,T) = T\big(X^{-2}(\log T)^2 + \log X\big)\Big(\frac{1}{2\pi} + O\big((\log T)^{-1/2}(\log\log T)^{1/2}\big)\Big) \tag{24}$$

uniformly for $1 \leqslant X \leqslant T$.

Proof. The first assertion is an immediate consequence of either of the two identities

$$F(X,T) = 2\pi \int_{-\infty}^{+\infty} e^{-4\pi|u|} \left| \sum_{0<\gamma\leq T} X^{i\gamma} e(\gamma u) \right|^2 du, \tag{25}$$

or

$$F(X,T) = \frac{2}{\pi} \int_{-\infty}^{+\infty} \left| \sum_{0<\gamma\leq T} \frac{X^{i\gamma}}{1+(t-\gamma)^2} \right|^2 dt.$$

The observation that F is non-negative has also been made by Mueller (unpublished). The second assertion is obvious from the definition of F. The estimate (24) is substantially due to Goldston [6, Lemma B], and may be proved by substituting an appeal to Lemma 7 in the argument of Montgomery [13].

Lemma 9. *If* $0 \leq h \leq T$ *then*

$$\#\{(\gamma,\gamma') : 0 \leq \gamma \leq T,\ |\gamma - \gamma'| \leq h\} \ll (1 + h \log T)T \log T . \tag{27}$$

Proof. We argue unconditionally, although if RH is assumed then the above follows easily from Lemma 8 (see (6) of Montgomery [13]). Let $N(T) = \#\{\gamma: 0 < \gamma \leq T\}$. Following Selberg, Fujii [2] showed that

$$\int_0^T \left(N(t+h) - N(t) - \frac{1}{2\pi} h \log t\right)^2 dt \ll T \log(2 + h \log T)$$

for $0 \leq h \leq 1$. Hence

$$\int_0^T \left(N(t+h) - N(t)\right)^2 dt \ll h^2\ T(\log T)^2$$

for $(\log T)^{-1} \leq h \leq 1$. This gives (27) in this case. To derive (27) when $0 \leq h \leq (\log T)^{-1}$, it suffices to consider $h = (\log T)^{-1}$. As for the range $1 \leq h \leq T$, it suffices to use the bound

$$N(T + 1) - N(T) \ll \log T \tag{28}$$

(see Titchmarch [16,p. 178]) .

Lemma 10. *For* $0 < \delta \leqslant 1$ *let*

$$a(s) = ((1 + \delta)^s - 1)/s . \tag{29}$$

If $|c(\gamma)| \leqslant 1$ *for all* γ *then*

$$\int_{-\infty}^{+\infty} |a(it)|^2 \, \Big| \sum_{\gamma} \frac{c(\gamma)}{1 + (t-\gamma)^2}\Big|^2 \, dt = \int_{-\infty}^{+\infty} \Big| \sum_{|\gamma| \leqslant Z} \frac{a(1/2 + i\gamma)c(\gamma)}{1 + (t-\gamma)^2}\Big|^2 \, dt$$

$$+ O(\delta^2(\log 2/\delta)^3) + O(Z^{-1}(\log Z)^3) \tag{30}$$

provide that $Z \geqslant 1/\delta$.

Proof. By (28), the sum that occurs in the integral on the left is $\ll \log (2 + |t|)$. Since

$$a(s) \ll \min(\delta, |s|^{-1}) \tag{31}$$

in the strip $|\sigma| \leqslant 1/\delta$, it follows by Cauchy's formula or by direct calculation that

$$a'(s) \ll \min(\delta^2 , \delta/|s|) \tag{32}$$

for $|\sigma| \leqslant (2\delta)^{-1}$. Hence in particular,

$$a(it) - a(\tfrac{1}{2} + it) \ll \min (\delta^2 , \delta/|t|) ,$$

and consequently

$$|a(it)|^2 - |a(\tfrac{1}{2} + it)|^2 \ll \min(\delta^3, \delta/t^2) .$$

Let I denote the integral on the left in (30), and J the corresponding integral with a(it) replaced by $a(\frac{1}{2} + it)$. Then

$$I - J \ll \int \min(\delta^3, \delta/t^2)(\log(2 + |t|))^2 \, dt \ll \delta^2 (\log 2/\delta)^2 .$$

Write J in the form $J = \int |A|^2$. From (28) and (31) we see that

$$A \ll \min(\delta, |t|^{-1}) \log(2 + |t|) \qquad (33)$$

Now let K be the integral with $a(1/2 + it)$ replaced by $a(1/2 + i\gamma)$, and write $K = \int |B|^2$. Then B also satisfies the estimate (33). From (31) and (32) we see that

$$a(1/2 + i\gamma) - a(1/2 + it) \ll |t - \gamma| \min(\delta^2, \delta/|t|) .$$

Thus

$$A - B \ll \min(\delta^2, \delta/|t|)(\log(2/\delta + |t|))^2,$$

so that

$$|A|^2 - |B|^2 \ll \min(\delta^3, \delta/t^2)(\log(2/\delta + |t|))^3 ,$$

and hence

$$J - K \ll \delta^2(\log 2/\delta)^3 .$$

Finally, let $L = \int |C|^2$ be the integral on the right in (30). We note that C also satisfies the estimate (33). Since

$$B - C \ll \min(Z^{-1}, |t|^{-1}) \log(2Z + |t|),$$

we find that

$$|B|^2 - |C|^2 \ll \min(Z^{-1}(1 + |t|)^{-1}, t^{-2})(\log(2Z + |t|))^2 .$$

Thus

$$K - L \ll Z^{-1} (\log 2Z)^3 ,$$

and the proof is complete.

4. Proof of Theorem 1.

Although we arrange the technical details differently, the ideas are entirely the same as in Selberg's paper. If $\delta X \leq 1$ then there is at most one prime power in the interval $(x, (1 + \delta)x]$, so that our integral is

$$\ll \delta \sum_{n\leq X} \Lambda(n)^2/n + \delta^2 X \ll \delta(\log X)^2 ,$$

which suffices. We now suppose that $\delta X > 1$. By the above argument we see that

$$\int_0^{1/\delta} \ldots \ll \delta(\log 2/\delta)^2.$$

Thus it suffices to consider the range $1/\delta \leq x \leq X$. Here we apply the explicit formula for $\psi(x)$ (see Davenport [1, 17]), which gives

$$\begin{aligned}\psi((1+\delta)x) - \psi(x) - \delta x = -\sum_{|\rho|\leq Z} a(\rho)x^{\rho} &\qquad (34)\\ + O\Big((\log x)\min\Big(1, \frac{x}{Z\|x\|}\Big)\Big)&\\ + O\Big((\log x)\min\Big(1, \frac{x}{Z\|(1+\delta)x\|}\Big)\Big)&\\ + O(x Z^{-1}(\log xZ)^2)&\end{aligned}$$

where $a(s)$ is given in (29), and $\|\theta\| = \min_n \|\theta - n\|$ is the distance from θ to the nearest integer. The error terms contribute a negligible amount if we take $Z = X(\log X)^2$. Writing $\rho = \frac{1}{2} + i\gamma$, $x = e^y$, $Y = \log X$, we see that it remains to show that

$$\int_{\log 1/\delta}^{Y} \Big|\sum_{|\gamma|\leq Z} a(\rho)\, e^{i\gamma y}\Big|^2 dy \ll \delta Y \log 2/\delta. \qquad (35)$$

By Lemma 5 we see that this integral is

$$\ll Y^2 \int_{-\infty}^{\infty} \Big(\sum_{\substack{|\gamma|\leq Z\\ |\gamma-2\pi u|\leq 2/Y}} a(\rho)^2\Big)\, du \ll Y \sum_{\substack{|\gamma|\leq Z\\ |\gamma'|\leq Z\\ |\gamma-\gamma'|\leq 4/Y}} |a(\rho)a(\rho')| .$$

By (31) and Lemma 9 this gives (35), and the proof is complete.

5. Proof of Theorem 2.

We first assume (8) as needed, and derive (10). Let

$$J(T) = J(X,T) = 4 \int_0^T \Big| \sum_\gamma \frac{X^{i\gamma}}{1+(t-\gamma)^2} \Big|^2 dt \quad .$$

Montgomery [13] (see his (26), but beware of the changes in notation) used (28) to show that

$$J(X,T) = 2\pi\, F(X,T) + O\big((\log T)^3\big) \; .$$

Thus (8) is equivalent to

$$J(X,T) = \big(1 + o(1)\big) T \log T \quad . \tag{36}$$

With a(s) defined in (29), we note that

$$|a(it)|^2 = 4\Big(\frac{\sin \kappa t}{t}\Big)^2$$

where $\kappa = \frac{1}{2}\log(1+\delta)$. Then by Lemma 2 we deduce that

$$\int_0^\infty |a(it)|^2 \Big|\sum_\gamma \frac{X^{i\gamma}}{1+(t-\gamma)^2}\Big|^2 dt = (\pi/2 + o(1))\kappa \log 1/\kappa$$

$$= (\pi/4 + o(1))\delta \log 1/\delta \; . \tag{37}$$

The values of T for which we have used (8) lie in the range

$$\delta^{-1}(\log 1/\delta)^{-2} \leqslant T \leqslant 3\delta^{-1}(\log 1/\delta)^2 \; . \tag{38}$$

The integrand is even, so that the value is doubled if we integrate over negative values of t as well. Then by Lemma 10

$$\int_{-\infty}^{+\infty} \Big|\sum_{|\gamma| \leqslant Z} \frac{a(\rho)\, X^{i\gamma}}{1+(t-\gamma)^2}\Big|^2 dt = \big(\pi/2 + o(1)\big)\delta \log 1/\delta$$

provided that $Z \geqslant \delta^{-1}(\log 1/\delta)^3$. Let S(t) denote the above sum over γ. Its Fourier transform is

$$\hat{S}(u) = \int_{-\infty}^{+\infty} S(t)\, e(-tu)dt = \pi \sum_{|\gamma| \leqslant Z} a(\rho)\, X^{i\gamma}\, e(-\gamma u) e^{-2\pi|u|} \; .$$

Hence by Plancherel's identity the integral above is

$$= \pi^2 \int_{-\infty}^{+\infty} \Big| \sum_{|\gamma| \leqslant Z} a(\rho) X^{i\gamma}\, e(-\gamma u)\Big|^2 e^{-4\pi|u|}\, du \; .$$

On writing $Y = \log X$, $-2\pi u = y$, we find that

$$\int_{-\infty}^{+\infty} \Big| \sum_{|\gamma| \leqslant Z} a(\rho)\, e^{i\gamma(Y+y)} \Big|^2 e^{-2|y|}\, dy = (1 + o(1))\delta \log 1/\delta \,. \tag{39}$$

In Lemma 1 we take

$$R(y) = \begin{cases} e^{2y} & 0 \leqslant y \leqslant \log 2, \\ 0 & \text{otherwise}\,. \end{cases}$$

On making the change of variable $x = e^{Y+y}$ we deduce that

$$\int_X^{2X} \Big| \sum_{|\gamma| \leqslant Z} a(\rho) x^{\rho} \Big|^2 dx = \big(3/2 + o(1)\big)\, \delta X^2 \log 1/\delta \,.$$

We replace X by $X2^{-k}$, sum over k, $1 \leqslant k \leqslant K$, and use the explicit formula (34) with $Z = X(\log X)^3$ to see that

$$\int_{X2^{-K}}^{X} \big(\psi((1+\delta)x) - \psi(x) - \delta x\big)^2 dx = \frac{1}{2}\big(1-2^{-2K} + o(1)\big)\, \delta X^2 \log 1/\delta.$$

We take $K = [\log\log X]$, and note that it suffices to have (8) in the range (11). To bound the contribution of the range $1 \leqslant x \leqslant X2^{-K}$, we appeal to (3) with X replaced by $X2^{-K}$. Thus we have (10) .

We now deduce (8) from (10). By integrating (10) by parts from X_1 to $X_2 = X_1(\log X_1)^{2/3}$, we find that

$$\int_{X_1}^{X_2} \big(\psi((1+\delta)x) - \psi(x) - \delta x\big)^2 x^{-4} dx = \Big(\frac{1}{2} + o(1)\Big)\delta(\log 1/\delta)X_1^{-2}.$$

From (3) we similarly deduce that

$$\int_{X_2}^{\infty} \big(\psi((1+\delta)x) - \psi(x) - \delta x\big)^2 x^{-4}\, dx \ll \delta\, (\log 1/\delta)^2\, X_2^{-2}$$

$$= O\big(\delta\, (\log 1/\delta)\, X_1^{-2} \big)\,.$$

We add these relations, and multiply through by X_1^2 . By making a further appeal to (10) with $X = X_1$ we deduce that

$$\int_0^{\infty} \min(x^2/X_1^2,\, X_1^2/x^2)\big(\psi((1+\delta)x) - \psi(x) - \delta x\big)^2 x^{-2}\, dx$$

$$= (1 + o(1))\delta \log 1/\delta \,.$$

We write X for X_1, put $Y = \log X$, $x = e^{Y+y}$, and appeal to the explicit formula (34) with $Z = X(\log X)^3$, and we find that we have (39). Retracing our steps, we find that we have (37). Then by Lemma 4 we obtain (36), and hence (8). The values of δ and X for which we have used (10) also satisfy (12).

6. Proof of the Corollary.

We note that Lemma 8 gives (8) when

$$X(\log X)^{-3} \leqslant T \leqslant X,$$

and that (10) is trivial when

$$X^{-1}(\log X)^{-3} \leqslant \delta \leqslant X^{-1}.$$

Thus the equivalence of (a) and (b) follows immediately from Theorem 2.

We now show that (b) implies (c). We suppress the converse argument, which is similar. The method here is that of Saffari and Vaughan [14]. Our first goal is to deduce from (b) that

$$\int_0^H \int_0^X (\psi(x+h) - \psi(x) - h)^2 \, dxdh \sim \frac{1}{2} H^2 X \log X/H \qquad (40)$$

uniformly for $1 \leqslant H \leqslant X^{1-\varepsilon}$. To this end it suffices to show that

$$\int_{\frac{1}{2}X}^X \int_0^H (\psi(x+h) - \psi(x) - h)^2 \, dh \, dx \sim \frac{1}{4} H^2 X \log X/H \qquad (41)$$

In this integral we replace h by $\delta = h/x$, and invert the order of integration. Thus the left hand side above is

$$\int_0^{H/X} \int_{\frac{1}{2}X}^X f(x,\delta x)^2 \, x \, dx \, d\delta + \int_{H/X}^{2H/X} \int_{\frac{1}{2}X}^{H/\delta} f(x,\delta x)^2 \, x \, dx \, d\delta$$

where $f(x,y) = \psi(x+y) - \psi(x) - y$. By integrating by parts, we see from (b) that if $A \approx B \approx X$ then

$$\int_A^B f(x,\delta x)^2 \, x \, dx = \frac{1}{3} (B^3 - A^3) \, \delta \log 1/\delta + O(X^3 \delta \log 1/\delta).$$

This yields (41) . Then (40) follows by replacing X by $X2^{-k}$ in (41), summing over $0 \leqslant k \leqslant K = [2 \log\log X]$, and by appealing to (4) with X replaced by $X2^{-K-1}$.

We now deduce (c) from (40). Suppose that $0 < \eta < 1$. By differencing in (40) we see that

$$\int_H^{(1+\eta)H} \int_0^X f(x,h)^2 \, dx \, dh = (\eta + \tfrac{1}{2}\eta^2 + o(1)) XH^2 \log X/H \quad .$$

Let $g(x,h) = f(x,H)$. From the identity

$$f^2 - g^2 = 2f(f-g) - (f-g)^2$$

and the Cauchy-Schwartz inequality we find that

$$\iint f^2 - g^2 \ll \left(\iint f^2\right)^{1/2} \left(\iint (f-g)^2\right)^{1/2} + \iint (f-g)^2 \quad .$$

But $f(x,h) - g(x,h) = f(x+H, h-H)$, so that

$$\iint (f-g)^2 = \int_0^{\eta H} \int_H^{X+H} f(x,h)^2 \, dx \, dh$$

$$\ll \eta^2 H^2 X \log X/H$$

by (40). Hence we see that

$$\eta H \int_0^X (\psi(x+H) - \psi(x) - H)^2 \, dx = \iint g^2$$

$$= \iint f^2 + O(\eta^{3/2} XH^2 \log X/H)$$

$$= \left(\eta + O(\eta^{3/2}) + o(1)\right) XH^2 \log X/H \ .$$

We now divide both sides by ηH, and obtain the desired result by letting $\eta \to 0^+$ sufficiently slowly.

References.

1. H. Davenport, Multiplicative Number Theory, Second Edition, Springer-Verlag, 1980.

2. A. Fujii, On the zeros of Dirichlet L-functions, I, *Trans. Amer. Math. Soc.* **196** (1974), 225-235. (Corrections to this paper are noted in *Trans. Amer. Math. Soc.* **267** (1981), pp 38-39, and in [5; pp. 219-220].)

3. P.X. Gallagher, A large sieve density estimate near $\sigma = 1$, *Invent. Math.* **11** (1970), 329-339.

4. P.X. Gallagher, On the distribution of primes in short intervals, *Mathematika* **23** (1976), 4-9.

5. P.X. Gallagher and Julia H. Mueller, Primes and zeros in short intervals, *J. Reine Agnew. Math.* **303/304** (1978), 205-220.

6. Daniel A. Goldston, Large differences between consecutive prime numbers, Thesis, University of California Berkeley, 1981.

7. H.Halberstam and H.-E. Richert, Sieve Methods. Academic Press, London, 1974.

8. G.H. Hardy, Divergent Series, Oxford University Press, 1963.

9. G.H. Hardy and W.W. Rogosinski, Notes on Fourier series (I): On sine series with positive coefficients, *J. London Math. Soc.* **18** (1943), 50-57.

10. D.R. Heath-Brown, Gaps between primes, and the pair correlation of zeros of the zeta-function, *Acta Arith.* **41** (1982), 85-99.

11. D.R. Heath-Brown and D.A. Goldston, A note on the difference between consecutive primes, *Math. Ann.* **266** (1984), 317-320.

12. Julia Huang (=J.H. Mueller), Primes and zeros in short intervals, Thesis, Columbia University, 1976.

13. H.L Montgomery, The pair correlation of zeros of the zeta function, *Proc. Sympos. Pure Math.* **24** (1973), 181-193.

14. B. Saffari and R.C. Vaughan, On the fractional Parts of x/n and related sequences II, *Ann. Inst. Fourier* (Grenoble) **27**, no. 2, (1977), 1-30.

15. A. Selberg, On the normal density of primes in small intervals, and the difference between consecutive primes, *Arch. Math. Naturvid.* **47**, no. 6, (1943), 87-105.

16. E.C. Titchmarsh, The theory of the Riemann zeta-function, Oxford University Press, 1951.

17. J.D. Vaaler, Some extremal functions in Fourier analysis, *Bull. Amer. Math. Soc.*, **12**, No. 2, (1985), 183-216.

D. A. Goldston
San Jose State University,
San Jose, CA 95192,
U.S.A.

H. L. Montgomery
University of Michigan,
Ann Arbor, MI 48109,
U.S.A.

ONE AND TWO DIMENSIONAL EXPONENTIAL SUMS

S. W. Graham and G. Kolesnik

1. Introduction

In number theory, one often encounters sums of the form

$$\sum_{(n_1,\ldots,n_k) \in \mathcal{D}} \sum e\big(f(n_1,\ldots,n_k)\big), \tag{1}$$

where $\mathcal{D}$ is a bounded domain in $\mathbf{R}^k$ and $e(w) = e^{2\pi iw}$. We shall refer to the case $k = 1$ as the one-dimensional case, $k = 2$ as the two-dimensional case, etc. Our objective here is to give an exposition of van der Corput's method for estimating the sums in (1). The one-dimensional case is well understood. Our knowledge of the two-dimensional case is fragmentary, and dimensions higher than two are *terra incognita*. We shall review the one-dimensional case in Section 2. In Section 3 we will give an outline of what is known and what is conjectured about the two-dimensional case.

2. The one-dimensional case

Let N be a large positive integer, I a subinterval of $(N, 2N]$, and $f : I \to \mathbf{R}$. We wish to get an upper bound for $S := \sum_n e\big(f(n)\big)$. Since $|e(f(n))| = 1$, we have the trivial upper bound $|S| \leqslant N$. Moreover, this upper bound is attained when $f(n) = an + b$, a is an integer, and $I = (N, 2N]$. A non-trivial upper bound thus requires some conditions on f. Usually these conditions are hypotheses about the derivatives of f. One example is

Theorem 1. (Kusmin-Landau inequality)

Assume that f' *is monotonic and that* $\|f'\| \geqslant \lambda$ on I, *where*

$$\|x\| := \min_{n \in \mathbf{Z}} |x - n|.$$

Then

$$\sum_{n \in I} e(f(n)) \ll \lambda^{-1}.$$

This inequality is implicit in Lemmas 4.8 and 4.2 of Titchmarsh's book [20]. An elementary proof can be found in Herzog-Piranian [6].

The condition that $\|f'\| \geqslant \lambda$ is too restrictive for most applications. Van der Corput's method applies to a much wider class of functions. It depends upon two processes, which have become known as the A-process and the B-process. The A-process may be formulated as

Lemma 1. *Let* I *and* f *be as before. Then*

$$\Big|\sum_{n \in I} e(f(n))\Big|^2 \leqslant \frac{|I| + Q}{Q} \sum_{|q|<Q} \left(1 - \frac{|q|}{Q}\right) \sum_{\substack{n \in I \\ n+q \in I}} e\big(f(n+q) - f(n)\big)$$

The proof of Lemma 1 uses the Cauchy-Schwarz inequality on the sum

$$\sum_{q=1}^{Q} \sum_{\substack{n \in I \\ n+q \in I}} e\big(f(n+q)\big) ,$$

see Titchmarsh [20] for details.
In most applications, the following variant of Lemma 1 is used.

Lemma 1A. *Let* I *and* f *be as before, let*

$$S = \sum_{n \in I} e\big(f(n)\big) \text{ and } S_q = \sum_{\substack{n \in I \\ n+q \in I}} e\big(f(n+q) - f(n)\big).$$

If $Q \leqslant |I|$ *then*

$$|S|^2 \ll \frac{|I|^2}{Q} + \frac{|I|}{Q} \sum_{0<|q|<Q} |S_q|. \qquad (2)$$

Of course, $|I|$ on the right-hand side of (2) may be replaced by N. But there are occassions when one needs to use the fact that I is short, and it is important to have $|I|$ in (2).

The B-process is a combination of the Poisson summation formula and the saddle point method. One possible formulation is

Lemma 2. *Let* $I = [a, b] \subset [N, 2N]$. *Assume* f *has four continuous derivatives and that* $f''(x) < 0$ *on* I. *Assume further that*

$$m_2 \leqslant |f''(x)| \leqslant Cm_2 , \quad |f^{(3)}(x)| \ll m_3 , \quad |f^{(4)}(x)| \ll m_4 ,$$

and that $m_3^2 = m_2 m_4$. *Let* $f'(b) = \alpha$, $f'(a) = \beta$, *and let* n_ν *be such that* $f'(n_\nu) = \nu$ *for* $\alpha < \nu < \beta$. *Then*

$$\sum_{n \in I} e\big(f(n)\big) = \sum_{\alpha<\nu\leq\beta} e\big(f(n_\nu) - \nu n_\nu - 1/8\big)|f''(n_\nu)|^{-1/2}$$
$$+ O\big(m_2^{-1/2} + |I|m_3 + \log(2 + |I|m_2)\big).$$

This is Lemma 3 of Phillips [13]. Heath-Brown [5] and Atkinson [1] give other versions of this lemma in which f is assumed to be analytic in some appropriate domain; this hypothesis naturally leads to strong error terms.

The efforts of the A and B processes can be explained succinctly by the theory of exponent pairs. In this theory, we deal with functions f satisfying the following conditions:

(3.1) f has infinitely many derivatives on I,

(3.2) there exists $y > 0$, $s > 0$, and d, $0 < d < 1/2$, such that for all integers $p \geqslant 0$ and all $x \in I$,

$$|f^{(p+1)}(x) - (-1)^p (s)_p y x^{-s-p}| < d(s)_p y x^{-s-p},$$

(3.3) $z := ya^{-s} \geqslant 1/2$.

The symbol $(s)_p$ in condition (3.2) is defined by $(s)_o = 1$ and $(s)_p = s(s+1)\cdots(s+p-1)$ if $p \geqslant 1$. Condition (3.2) states that f is, in an appropriate sense, well approximated by yx^{-s}. In condition (3.3), z is effectively $f'(a)$. The condition $z \geqslant 1/2$ is motivated by the fact that we can apply the Kusmin-Landau inequality in the contrary case.

Definition. The ordered pair (k,ℓ) is an *exponent pair* if $0 \leqslant k \leqslant 1/2 \leqslant \ell \leqslant 1$, and if for all f satsifying (3.1)-(3.3), the estimate

$$\sum_{n \in I} e\big(f(n)\big) \ll z^k N^\ell$$

holds.

The trivial estimate shows that (0,1) is an exponent pair. By application of Lemma 1A, one can prove that if (k,ℓ) is an exponent pair then

$$A(k,\ell) = \left(\frac{k}{2k + 2} , \frac{k + \ell + 1}{2k + 2} \right)$$

is also an exponent pair. By application of Lemma 2, one can prove that if (k,ℓ) is an exponent pair and if $k + 2\ell \geqslant 3/2$ then

$$B(k,\ell) = (\ell - 1/2, k + 1/2)$$

is also an exponent pair. Proofs of these results can be found in Phillips [13]. The restriction $k + 2\ell \geqslant 3/2$ in the B-process can be removed by appealing to the stronger versions of Lemma 2 previously mentioned. Moreover, this condition is satisfied by every exponent pair that arises from the A-process. There is no point in applying B to an exponent pair arising from the B-process since $B^2(k,\ell) = (k,\ell)$.

For computational purposes, it is convenient to think of A and B as linear transformations on projective space. Let

$$A = \begin{bmatrix} 1 & 0 & 0 \\ 1 & 1 & 1 \\ 2 & 0 & 2 \end{bmatrix} \quad \text{and} \quad B = \begin{bmatrix} 0 & 2 & -1 \\ 2 & 0 & 1 \\ 0 & 0 & 2 \end{bmatrix} .$$

Then

$$A \begin{bmatrix} k \\ \ell \\ 1 \end{bmatrix} = \begin{bmatrix} k \\ k + \ell + 1 \\ 2k + 2 \end{bmatrix}$$

In projective space this is equal to

$$\begin{bmatrix} k/(2k + 2) \\ (k + \ell + 1)/(2k + 2) \\ 1 \end{bmatrix} = \begin{bmatrix} \kappa \\ \lambda \\ 1 \end{bmatrix} ,$$

where $(\kappa,\lambda) = A(k,\ell)$. The B matrix has an analogous effect. We are, of course, abusing notation by using the same letters in two different senses, but the intended meaning will be clear from the context.

As we noted before, B is an involution. Moreover, $A(0,1) = (0,1)$. It follows that any exponent pair obtainable from the A and B processes can be written either in the form

$$A^{q_1}BA^{q_2}B \cdots A^{q_r}B(0,1) \tag{4}$$

or in the form

$$BA^{q_1}BA^{q_2}B \cdots A^{q_r}B(0,1), \tag{5}$$

where $q_1,\ldots,q_r$ are non-negative integers. (When A and B are thought of as functions on $\mathbf{R}^2$, $A^{q_1}BA^{q_2}B \cdots A^{q_r}B$ is a composition of functions. Thus $AB(0,1) = A(B(0,1)) = A(1/2,1/2) = (1/6,2/3)$. When A and B are thought of as matrices, $A^{q_1}B \cdots A^{q_r}B$ is a matrix multiplication.)

We use $\mathcal{P}$ to denote the set of all exponent pairs obtainable from (0,1) by A and B. Exponent pairs of the form (4) are in the set $A\mathcal{P}$; those of the form (5) are in the set $BA\mathcal{P}$. Note that $(0,1) \in A\mathcal{P}$ since $A(0,1) = (0,1)$.

Exponent pairs enjoy a convexity property. From the inequality

$$\min(X,Y) \leqslant X^{\alpha}Y^{1-\alpha} \qquad (0 \leqslant \alpha \leqslant 1) \tag{6}$$

we see that if (k_1,ℓ_1) and (k_2,ℓ_2) are exponent pairs, then so is

$$(\alpha k_1 + (1 - \alpha)k_2,\ \alpha\ell_1 + (1 - \alpha)\ell_2)$$

for any α, $0 \leqslant \alpha \leqslant 1$. Consequently, $\overline{\mathcal{P}}$ - the convex hull of $\mathcal{P}$ - is a set of exponent pairs. In fact all known exponent pairs are in $\overline{\mathcal{P}}$. However, it is possible that there are other exponent pairs. For example, it has been conjectured that $(\varepsilon, 1/2 + \varepsilon)$ is an exponent pair for every $\varepsilon > 0$.

In applications, it is usually desirable to minimize some

function on P. We illustrate this with the following examples. Let $\zeta(s)$ denote Riemann's zeta function, let $d(n)$ be the number of divisors of n, and let $r(n)$ be the number of ways of writing n as a sum of two squares. Set

$$\Delta(x) = \sum_{n \leq x} d(n) - x(\log x + 2\gamma - 1),$$

and

$$R(x) = \sum_{n \leq x} r(n) - \pi x .$$

It can be proved that if (k,ℓ) is an exponent pair and if $\Theta(k,\ell) = k + \ell - 1/2$ then

$$\Delta(x) \ll x^{\Theta} \log x + x^{1/4} \log x ,$$

$$R(x) \ll x^{\Theta} \log x + x^{1/4} \log x,$$

and

$$\zeta(1/2 + it) \ll t^{\Theta/2} \log t.$$

This motivates the problem of finding

$$\inf_{(k,\ell) \in P} (k+\ell). \tag{7}$$

In 1945, Rankin [14] found an algorithm for computing (7). His work was published ten years leater, but he did not give the details of his method since they involved much heavy algebra. Recently, one of us (Graham) has found an algorithm for computing

$$\inf_{(k,\ell)\in P} \frac{ak + b + c}{dk + e + f} ; \tag{8}$$

the algebra can be considerably lightened by appealing to matrix notation.

The algorithm yields a sequence of exponent pairs which provide approximations to the desired infimum. The rth term in this sequence has the form $A^{q_1}BA^{q_2} \cdots A^{q_r}B(0, 1)$, where all the q_i's are non-negative integers, and only q_1 can be zero. The sequence $(q_1, q_2, \ldots)$ is called the q-sequence. It is unusual to have $q_i \geq 10$, so it is convenient to use baseball notation and write the q-sequence as $q_1q_2q_3q_4q_5 \quad q_6q_7q_8q_9q_{10} \cdots$. For example, in the problem of finding (7) the optimal q-sequence is

$$13211 \qquad 21122 \qquad 12221 \qquad 21122 \qquad 11213 \qquad \ldots \tag{9}$$

This means that the sequence of exponent pairs leading to the infimum is

$$AB(0,\ 1) = (1/6,\ 2/3),$$

$$ABA^3B(0,\ 1) = (11/82,\ 57/82),$$

$$ABA^3BA^2B(0,\ 1) = (33/234),\ 161/234),$$

etc. Using a Casio FX-700P programmable calculator, we have carried the sequence in (9) out to 100 terms. Glen Ierley and his IBM PC-XT have shown that this gives $\inf(k + \ell)$ to 85 decimal places. To 30 places, the answer is

$$\inf_{(k,\ell) \in P} (k + \ell) = .82902 \ 13568 \ 59133 \ 59240 \ 92397 \ 77283.$$

The details of the above mentioned algorithm will appear later, but we can give a short sketch of it here. Let

$$\Theta(k,\ell) = \frac{ak + b \quad + c}{dk + e \quad + f} .$$

It is necessary to assume that $dk + e\ell + f > 0$ for all $(k,\ell) \in P$. In practice, this requires checking only the points (0,1), (1/2,1/2) and (0,1/2), for P is contained inside the triangle determined by these points.

We may also regard Θ as a matrix, i.e.

$$\Theta = \begin{bmatrix} a & b & c \\ d & e & f \end{bmatrix} .$$

Let u, v, and w denote tha 2×2 sub-determinants of Θ, so that

$$u = \begin{vmatrix} b & c \\ e & f \end{vmatrix}, \quad v = \begin{vmatrix} a & c \\ d & f \end{vmatrix}, \text{ and } w = \begin{vmatrix} a & b \\ d & e \end{vmatrix} .$$

The algorithm is based on

Lemma 3. *If* $(k,\ell) \in AP$, *then* $\Theta B(k,\ell) - \Theta(k,\ell)$ *has the sign as*

$$w(k + \ell) + v - u.$$

We then apply this lemma as follows. Let

$$r = \inf_{(k,\ell) \in P} (k + \ell) = .82902\ 13568\ 59133 \ldots ,$$

$$Y = \max \{w + v - u,\ wr + v - u\},$$

and

$$Z = \min \{w + v - u,\ wr + v - u\}.$$

The analysis then breaks into three cases.

Case 1. $Z \geqslant 0$. Then $\Theta B(k,\ell) \geqslant \Theta(k,\ell)$ for all (k,ℓ) in P . Consequently,

$$\inf_{(k,\ell) \in P} \Theta(k,\ell) = \inf_{(k,\ell) \in P} \Theta A(k,\ell).$$

We let $\Theta_1 = \Theta A$, and we repeat the analysis.

Case 2. $Y \leqslant 0$. Then $B(k,\ell) \leqslant \Theta(k,\ell)$ for all (k,ℓ) in P. Consequently,

$$\inf_{(k,\ell) \in P} \Theta(k,\ell) = \inf_{(k,\ell) \in P} \Theta BA(k,\ell).$$

We let $\Theta_1 = \Theta BA$, and we repeat the analysis.

Case 3. $Z < 0 < Y$. In this case, the algorithm branches. We pursue each branch until one of them can be shown to be superior.

3. Two dimensional sums.

Let $\mathcal{D} \subset [X, 2X] \times [Y, 2Y]$ and let $f\colon \mathcal{D} \to \mathbf{R}$. Define

$$S = \sum_{(m,n) \in \mathcal{D}} e\big(f(m,n)\big) .$$

In analogy with the one-dimensional theory of exponent pairs, it is appropriate to assume that

$$D_{x^i y^j} f(x,y) = D_{x^i y^j} (Ax^{-\alpha} y^{-\beta})\{1 + O(\Delta)\}$$

where

$$D_{x^i y^j} = \frac{\partial^{i+j}}{\partial x^i \partial y^j},$$

A is a non-zero real constant,

$\alpha < 1$, $\beta < 1$, $\alpha\beta \neq 0$, and

$\Delta = \Delta(X, Y) \to 0$ as $X \to \infty$ and $Y \to \infty$.

The primary tools for estimating S are two dimensional analogues of the Poisson summation formula and the Weyl-van der Corput inequality.

First, let us consider the Poisson summation formula. Recall that in Lemma 2, terms of the form $|f''(x_\nu)|^{-1/2}$ appear, so that the usefulness of that lemma is lessened when f" becomes small. In two dimensional sums, the Hessian of f plays a similar role. The Hessian of f is defined by

$$Hf = \det \begin{vmatrix} D_{xx}f & D_{xy}f \\ D_{xy}f & D_{yy}f \end{vmatrix} .$$

A precise version of the two dimensional Possion summation formula is complicated to state; see [17], Lemma 4 or [9], Lemma 2. We will mention only that under suitable conditions on f and $\mathcal{D}$, we have

$$\sum_{(m,n)\in\mathcal{D}} e\big(f(m,n)\big)$$

$$\ll M^{-1/2} \Big| \sum_{(u,\nu)\in\Delta'} e\big(f(\xi,\eta) - \mu\xi - \nu\eta\big)\Big| + \text{Error terms.}$$

Here, it is understood that

(i) M satisfies $M \ll Hf < M$,

(ii) Δ is the image of $\mathcal{D}$ under $\mu = D_x f$, $\nu = D_y f$,

(iii) Δ' is some subset of Δ,

(iv) $\xi = \xi(\mu,\nu)$ and $\eta = \eta(\mu,\nu)$ are defined by $D_x f(\xi,\eta) = \mu$ and $D_y f(\xi,\eta) = \nu$.

The two dimensional Weyl-van der Corput inequality can be expressed as

Lemma 4. *If* $Q \leq X$ *and* $R \leq Y$ *then*

$$|S|^2 \ll \frac{X^2Y^2}{QR} + \frac{XY}{QR} \sum_{\substack{|q| < Q \quad |r| < R \\ (q,\, r) \neq (0,\, 0)}} \sum |S_1(q, r)|,$$

where

$$S_1(q, r) = \sum_{(m,n) \in \mathcal{D}_1(q,r)} e\big(f_1(m,n;q,r)\big),$$

with

$$f_1(m,n;q,r) = f(m + q, n + r) - f(m, n)$$

$$= \int_0^1 \frac{\partial}{\partial t} f(m + qt, n + rt)\, dt,$$

and

$$\mathcal{D}_1(q,r) = \big\{(m,n) : (m + qt, n + rt) \in \mathcal{D} \text{ for } t = 0, 1 \big\}.$$

In analogy with the one-dimensional case, we can hope to prove an estimate of the form

$$S \ll L_1^{k_1} X^{\ell_1} L_2^{k_2} Y^{\ell_2}, \tag{10}$$

where $L_1 = |A|X^{-\alpha-1}Y^{-\beta}$ and $L_2 = |A|X^{-\alpha}Y^{-\beta-1}$. Note that $L_1 \approx D_x f$ and $L_2 \approx D_y f$. If we can prove an estimate of the form (10) under appropriate assumptions on f and $\mathcal{D}$, we say that $(k_1,\ell_1;\, k_2,\ell_2)$ is an exponent quadruple. Note that since

$$|S| \leq \sum_m \Big| \sum_n e\big(f(m,n)\big)\Big|, \tag{11}$$

$(0,1;\, k,\ell)$ is an exponent quadruple whenever (k,ℓ) is an exponent pair. Similarly, $(k,\ell;\, 0,1)$ is an exponent quadruple.

Unfortunately, the application of Lemma 4 and the Poisson summation formula is not as straightforward as it is in the one-dimensional case. To illustrate why this is so, we consider

$$f(m,n) = Am^{-\alpha}n^{-\beta}.$$

After applying Lemma 4, we encounter functions of the form

$$f_1(m,n;\, q,r) = \int_0^1 \frac{d}{dt} A(m + qt)^{-\alpha}(n + rt)^{-\beta} dt$$

$$\sim -Am^{-\alpha}n^{-\beta}(\alpha q m^{-1} + \beta r n^{-1}) .$$

If we then apply the Poisson summation formula, we must first compute Hf_1. Now

$$Hf_1 \sim A^2 m^{-2\alpha} n^{-2\beta} \alpha\beta(\alpha + \beta + 2)P,$$

where

$$P = \frac{(\alpha)_2 q^2}{m^2} + \frac{2(\alpha + 1)(\beta + 1)qr}{mn} + \frac{(\beta)_2 r^2}{n^2} .$$

For some values of the parameters, the expression P will vanish, or it will be inconveniently small. The effect of this is that the Poisson summation formula cannot be applied directly. Instead, we subdivide $\mathcal{D}$ into a region where P is small and another region where P is large. In the latter region, we can apply the Poisson summation formula. In the former region, we use some other estimate such as (11). There are considerable technical difficulties in carrying this out, and the difficulties become even more pronounced when Lemma 4 is used more that once. Here we shall ignore these difficulties and argue heuristically. By Lemma 4,

$$S^2 \ll \frac{X^2Y^2}{QR} + \frac{XY}{QR} \sum_{|q|<Q} \sum_{\substack{|r|<R \\ (q,r) \neq (0,0)}} |S_1(q,r)| .$$

Now

$$S_1(q,r) = \sum_{(m,n) \in \mathcal{D}_1(q,r)} e\big(f_1(m,n;\ q,r)\big) ,$$

and

$$f_1(m,n;\ q,r) = \int_0^1 \frac{d}{dt} f(m + at,\ n + rt)dt$$

$$\approx \frac{qF}{X} + \frac{rF}{Y} \approx \rho F$$

where $F = |A|X^{-\alpha}Y^{-\beta}$ and $\rho = \max(|q|X^{-1}, |r|Y^{-1})$. If $(k_1,\ell_1;\ k_2,\ell_2)$ is an exponent quadruple, then

$$S_1(q,r) \ll (F\rho)^{k_1+k_2}\, X^{\ell_1-k_1}\, Y^{\ell_2-k_2}.$$

Now assume that Q and R are chosen so that $QX^{-1} = RY^{-1}$. If we set $Z = Q^2YX^{-1} = R^2XY^{-1}$, then

$$\frac{1}{QR} \sum_{|q|<Q} \sum_{|r|<R} \rho^{k_1+k_2} \ll \left(\frac{Z}{XY}\right)^{(k_1+k_2)/2} .$$

It follows that

$$S^4 \ll X^4Y^4Z^{-2} + F^{2k_1+2k_2}X^{2+2\ell_1-3k_1-k_2}Y^{2+2\ell_2-k_1-3k_2}Z^{k_1+k_2}.$$

Choose Z so that the two terms on the right-hand side are equal. Then

$$S^{2k_1+2k_2+4} \ll \left(\frac{F}{X}\right)^{2k_1} X^{k_1+k_2+2\ell_1+2} \left(\frac{F}{X}\right)^{2k_2} Y^{k_1+k_2+2\ell_2+2} .$$

Thus we see heuristically, that if $(k_1,\ell_1;\ k_2,\ell_2)$ is an exponent quadruple then so is

$$A(k_1,\ell_1;\ k_2,\ell_2) = \frac{1}{2k_1+2k_2+4}(2k_1, k_1+k_2+2\ell_1+2;\ 2k_2,\ k_1+k_2+2\ell_2+2).$$

Similarly, a heuristic argument with the Poisson summation formula yields the exponent quadruple

$$B(k_1,\ell_1;\ k_2,\ell_2) = (\ell_1 - 1/2,\ k_1 + 1/2,\ \ell_2 - 1/2,\ k_2 + 1/2).$$

One way of avoiding the difficulties implicit in Lemma 4 is to apply it with Q = 1 or R = 1. Classical scholars will recall that Titchmarch [18] used this approach. In his notation, Lemma 4 is Lemma β, and Lemma 4 with R = 1 is Lemma β'. By taking R = 1 and arguing heuristically, we see that this approach should lead to the exponent quadruple

$$A_1(k_1,\ell_1;\ k_2,\ell_2) = \frac{1}{2k_1+2k_2+2}(k_1,\ k_1+\ell_1+k_2+1;\ k_2,\ 2k_1+2k_2+\ell_2+1).$$

Similarly, with Q = 1 one gets

$$A_2(k_1,\ell_1;\ k_2,\ell_2) = \frac{1}{2k_1+2k_2+2}(k_1,\ 2k_1+\ell_1+2k_2+1;\ k_2,\ k_1+k_2+\ell_2+1).$$

We may use (6) and take the average of A_1 and A_2 to get the exponent quadruple

$A_s(k_1,\ell_1;\ k_2,\ell_2)$

$$= \frac{1}{4k_1+4k_2+4}\,(2k_1,\ 3k_1+2\ell_1+3k_2+\ell_2;\ 2k_2,\ 3k_1+\ell_1+3k_2+\ell_2).$$

The "s" here stands for Srinivasan, who used essentially this operation in his method of exponent pairs [17]. We shall say more about this later.

It is also possible to apply the Poisson summation formula to one variable at a time and get the exponent quadruples

$$B_1(k_1,\ell_1;\ k_2,\ell_2) = (\ell_1 - 1/2,\ k_1 + 1/2;\ k_2,\ \ell_2)$$

and

$$B_2(k_1,\ell_1;\ k_2,\ell_2) = (k_1,\ \ell_1;\ \ell_2-1/2,\ k_2 + 1/2).$$

In some applications, the critical cases for estimating S occur when $X \approx Y$. In such a case, it is desirable to have $k_1 = k_2$ and $\ell_1 = \ell_2$. Note that

$$A(k,\ell;\ k,\ell) = \left(\frac{k}{2k+2},\ \frac{k+\ell+1}{2k+2};\ \frac{k}{2k+2},\ \frac{k+\ell+1}{2k+2}\right)$$

and

$$B(k,\ell;\ k,\ell) = (\ell - 1/2,\ k + 1/2;\ \ell - 1/2,\ k + 1/2).$$

We thus have the following

Conjecture. If (k,ℓ) is an exponent pair, then $(k,\ell;\ k,\ell)$ is an exponent quadruple.

The conjecture is known in the following special cases.

1. $f(x,y) = g(x) + h(y)$ and $\mathcal{D}$ is a rectangle. In this case,

$$S = \sum_{(m,n)\in\mathcal{D}} e\big(g(m) + h(n)\big) = \sum_m e\big(g(m)\big) \sum_n e\big(h(n)\big)\ ,$$

and the result follows immediately.

2. $(k,\ell) = (0,1)$. This is the trivial estimate.

3. $(k,\ell) = B(0,1) = (1/2,1/2)$. This has been proved by several authors independently; see [3], [5], and [16].

4. $(k,\ell) = AB(0,1) = (1/6,2/3)$. See Theorem 1 of Kolesnik [10].

5. $(k,\ell) = A^qB(0,1)$ for any $q \geqslant 0$. This is a result of the authors which is in preparation.

Srinivasan [17] has used A_s to develop a theory of exponent quadruples. Roughly stated, his theory is as follows. Let P_s be the set of all pairs obtained from (0,1) by

$$A_s(k,\ell) = \left(\frac{k}{4k+2}, \frac{3k+\ell+1}{4k+2}\right).$$

and

$$B(k,\ell) = (\ell - 1/2,\ k + 1/2).$$

If $(k,\ell) \in P_s$, then $(k,\ell;\ k,\ell)$ is an exponent quadruple.

It should be noted that Srinivasan's notation is different from ours; he says that (k,ℓ) is a two-dimensional exponent pair if

$$S \ll L_1^k X^{1-\ell} L_2^k Y^{1-\ell}.$$

The applications mentioned in Section 1 can be done with two dimensional sums. Assume that $(k,\ell;\ k,\ell)$ is an exponent quadruple, and let

$$\Theta = \Theta(k,\ell) = \frac{2k + 2\ell - 1}{4\ell - 1}.$$

Then for some constant $C > 0$, we have

$$\zeta(1/2 + it) \ll t^{\Theta/2} \log^C t, \tag{12.1}$$

$$\Delta(x) \ll x^{\Theta} \log^C x + x^{1/4} \log x, \tag{12.2}$$

$$E(x) \ll x^{\Theta} \log^C x + x^{1/4} \log x. \tag{12.3}$$

Here is a historical survey of the results of this type that have appeared in the literature.

1. $(k,\ell) = A_s^3B(0,1)$; $\Theta = 19/58$. This was done by Titchmarsh [19] for $\zeta(1/2 + it)$.

2. $(k,\ell) = A_s^2AB(0,1)$; $\Theta = 15/46$. This was done by Titchmarsh

[18] for $E(x)$, by Min [12] for $\zeta(1/2 + it)$, and by Richert [15] for $\Delta(x)$.

3. $(k,\ell) = A_s A^2 B(0,1)$; $\Theta = 13/40$. This was done by Hua [7] for $E(x)$.

4. $(k,\ell) = A^3 B(0,1)$; $\Theta = 12/37$. This was done by Haneke [4] for $\zeta(1/2 + it)$, by Chen [2] for $E(x)$, and by Kolesnik [8] for $\Delta(x)$.

5. $(k,\ell) = A^3 B A_s^3 B(0,1)$; $\Theta = 35/108$. This was done by Kolesnik [11] for $\zeta(1/2 + it)$ and $\Delta(x)$.

Note that

$$\frac{35}{108} = .324\ \overline{074}\ .$$

If our conjecture is true for $(k,\ell) = A^3 B A^3 B(0,1)$, then we could prove (12.1), (12.2), and (12.3) with

$$\Theta = \frac{23}{71} = .32394\ \ 36619\ \ldots\ .$$

If we assume the conjecture for all (k,ℓ) and apply the algorithm mentioned in Section 1, we find that the optimal q-sequence is

32122 11121 21211 21121 11122 11111 ...

and the limiting value for Θ is

.32392 47503 76239 83494 00175 84916 ... .

We would like to mention two more applications. We can apply Lemma 4 with $Q = 1$ to estimate sums of the form

$$\sum_{(m,n)\in\mathcal{D}} a(m)\ e\big(f(m,n)\big).$$

An example of this is given in Lemma 4 of [3]. By making some slight modifications of that lemma, we can prove that if $\sum_m |a(m)|^2 \ll X$ and $\beta > 0$ then

$$\sum_{X<m<2X}\ \sum_{Y<n<2Y} a(m)\ e\big(xm^{-\beta}n^{-\beta}\big)$$

$$\ll F^{1/4}\, X^{3/4}\, Y^{1/2} + X^{5/6}\, Y^{5/6} + F^{-1/4}\, XY + XY^{1/2},$$

where $F = xX^{-\beta}Y^{-\beta}$. The first term may be written as

$$\left(\frac{F}{X}\right)^{1/8} X^{7/8} \left(\frac{F}{Y}\right)^{1/8} Y^{5/8} .$$

Note that $A_2B(0,1;\ 0,1) = (1/8,7/8;\ 1/8,5/8)$.

In our final application, we let $d_3(n)$ be the number of ways of writing n as a product of three factors, and we define

$$\Delta_3(x) = \sum_{n \leqslant x} d_3(n) - xf_3(\log x),$$

where $f_3(\log x)$ is the residue of $\zeta^3(s)x^s/s$ at $s = 1$. An examination of Kolesnik's arguments in [9] shows that if $(k_1,\ell_1;\ k_2,\ell_2)$ is an exponent quadruple and if

$$\Theta = \frac{2k_1 + 12\ell_1 + 10k_2 + 4\ell_2 - 5}{6(4\ell_1 + 3k_2 + \ell_2 - 1)}$$

then

$$\Delta_3(x) \ll x^{\Theta+\varepsilon}. \tag{13}$$

Kolesnik takes $(k_1,\ell_1;\ k_2,\ell_2) = ABA_1B(0,1;\ 0,1) = (1/20,15/20;\ 3/20,15/20)$ to get $\Theta = 43/96 = .447916 \cdots$. If we take $k_1 = k_2 = k$ and $\ell_1 = \ell_2 = \ell$ then

$$\Theta = \frac{12k + 16\ell - 5}{18k + 30\ell - 6} .$$

For this Θ, the optimal q-sequence is

11112 22121 21211 23321 11221 11111 ··· .

Our conjecture would therefore imply (13) with

$$\Theta = .44607\ 41756\ 73843\ 37652.$$

Acknowledgements.

We had the opportunity to speak on this material at the Mathematisches Forschingstitut of Oberwohlfach, at Oklahoma State University, and the University of Michigan. We wish to thank those institutions for their hospitality.

References.

1. F. V. Atkinson, The mean value of the Riemann zeta-function, *Acta. Math.* **81** (1949), 353-376.

2. Chen Jing-Run, The lattice points in a circle, *Sci. Sinica* **12** (1963),633-649.

3. S. W. Graham, The distribution of squarefree numbers, *J. London Math. Soc.* (2) **24** (1981), 54-64.

4. W. Haneke, Verscharfung der Abschätzung von $\zeta(1/2 + it)$, *Acta Arith.* **8** (1963), 357-430.

5. D. R. Heath-Brown, The Pjateckii-Sapiro prime number theorem, *J. No. Theory* **16** (1963), 242-266.

6. F. Herzog and G. Piranian, Sets of convergence of Taylor Series I, *Duke Math. Jnl.* **16** (1949) 529-534.

7. L. K. Hua, The lattice points in a circle, *Quart. J. Math.* (Oxford) **12** (1941), 193-200.

8. G. Kolesnik, Improvement of remainder term for the divisors problem, *Math. Zametki* **6** (1969), 545-554.

9. ___________, On the estimation of multiple exponential sums, Recent Progress in Analytic Number Theory, Vol. 1 (eds. H. Halberstam and C. Hooley, Academic Press, New York, 1981) 247-256.

10. ___________, On the number of abelian groups of a given order, *J. Reine Angew. Math.* **329** (1981), 164-175.

11. ___________, On the order of $\zeta(1/2 + it)$ and $\Delta(R)$, *Pac. Jnl. of Math.*, **98** (1982) 107-122.

12. S. H. Min, On the order of $\zeta(1/2 + it)$, *Trans. Amer. Math.*

Soc. **65** (1949) 448-472.

13. E. Phillips, The zeta-function of Riemann; further developments of van der Corput's method, *Quart. J. Math.* (Oxford) **4** (1933) 209-225.

14. R. A. Rankin, Van der Corput's method and the theory of exponent pairs, *Quart. J. Math. Oxford* (2), **6** (1955) 147-153.

15. H. E. Richert, Verscharfüng der Abschatzüng beim Dirichletschen Teilerproblem, *Math. Z.* **58** (1953) 204-218.

16. P. G. Schmidt, Zur Anzahl Abelscher Gruppen gegebner Ordnung I, *Acta Arith.* **13** (1968) 405-417.

17. B. R. Srinivasan, The lattice point problem in many dimensional hyperboloids, III, *Math. Ann.* **160** (1965) 280-311.

18. E. C. Titchmarsh, The lattice points in a circle, *Proc. London Math.Soc.* (2) **38** (1934) 96-155; see also "Corrigendum", op. cit. **55** (1935).

19. ___________, On the order of $\zeta(1/2 + it)$, *Quart. J. Math.* (Oxford) **13** (1942) 11-17.

20. ___________, The theory of the Riemann-zeta function, Clarendon Press, Oxford 1951.

S.W Graham
Michigan Technology University
Houghton, Michigan 49931 USA

G. Kolesnik
California State University - Los Angeles
Los Angeles, CA 90032 USA

NON-VANISHING OF CERTAIN VALUES OF *L*-FUNCTIONS*

Ralph Greenberg

1. Let K be an imaginary quadratic field. The L-functions that we will consider are defined by

$$L(\chi,s) = \sum_{\mathfrak{a}} \frac{\chi(\mathfrak{a})}{N(\mathfrak{a})^s}$$

where the sum is over the nonzero ideals of the ring of integers O_K of K. Here χ is a grossencharacter of K of type A_0. That is, χ is a complex-valued multiplicative function on the ideals of O_K such that $\chi((\alpha)) = \alpha^n \bar{\alpha}^m$ for all α O_K, $\alpha \equiv 1 \pmod{f_\chi}$, where $n, m \in Z$ and f_χ is an ideal of O_K (the conductor of χ). We call (n,m) the infinity type of χ. The above series defines an analytic function for $Re(s)$ sufficiently large which can be analytically continued to the entire complex plane and satisfies a functional equation. By translating s or applying complex conjugation, we can clearly assume that χ has infinity type $(n,0)$ with $n = n_\chi \geqslant 0$, as we will from here on. The functional equation is then as follows. Let

$$\Lambda(\chi,s) = A^{-s}\Gamma(s)L(\chi,s)$$

where $A = 2\pi/\sqrt{N_\chi}$ and $N_\chi = |\mathrm{disc}(K)|N(f_\chi)$. Then

$$\Lambda(\chi,n+1-s) = w_\chi \, \Lambda(\bar{\chi},s).$$

Here the root number w_χ is a complex number of absolute value 1 which can be computed in terms of Gauss sums. Now

---

*Supported in part by a National Science Foundation grant.

$$L(\overline{\chi},s) = \sum_{a} \frac{\overline{\chi(a)}}{N(a)^s} = \sum_{a} \frac{\overline{\chi(\overline{a})}}{N(\overline{a})^s} = \sum_{a} \frac{\overline{\chi}\circ c(a)}{N(a)^s} = L(\overline{\chi}\circ c,s),$$

since complex conjugation simply permutes the ideals of O_K. Here c denotes complex conjugation (in $\mathrm{Gal}(K/\mathbf{Q})$). The above functional equation becomes

$$\Lambda(\chi,n+1-s) = w_{\chi}\ \Lambda(\overline{\chi}\circ c,s) .$$

Note that $\overline{\chi}\circ c$ and χ have the same infinity type. If $\overline{\chi}\circ c = \chi$, then clearly $w_{\chi} = \pm 1$. In the case $w_{\chi} = -1$, the functional equation then implies that $L(\chi, \frac{n+1}{2}) = 0$. We will assume from now on that n is odd so that the point of symmetry $s = \frac{n+1}{2}$ in the functional equation is an integer. If $\overline{\chi}\circ c = \chi$, $w_{\chi} = -1$, and n is odd, then we will call the zero of $L(\chi,s)$ at $s = (n+1)/2$ a "trivial critical zero". The following theorem concerns the cases where either $\overline{\chi}\circ c = \chi$ and $w = +1$ or $\overline{\chi}\circ c \neq \chi$. It is proved in [2].

Theorem 1. *Let* $B > 0$. *Excluding the trivial critical zeros,* $L(\chi, \frac{n+1}{2})$ *vanishes for only finitely many grossencharacters* χ *such that* $N_{\chi} < B$.

As an example, consider an elliptic curve E defined over $\mathbf{Q}$ and with complex multiplication by O_K. The Hasse-Weil L-function for E over $\mathbf{Q}$ turns out to be $L(\psi,s)$ for a certain grossencharacter $\psi = \psi_E$ for K (proved by Deuring). The infinity type of ψ is (1,0). The assumption that E is defined over $\mathbf{Q}$ is equivalent to the equality $\overline{\psi}\circ c = \psi$. (See [3].) The grossencharacters $\chi = \psi^{2k+1}$ for $k \geqslant 0$ have infinity type $(2k+1,0)$ and clearly satisfy $\overline{\chi}\circ c = \chi$. Also N_{χ} is bounded (by N_{ψ}). It is not hard to compute w_{χ} (see [1]). If $K \neq \mathbf{Q}(\sqrt{-1})$ or $\mathbf{Q}(\sqrt{-3})$, then $w_{\chi} = (-1)^k w_{\psi}$. Thus $L(\psi^{2k+1},s)$ has a trivial critical zero at $s = k+1$ for half of the k's. According to the above theorem, for the remaining k's, only finitely many of the values $L(\psi^{2k+1},k+1)$ are zero. (Actually this special case of the theorem was proved earlier, in [1].) There can in fact be zeros among these remaining values. If the Mordell-Weil group $E(\mathbf{Q})$ is infinite and of even rank, then the Birch and Swinnerton-Dyer conjecture would imply that $L(\psi,s)$ has an even order zero at $s = 1$

(so that $w_\psi = +1$) and this of course is true for many elliptic curves E. Also, Nelson Stephens has found a number of examples where $L(\psi^{2k+1}, s)$ vanishes to even order at $s = k+1$ for small values of $k > 0$.

Rohrlich has proved other non-vanishing results, which we combine in the following theorem. Here $\psi = \psi_E$ is the grossencharacter attached to an elliptic curve E as above.

Theorem 2.(Rohrlich) *Let S be a finite set of primes. Let ϕ vary over all Hecke characters of finite order for K such that $N(f_\phi)$ is divisible only by primes in S and either (i) $\phi \circ c = \phi^{-1}$ and $w_{\psi\phi} = +1$ or (ii) $\phi \circ c = \phi$. Then $L(\psi\phi, 1)$ vanishes for only finitely many such ϕ's. If (iii) $\phi \circ c = \phi^{-1}$ and $w_{\psi\phi} = -1$ and if the conductor of ϕ is restricted as above, then $L'(\psi\phi, 1)$ vanishes for only finitely many such ϕ's.*

Cases (i) and (iii) in this theorem are proved in [6]. Note that if $\chi = \psi\phi$ where ϕ is of finite order and satisfies $\phi \circ c = \phi^{-1}$, then $\overline{\chi} \circ c = (\overline{\psi} \circ c)(\overline{\phi^{-1}} \circ c) = \psi\phi = \chi$. The infinity type of χ is $(1,0)$. One intriguing connection between the proof of Theorem 1 in [2] and Rohrlich's arguments in [6] is that we both use Roth's theorem on approximating algebraic numbers by rational numbers in a crucial way. Although in [2] we use the archimedean version and in [6] Rohrlich uses the nonarchimedean version, there is a certain similiarity to how Roth's theorem comes into the arguments which we will explain later. Case (ii) of the above theorem is proved in [7]. Actually Rohrlich considers the more general L-functions attached to the twists by ϕ of the L-series for modular forms of weight 2. If $\chi = \psi\phi$ and $\phi \circ c = \phi$, then $\overline{\chi} \circ c = \psi\phi^{-1}$. Except for the finitely many ϕ's with $\phi = \phi^{-1}$ (and conductor restricted as above), we have $\overline{\chi} \circ c \neq \chi$. Theorems 1 and 2 would obviously be consequences of the following conjecture.

Conjecture 1. *Let S be a finite set of primes. Let χ vary over all grossencharacters of K such that $N(f_\chi)$ is divisible only by primes in S (and of infinity type $(n,0)$ with n odd, positive, but not fixed). Excluding the trivial critical zeros, $L(\chi, \frac{n+1}{2})$ is nonzero with at most finitely many exceptions. The trivial critical* *zeros are simple with at most finitely many exceptions.*

One could also consider the following more general questions. Let f be a cusp form of weight k which is an eigenform for the Hecke operators and a new form of level N_f. The corresponding L-function satisfies the functional equation

$$\Lambda(f,k-s) = w_f \Lambda(\bar{f},s)$$

where

$$\Lambda(f,s) = A^{-s}\Gamma(s)L(f,s), \qquad A = 2\pi/\sqrt{N_f}.$$

Here $\bar{f}$ is obtained by applying complex conjugation to the coefficients in the q-expansion of f. If $\bar{f} = f$ and $w_f = -1$, then clearly $L(f,k/2) = 0$. If N_f is divisible only by primes in some finite set S but k is not restricted (except perhaps to be even), will these zeros forced by the functional equation account for all but finitely many of the values $L(f,k/2)$ which vanish? Will the zeros forced by the functional equation be simple with at most finitely many exceptions? The L-function $L(\chi,s)$ attached to a grossencharacter χ of K corresponds to a modular form of weight $n_\chi + 1$ and level N_χ. The above condition on N_χ would limit K to finitely many imaginary quadratic fields and would limit $N(f_\chi)$ to be divisible only by primes in S.

2. We now want to discuss the connection of the above nonvanishing results to the arithmetic of elliptic curves. As before, let E be an elliptic curve defined over $\mathbf{Q}$ and with complex multiplication by O_K. If p is any prime, we will consider towers of fields $K = K_0 \subset K_1 \subset \dots \subset K_n \subset \dots$ with K_n a cyclic extension of K of degree p^n. The field $K_\infty = \bigcup_{n \geq 0} K_n$ is then a Galois extension of K with $\mathrm{Gal}(K_\infty/K) \cong \varprojlim(\mathbf{Z}/p^n\mathbf{Z}) \cong \mathbf{Z}_p$, the additive group of p-adic integers. K_∞ is a so-called $\mathbf{Z}_p$-extension of K. The question of how the rank of $E(K_n)$ behaves as $n \to \infty$ (and related questions) was first discussed by Mazur (see [5]).

Now the Birch and Swinnerton-Dyer conjecture states that, if F is any number field, then the rank of $E(F)$ should equal the order of vanishing of the Hasse-Weil L-function $L_F(E,s)$ for E over F at s=1. If F is abelian over K, we have the following essentially formal identity:

$$L_F(E,s) = \prod_{\phi} L(\psi\phi,s)^2.$$

Here ϕ runs over the characters of Gal(F/K) (which can be identified with Hecke characters of finite order for K by class field theory). Also $\psi = \psi_E$ as before. Note that the fact that $L_F(E,s)$ has even order at s=1 agrees with the fact that $E(F)$ is an O_K-module and $\text{rank}_{\mathbf{Z}}(E(F)) = 2\ \text{rank}_{O_K}(E(F))$. Conjecturally, the behavior of the rank (over $\mathbf{Z}$) of $E(K_n)$ as $n \to \infty$ should be related to the vanishing of $L(\psi\phi,s)$ at s=1 as ϕ varies over the characters of Gal(K_∞/K) of finite order (each of which factors through Gal(K_n/K) for some n). It is easy to show that only primes of K dividing p can ramify in a $\mathbf{Z}_p$-extension K_∞/K. Hence the conductor of the grossencharacter $\chi = \psi\phi$ will be divisible only by primes in some finite set.

We will single out two special $\mathbf{Z}_p$-extensions K_∞^+ and K_∞^- of K. Both are Galois extensions of $\mathbf{Q}$. The element c in Gal$(K/\mathbf{Q})$ acts (as an inner automorphism) on Gal(K_∞^+/K) trivially and on Gal(K_∞^-/K) by multiplication by -1. Thus the n-th level K_n^+ of K_∞^+ is abelian over $\mathbf{Q}$ of degree $2p^n$. The n-th level K_n^- of K_∞^- is a dihedral extension of $\mathbf{Q}$, also of degree $2p^n$. If ϕ is a character of Gal(K_n^+/K) or Gal(K_n^-/K), then (identifying ϕ with a Hecke character for K) one finds that $\phi \circ c = \phi$ or ϕ^{-1}, respectively. The existence of these $\mathbf{Z}_p$-extensions can be proven by class field theory. Actually K_∞^+ is easily described explicitly. It is a subfield of $K(\mu_{p^\infty})$ where μ_{p^∞} denotes the p-power roots of unity, and is called the cyclotomic $\mathbf{Z}_p$-extension of K for that reason. K_∞^- is often called the anticyclotomic $\mathbf{Z}_p$-extension of K. It could also be described explicitly as a subfield of the field obtained by adjoining certain values of the j-function to K. By class field theory, one can show that every $\mathbf{Z}_p$-extension of K is contained in $\mathbf{K}_\infty = K_\infty^+ K_\infty^-$. Also Gal$(\mathbf{K}_\infty/K) \cong \mathbf{Z}_p^2$ and so obviously K has infinitely many distinct $\mathbf{Z}_p$-extensions.

Consider first the anti-cyclotomic $\mathbf{Z}_p$-extension. If ϕ is a character of $\mathrm{Gal}(K_\infty^-/K)$, then $\chi = \psi\phi$ satisfies $\bar{\chi} \circ c = \chi$. The root numbers w_χ behave as follows (see [1]). We assume E has good reduction at p. If p splits in K, then $w_{\psi\phi} = w_\psi$. In particular, if $w_\psi = +1$ (i.e. if $L_{\mathbf{Q}}(E,s)$ has an even order zero), then Rohrlich's theorem implies that $L(\psi\phi,1) \neq 0$ for all but finitely many such ϕ's. Rubin's generalization of the Coates-Wiles theorem then shows that the rank of $E(K_n^-)$ becomes constant for n sufficiently large. If $w_\psi = -1$, then $L(\psi\phi,1) = 0$ for all ϕ. If p remains prime in K, then $w_{\psi\phi} = \pm w_\psi$ and both signs occur depending just on whether the order of ϕ is an even or odd power of p. Thus $L(\psi\phi,1) = 0$ for infinitely many ϕ's. But Rohrlich proves that these zeros are mostly simple. This result together with a recent theorem of Gross and Zagier (which connects the heights of certain "Heegner points" on E with the values $L'(\chi,1)$) shows that $\mathrm{rank}(E(K_n^-)) \to \infty$ as $n \to \infty$ if either p splits in K and $w_\psi = -1$ or if p remains prime in K. In the first case, $\mathrm{rank}(E(K_n^-)) > 2p^n - c$ for all n, where c is some constant. The Birch and Swinnerton-Dyer conjecture would imply the more precise statement that $\mathrm{rank}(E(K_n^-)) - 2p^n$ becomes constant for $n \gg 0$. In the case where p remains prime in K, the growth of $E(K_n^-)$ is less regular. For $n \gg 0$, the rank of $E(K_n^-)$ increases only for either the even or odd n's. We still have an inequality $\mathrm{rank}(E(K_n^-)) > ap^n$ for $n \gg 0$, where a is some positive constant.

If ϕ factors through $\mathrm{Gal}(K_\infty^+/K)$, then for $\chi = \psi\phi$, we have $\bar{\chi} \circ c \neq \chi$ (except if ϕ has order 2). Again Rohrlich's result together with Rubin's theorem implies that $\mathrm{rank}(E(K_n^+))$ becomes constant for large enough n. More generally, consider any $\mathbf{Z}_p$-extension K_∞ of K other than the anti-cyclotomic one. If ϕ is a character of $\mathrm{Gal}(K_\infty/K)$, then one sees easily that $\phi \circ c \neq \phi^{-1}$ except possibly for finitely many such ϕ's. Again, for $\chi = \psi\phi$, we will usually have $\bar{\chi} \circ c \neq \chi$. The argument given in [7] can be adapted (with some difficulties) to prove the following result (suggested by the conjecture stated in Section 1).

Theorem 3. *Let* $K_\infty = \cup K_n$ *be any* $\mathbf{Z}_p$*-extension of* K, $K_\infty \neq K_\infty^-$. *Then* $\mathrm{rank}(E(K_n))$ *is bounded as* $n \to \infty$.

A stronger result should be true. Conjecture 1 actually would imply the following conjecture. Let F be any finite abelian extension of K and let $F_\infty = FK_\infty$. Let F_∞^* be the largest subfield of F_∞ such that the characters ϕ of $\mathrm{Gal}(F_\infty^*/K)$ of finite order all have the property that $\phi \circ c = \phi^{-1}$. The field F_∞^* is a finite extension of K_∞^-. If $F_\infty = K_\infty$ and if p is odd, the field F_∞^* is K_∞^-. For any field L, we let $\mathbf{E}(L) = E(L)/E(L)_{\mathrm{torsion}}$.

Conjecture 2. $\mathbf{E}(F_\infty)/\mathbf{E}(F_\infty^*)$ *is finitely generated.*

It is tempting to speculate in a somewhat different direction. Let $\mathcal{F}$ be a Galois extension of $\mathbf{Q}$ such that $G = \mathrm{Gal}(\mathcal{F}/\mathbf{Q}) \cong \mathrm{GL}_2(\mathbf{Z}_p)$ for some prime p. Let E by any elliptic curve defined over $\mathbf{Q}$. Assume that Weil's conjecture is valid for E. That is, $L_{\mathbf{Q}}(E,s) = L(f,s)$, where f is a modular form of weight 2. Let F be any finite Galois extension of Q contained in $\mathcal{F}$. The Hasse-Weil L-function $L_F(E,s)$ is formally a product of L-functions $L(f,\phi,s)$, where ϕ is an irreducible character of $\mathrm{Gal}(F/Q)$. Each L-function occurs d_ϕ times in this product, where d_ϕ is the degree of the character. The function $L(f,\phi,s)$ is defined (for Re(s) > 3/2) by an Euler product whose factors are (mostly) of degree $2d_\phi$ and which are easily described from the Euler factors for $L(f,s)$ and those for the Artin L-function $L(\phi,s)$. The properties of these L-functions don't seem to be known in general, but it seems reasonable to believe that they have analytic continuations with a functional equation relating $L(f,\phi,2-s)$ to $L(f,\overline{\phi},s)$. (The modular form f here would satisfy $\overline{f} = f$.) If $\overline{\phi} = \phi$ and if the root number $w_{f,\phi}$ occuring in the functional equation is -1, then $L(f,\phi,1)$ would be forced to vanish. If $\overline{\phi} \neq \phi$, one might believe that $L(f,\phi,1)$ should be nonzero with at most finitely many exceptions as ϕ varies over all such irredcible characters of G. (Perhaps we should assume here that only finitely many primes of Q are ramified in $\mathcal{F}$).

Now it is easy to show that in the group $G^* = \mathrm{PGL}_2(\mathbf{Z}_p) = G/\mathbf{Z}_p^\times$, every element is conjugate to its inverse. Thus every character of G^* is real-valued. Also every real-valued irreducible character of G factors through $G^{**} = G/(\mathbf{Z}_p^\times)^2$. Let $\mathcal{F}^*$ and $\mathcal{F}^{**}$ denote the corresponding subfields of $\mathcal{F}$. Thus $\mathrm{Gal}(\mathcal{F}^*/Q) \cong \mathrm{PGL}_2(Z_p)$ and $\mathcal{F}^{**}$ is a finite (quadratic if p ≠ 2) extension of $\mathcal{F}^*$. In analogy with Conjecture 2, it may be reasonable to believe that $\mathbf{E}(\mathcal{F})/\mathbf{E}(\mathcal{F}^{**})$ is finitely generated in general. Also, under certain assumptions, M. Harris [4] has shown that an elliptic curve can have unbounded rank in a $PGL_2(\mathbf{Z}_p)$-extension of some number field. A calculation of what the root numbers $w_{\mathfrak{f},\phi}$ should be would give some idea of what to expect in general. Such calculations can be done if the elliptic curve E has good reduction at all primes ramified in $F^*/\mathbf{Q}$. Assume $p > 2$. There is a unique character $\varepsilon : G^* \to \pm 1$. Let N_E denote the conductor of E. If $\varepsilon(-N_E) = +1$, then all but finitely many of the $w_{\mathfrak{f},\phi}$'s turn out to be $+ 1$ when ϕ factors through G^*. Possibly E has bounded rank in F^* (and even F) in this case. If $\varepsilon(-N_E) = -1$, then infinitely many of the $w_{\mathfrak{f},\phi}$'s are -1 (namely for those ϕ's with ε as corresponding determinant). This suggests that the rank of E should be unbounded in F^*. There is a canonical tower of fields F_n^*, $n \geq 1$, with $\mathrm{Gal}(F_n^*/\mathbf{Q}) \cong PGL_2(\mathbf{Z}/(p^n))$ such that $F^* = \bigcup F_n^*$. We have $[F_n^* : \mathbf{Q}] \sim c(p^n)^3$ for some constant c. If $\varepsilon(-N_E) = -1$, it seems that $\mathrm{rank}(E(F_n^*))$ should be $> a(p^n)^2$ for some $a > 0$ when $n \gg 0$. This rate of growth is the most one could find by just root number calculations. A higher rate of growth would indicate that many of the L-functions $L(\mathfrak{f},\phi,s)$ have high order zeros at $s = 1$.

Now let E be an elliptic curve without complex multiplication and let F be the field generated by the coordinates of the p-power division points on E. For all but finitely many p, we will have $\mathrm{Gal}(F/\mathbf{Q}) \cong GL_2(\mathbf{Z}_p)$. It is this case that seems closest to the situation described earlier in this section. Although we haven't calculated root numbers, we suspect that $w_{\mathfrak{f},\phi} = -1$ for infinitely many characters ϕ factoring through G^* and hence that E has unbounded rank in F^*.

3. We want to say something about the proofs of Theorems 1 and 2. Since they are already in print, we will be very sketchy. Mainly, we will try to explain a certain similarity in how Roth's theorem occurs in the arguments.

We will simplify our discussion of Theorem 1 by restricting attention to the values $L(\psi^{2k+1}, k + 1)$ for $k \geq 0$, where ψ is the grossencharacter for an elliptic curve E as in Section 1. The root numbers $w_k = w(\psi^{2k+1})$ turn out to depend only on the residue class of k modulo m, where m is the number of roots of unity in K. Let m' be any multiple of m and let k' be a fixed integer such that $\omega_{k'} = +1$. The essential part of our proof is to show that the Abel average of the L-values over all $k \equiv k' \pmod{m'}$ is nonzero and so *infinitely many* of these L-values are also nonzero. One improves this to *only finitely many* by using the fact that these L-values are (up to a factor) certain special values of p-adic L-functions constructed by Katz. This role of p-adic L-functions in our argument is the reason our result is limited to grossencharacters with n_χ odd.

One can derive a convergent series for the L-values considered here by using the same integral representation for $L(\psi^{2k+1},s)$ which gives the analytic continuation and functional equation. The integrals can be evaluated when $s = k+1$. The result is that $L(\psi^{2k+1},k+1) = (1+\omega_k)G_k$, where

$$G_k = \sum_{\mathfrak{a}} \frac{\psi^{2k+1}(\mathfrak{a})}{N(\mathfrak{a})^{k+1}} e^{-AN(\mathfrak{a})} \sum_{j=0}^{k} \frac{(AN(\mathfrak{a}))^j}{j!}$$

$$= \sum_{\mathfrak{a}} \frac{\psi(\mathfrak{a})\phi_0(\mathfrak{a})^k}{N(\mathfrak{a})} e^{-AN(\mathfrak{a})} \sum_{j=0}^{k} \frac{(AN(\mathfrak{a}))^j}{j!}$$

Here $\phi_0(\mathfrak{a}) = \psi(\mathfrak{a})/\overline{\psi}(\mathfrak{a})$ and A is the same constant which appears in the functional equation. (A small difficulty occurs if $K = Q(\sqrt{-3})$. Then A might vary slightly with k and also the second series above will be different. We assume here A is constant.) The Abel average

$$\lim_{x\to 1} (1-x) \sum_{k=0}^{\infty} G_k x^k$$

can be evaluated. The terms in G_k for which $\phi_0(\mathfrak{a})=1$ (or equivalently $\mathfrak{a} = \overline{\mathfrak{a}}$) give a contribution of $\sum_{\mathfrak{a}=\overline{\mathfrak{a}}} \frac{\psi(\mathfrak{a})}{N(\mathfrak{a})}$ to this Abel average. The conductor of ψ is divisible by the ramified primes of K and so one need consider only the ideals $\mathfrak{a} = (a)$, where $a \in \mathbf{Z}$. Now $\psi((a)) = \xi(a)a$ for some Dirichlet character ξ. (It turns out that ξ is equivalent to the Dirichlet character for K, although usually nonprimitive.) Thus the above sum is just $L(\xi,1)$ and is certainly nonzero.

The terms in G_k for which $\phi_0(\mathfrak{a}) \neq 1$ give a contribution of zero to the Abel average. Also the Abel average of the sequence $G_k\zeta^k$ (where ζ is any m'-th root of unity, $\zeta \neq 1$) is zero. These facts immediately give the result stated earlier about the Abel average of our L-values over $k \equiv k' \pmod{m'}$. The estimates that are involved here are the most troublesome for those terms where $\beta = \phi_0(\mathfrak{a})$ (or $\beta = \phi_0(\mathfrak{a})\zeta$) is close to 1. One needs to show that $N(\mathfrak{a})$ increases rapidly for those terms. Now $\beta = \lambda/\bar{\lambda}$, where $\lambda = \psi(\mathfrak{a})$ (or $\psi(\mathfrak{a})\omega$ for some root of unity ω). The λ's which occur here belong to one of finitely many lattices $\mathbf{L}$ in the complex plane consisting of algebraic numbers. The most delicate estimates are needed for the terms where $Im(\lambda)$ is small. Let $\mathbf{L} = \mathbf{Z}\omega_1 + \mathbf{Z}\omega_2$. If $\lambda = a\omega_1 + b\omega_2$ is close to the real axis (and, say $b \neq 0$), then a/b is a good rational approximation to the algebraic number $Im(-\omega_2/\omega_1)$. Roth's theorem enters at this point in order to show a or b and so $\lambda\bar{\lambda} = N(\mathfrak{a})$ is large. One in fact needs Roth's theorem with an exponent $2 + \varepsilon$ for a rather small value of ε.

The values $L(\psi^{2k+1}, k+1)$ that we have considered can be written as $L(\psi\phi_0^k, 1)$. The grossencharacter $\phi = \phi_0^k$ satisfies $\phi \circ c = \phi^{-1}$, although of course ϕ is not of finite order if $k > 0$. Its infinity type is $(k,-k)$. In Rohrlich's theorem the analogous case is (1) and it is this case (and also case (3)) where Roth's theorem (the nonarchimedean version) plays a role. We will just consider case (1) and will assume that $S = \{p\}$, where p is an odd prime. For simplicity, we will restrict attention to Hecke characters ϕ such that the field K_ϕ which corresponds to ϕ by class field theory is a subfield of the field $\mathbf{K}_\infty$ defined in Section 2. The condition $\phi \circ c = \phi^{-1}$, means that $K_\phi = K_n^-$ for some n. Obviously the order of ϕ (denoted by $\mathrm{ord}(\phi)$) is p^n. Let $\Gamma = \mathrm{Gal}(\mathbf{K}_\infty/K)$. Then $\Gamma \cong \mathbf{Z}_p^2$ and c acts naturally on Γ (as an inner automorphism in $\mathrm{Gal}(\mathbf{K}_\infty/\mathbf{Q})$). This gives us a decomposition $\Gamma = \Gamma^+ \times \Gamma^-$, where Γ^+ and Γ^- can be identified with $\mathrm{Gal}(K_\infty^+/K)$ and $\mathrm{Gal}(K_\infty^-/K)$, respectively.

Rohrlich also uses an averaging argument. $\mathrm{Gal}(K(\text{values of } \phi)/K)$ acts on a character ϕ, giving a set of conjugate characters ϕ_i, $1 \leq i \leq e_\phi$, say. Denote by $L(\psi\phi_{av}, 1)$ the Galois average

$$L(\psi\phi_{av}, 1) = \frac{1}{e_\phi} \sum_{i=1}^{e_\phi} L(\psi\phi_i, 1).$$

Rohrlich shows that $\lim_{\phi} L(\psi\phi_{av},1)$ is nonzero as $\mathrm{ord}(\phi) \to \infty$ and ϕ varies as restricted above and such that $w(\psi\phi) = +1$. Now ψ has its values in K and so the grossencharacters $\chi_i = \psi\phi_i$ are all conjugate. The root numbers $w(\chi_i)$ are all $+1$ and a theorem of Shimura shows that either all or none of the values $L(\chi_i,1)$ are zero. Hence $L(\psi\phi,1)$ is nonzero if $\mathrm{ord}(\phi)$ is sufficiently large and $w(\psi\phi) = +1$.

We have the following convergent series for $L(\psi\phi,1)$:

$$L(\psi\phi,1) = \bigl(1 + w(\psi\phi)\bigr) \sum_{a} \frac{\psi\phi(a)}{N(a)} e^{-A_\phi N(a)}.$$

Here $A_\phi = 2\pi/\sqrt{N_{\psi\phi}}$, which is unchanged when ϕ is replaced by any of the ϕ_i's. One difficulty in handling these series is that $A_\phi \to 0$ as $\mathrm{ord}(\phi) \to \infty$. We will assume that $p \nmid N(f_\psi)$ so that $\psi\phi(a) = \psi(a)\phi(a)$ for all ideals a . Then we can replace ϕ by $\phi_{av} = e_\phi^{-1} \sum \phi_i$ in the above series, giving a convergent series for $L(\psi\phi_{av},1)$. When is $\phi_{av}(a) \neq 0$? If w is a p^n-th root of unity, then the sum of the conjugates of w will be zero unless $w^p = 1$. If $\mathrm{ord}(\phi) = p^n$, then we can regard ϕ as a character of $\mathrm{Gal}(K_n^-/K)$. Now $\phi(a) = \phi((\frac{K_n^-/K}{a}))$, and so $\phi_{av}(a) \neq 0$ implies that the Artin symbol $(\frac{K_n^-/K}{a})$ has order 1 or p. It must then fix K_{n-1}^- and so $(\frac{K_{n-1}^-/K}{a})$ must be trivial. As $\mathrm{ord}(\phi) \to \infty$, the terms that survive in the series for $L(\psi\phi_{av},1)$ are those for which $(\frac{K_\infty^-/K}{a})$ is trivial, i.e. those that correspond to ideals a such that $a = \overline{a}$. The contribution of those terms to the limit in question is

$$2 \sum_{a = \overline{a}} \frac{\psi(a)}{N(a)} ,$$ nonzero as before.

Let $\sigma_a = (\frac{K_\infty/K}{a})$. The condition $a = \overline{a}$ means that $\sigma_a \in \Gamma^+$. For a given ϕ such that $\mathrm{ord}(\phi) = p^n$, the remaining nontrivial terms in the series for $L(\psi\phi_{av},1)$ are those for which $a \neq \overline{a}$ and $\sigma_a |_{K_\infty^-} = \mathrm{proj}_{\Gamma^-}(\sigma_a)$ is in $(\Gamma^-)^{p^{n-1}}$. If $a = (\alpha)$, we can translate this into a statement about α. Class field theory gives a canonical isomophism $U/U_{\mathrm{torsion}} \cong \Gamma$, where U is the group of units in

$O_p = O_K \otimes_{\mathbf{Z}} Z_p$. (This ring is either the integers in the p-adic completion of K or the direct product of two copies of $\mathbf{Z}_p$, depending on whether p remains prime or splits in K.) The statement that $\mathrm{proj}_{\Gamma^-}(\sigma_{\mathfrak{a}})$ is in a small subgroup of Γ^- becomes equivalent to stating that $\alpha/\overline{\alpha}$ is close to some element ζ of the finite group U_{torsion}. In fact, ζ must be a global root of unity. One can write each ζ as $\zeta = \overline{\omega}/\omega$ where ω is the image of some algebraic number in O_p. Thus $\lambda = \alpha\omega$ belongs to one of finitely many "lattices" $\mathbf{L} = \omega\, O_K$ in O_p consisting of algebraic numbers and λ has the property that $\lambda/\overline{\lambda}$ is close to 1, that is, $\lambda - \overline{\lambda}$ is small. As before, but this time using a p-adic version of the theorem of Roth, one finds that $\lambda\overline{\lambda}$ (in $\mathbf{R}$ here) and so $N(\mathfrak{a}) = N(\alpha)$ is large. In this way, Rohrlich shows that the terms in the convergent series giving $L(\psi\phi_{av},1)$ for which $\mathfrak{a} \neq \overline{\mathfrak{a}}$ contribute zero to the limit.

References.

1. R. Greenberg, On the Birch and Swinnerton-Dyer conjecture. *Invent. Math.* **72**, 241-265 (1982).

2. R. Greenberg, On the critical values of Hecke L-functions for imaginary quadratic fields, *Invent. Math.* **79**, 79-94 (1985).

3. B. Gross, Arithmetic on Elliptic Curves with Complex Multiplications. Lecture Notes in Math. **776**.

4. M. Harris, Systematic growth of Mordell-Weil groups of abelian varieties in towers of number fields. *Invent. Math.* **51**, 123-141 (1979).

5. B. Mazur, Rational points of abelian varieties with values in towers of number field. *Invent. Math.* **18**, 183-226 (1972).

6. D. Rohrlich, On L-functions of elliptic curves and anticyclotimic towers. *Invent. Math.* **75**, 383-408 (1984).

7. D. Rohrlich, On L-functions of elliptic curves and cyclotomic towers. *Invent. Math.* **75**, 409-423 (1984).

R. Greenberg,
University of Washington,
Seattle, Washington 98195, U.S.A.

ON AVERAGES OF EXPONENTIAL SUMS OVER PRIMES

Glyn Harman

1. Introduction.

In this paper we shall be concerned with obtaining approximations to and estimates for the sum

$$S_N(\alpha) = \sum_{n\leqslant N} e(n\alpha)\Lambda(n) \tag{1}$$

where $e(x) = \exp(2\pi ix)$, α is real, and $\Lambda(n)$ is the von Mangoldt function. Although we are unable to establish the naturally conjectured results for this sum, we shall show how the introduction of averaging - in a form likely to occur in applications - can lead to substantial improvements.

To analyse the behaviour of $S_N(\alpha)$ we first need some information concerning diophantine approximations to α. If we suppose that

$$\alpha = \frac{a}{q} + \beta$$

where $|\beta| < q^{-2}$ and $(a,q) = 1$, then one expects that

$$S_N(\alpha) = \frac{\mu(q)}{\phi(q)} S_N(\beta) + E(N,q,\beta) \tag{2}$$

where $E(N,q,\beta)$ is some error which will be an increasing function of N, q and $|\beta|$. For small values of q, (2) would provide a good approximation to $S_N(\alpha)$ by a term which is $O(\min(N, |\beta|^{-1})/\phi(q))$ for certain values of the parameters. For some applications the exact form of the approximation is necessary (e.g. on the major arcs of the Hardy-Littlewood circle method, see [14]) and in other cases an upper bound suffices (e.g. section 7 of [1]). The fact that (2) holds on the Generalized Riemann Hypothesis is classical, with $E(N,q,\beta) \ll N^{1/2}q^{1/2}(1 + N|\beta|)^{1/2}(\log N)^2$. This analysis was fundamental to Hardy and Littlewood's conditional proof of the ternary Goldbach theorem [4] and the demonstration in [5] that the exceptional set in the binary Goldbach problem is $O(X^{1/2+\varepsilon})$ (actually they used a more general hypothesis and gave results depending on the width of the zero-free region). Ignoring powers of (log N) we note that for large q, (2) gives a bound $N^{1/2}q^{1/2}$, while for small q, if we only know $|\beta| < q^{-2}$, the upper estimate is $Nq^{-1/2}$.

Without any hypothesis one can only establish (2) with the current state of knowledge, for $q < (\log N)^A$ (any given A) and with

$$E(N,q,\beta) \ll N \exp\left(-c(A)(\log N)^{1/2}\right)(1 + N|\beta|)$$

(see for example, the proof of Lemma 3.1 in [14]). Vinogradov, however, proved the ternary Goldbach theorem unconditionally (see chapter 10 of [15]) by establishing a result of the form

$$S_N(\alpha) \ll (N^{4/5} + Nq^{-1/2} + N^{1/2}q^{1/2})(\log N)^{7/2} \tag{3}$$

The bound (3) in this form is due to R.C. Vaughan [12]. We note that for $q < N^{2/5}$ or $q > N^{3/5}$ and given only $|\beta| < q^{-2}$, this is only weaker than the result obtained on the GRH by a power of (log N). No stronger bounds are possible for small q when $|\beta|$ is substantially smaller than q^{-2}, however, by the Vinogradov-Vaughan method. Vaughan also established that

$$\sum_{h\leqslant H} |S_N(h\alpha)| \ll (\log N)^7(N^{3/4}H + (NHq)^{1/2} + NHq^{-2} + N^{4/5+\varepsilon}H^{3/5}), \tag{4}$$

which quickly leads to the result that, for α irrational, β arbitrary, there are infinitely many primes p such that

$$\|\alpha p + \beta\| < cp^{-1/4}(\log p)^7 \tag{5}$$

where c is an absolute constant. By sieve methods one can deduce a stronger result [6] but this sheds no light of $S_N(\alpha)$. On the GRH the exponent in (5) can be increased to 1/3 (I have not been able to locate this fact mentioned in the literature, but Prof. S. Graham remarked to me that he had proved it in an unpublished manuscript).

The Bombieri-Vinogradov theorem (chapter 28 of [2]) shows that, in some sense, the GRH is true on average. This leads one to hope that one could prove (2) to be true on average. Montgomery and Vaughan [9] effectively got such a result, drawing on some work of Gallager [3]. They proved that the integral

$$\int_M S_N(\alpha)^2 \, e(-n\alpha) \, d\alpha$$

where M is the union of major arcs, equals the value expected with a suitably small error plus some unpleasant terms coming from a possible 'exceptional' character (one whose L-function has a zero very close, in terms of n, to 1). In this use is being made of averaging over both numerator and denominator and the latter can take values up to a small power of n.

The following three theorems demontrate other average results on exponential sums over primes.

Theorem 1. *Let* $N \geq Q \geq 1$. *Suppose that, for* $Q \leq q \leq 2Q$ *we have* $\alpha_q - a(q)/q = \beta_q$ *with* $|\beta_q| < q^{-2}$, $\beta \leq |\beta_q| \leq 2\beta$ *and* $N^{-1} \leq \beta \leq N^{-3/5}$. *Then we have*

$$\sum_{Q\leq q\leq 2Q} |S_N(\alpha_q)| \ll (\log N)^5 \left(N^{7/8}\beta^{-1/8} + Q^{3/4}N\beta^{1/4} + Q^{3/2}N\beta^{1/2}\right) . \tag{6}$$

Theorem 2. *Given the hypotheses of Theorem 1 but with* $0 < \beta < N^{-1}\exp((\log N)^{1/2})$ *and* $Q < N^{1/3}\exp(-2(\log N)^{1/2})$, *then there exists an absolute constant* c *such that*

$$\sum_{Q\leq q\leq 2Q} \left|S_N(\alpha_q) - \frac{\mu(q)}{\phi(q)} S_N(\beta_q) + X_q\right| \ll N \exp\left(-c(\log N)^{1/2}\right), \tag{7}$$

where

$$X_q = \frac{\tau(\chi)}{\phi(q)} \chi(q/r)\mu(q/r) \sum_{n\leq N} e(n\beta_q) \, n^{1-\sigma},$$

this term occuring only if there is a modulus r *dividing* q *with a real primitive character* χ *whose L-function has a real zero* σ *with* $(1 - \sigma) < (\log N)^{-1/2}$. *(There can be at most one such* r *for a given*

N).

Theorem 3. *Suppose that* (a,q) = 1, *and* q, R, L, N ≥ 1. *Let* ε > 0 *be given. Then we have that*

$$\sum_{R\le r<2R}\ \sum_{L\le \ell<2L} \left|S_N\left(\frac{a\ell}{qr}\right)\right| \ll (Na)^{\varepsilon}\Big(RLN^{2/3} + R^2\big(q + N^{2/5}(1 + |\tfrac{q}{a}|)^{1/2}\big) \tag{8}$$

$$+ NLq^{-1/2} + N^{9/10}(RL)^{1/2} + RN^{4/5} + \Big(NLRq\big(1 + \frac{R^2}{|a|}\big)\Big)^{1/2}.$$

Alternatively the exponents 2/3, 2/5, 9/10, 4/5 *may be replaced by* 7/10, 3/5, 7/8, 3/4 *respectively.*

The author does not know of any applications at present for the first two theorems although they do imply a bound $O(N^{7/8}(\log N)^5 \min(N, \beta^{-1})^{1/8}/q)$ on average over q. Theorem 3 is, however, a stronger result than can be obtained by applying the GRH for each modulus qr, when the parameters are in certain ranges. For example, when $R = L = N^{1/3}$, $q = N^{1/2}$, $a < N$, the right hand side of (8) is $O(N^{4/3+\varepsilon})$ whereas applying the GRH (and not making use of the averaging over r) there is a term $(qLNR^3)^{1/2}$ which is of size $N^{4/3+1/12}$. Professor P.X. Gallagher has remarked that this may have some implications for the vertical distribution of zeros of L-functions. Theorem 3 is applied in [7] to prove that there are infinitely many solutions of $|\alpha p - P_3 + \beta| < p^{-1/300}$ where p is a prime, P_3 a number having no more than three prime factors, α is irrational, and β is arbitarary. This improves upon a result of Vaughan [11] who adapted his method in [10] which has a "GRH true on average" strength. Several variations on the above results are possible. We shall only briefly sketch the proofs of the results here.

2. Proofs of Theorems 1 and 2.

We shall adapt the argument of [10] to prove Theorem 1 and appeal to Theorem 7 of [3] in addition to establish Theorem 2. We donote by τ(χ) the usual Gauss sum. We note the well-known (Chapter 9 of [2]) results:

$$|\tau(\chi)| = q^{1/2} \text{ if } \chi \text{ is primitive mod } q$$

$$\leqslant |\tau(\chi_d)| \text{ if } \chi_d \text{ is the character mod } d \text{ which induces } \chi$$

$$\tau(\chi) = \mu(q) \text{ if } \chi \text{ is the principal character mod } q.$$

Let r(q) be the nearest integer to β_q^{-1}. Then $||\beta_q| - h/(hr(q) + 1)| < 3/r(q)^2$ for $h \geqslant 1$. It is elementary that the smallest integer in the arithmetic progression $hr(q) + 1$ which is coprime to q is $O(d(q)qr(q)/\phi(q))$. Hence for each q there exist integers t(q), k(q) with $(t(q),k(q)) = 1$, $1 \leqslant t(q) \ll d(q)r(q)\, q/\phi(q)$ and $|\beta_q - k(q)/t(q)| \ll \beta^2$. We write

$$\psi(N,\chi,\gamma) = \sum_{n\leqslant N} \Lambda(n)\chi(n)e(n\gamma) \quad \text{and} \quad \psi(y,\chi) = \psi(N,\chi,0).$$

Thus

$$\sum_{Q\leqslant q\leqslant 2Q} |S_N(n\alpha_q)| \leqslant \sum_{Q\leqslant q\leqslant 2Q} \frac{1}{\phi(q)} \sum_{\substack{\chi \\ \text{mod } q}} |\tau(\bar{\chi})||\psi(N,\chi,\beta_q)| + O\big(Q(\log N)^2\big). \tag{9}$$

We first assess the contribution to the right hand side of (9) arising from principal characters. In this case we use the bound (3) and obtain

$$\begin{aligned}\psi(N,\chi,\beta_q) &= S_N(\beta_q) + O(\log Q)\\ &\ll (N^{4/5} + Nr(q)^{-1/2} + N^{1/2}r(q)^{1/2})(\log N)^{7/2}\\ &\ll N^{7/8}\beta^{-1/8} (\log N)^{7/2}\end{aligned}$$

which is of a suitable size since $|\tau(\chi)| \leqslant 1$.

Now we must convert the remainder of the sum to one involving only primitive characters. Using * to denote summation over primitive characters only, the sum is

$$\ll \sum_{Q\leqslant q\leqslant 2Q} \frac{1}{\phi(q)} \sum_{m|q} \sideset{}{^*}\sum_{\chi \text{ mod } M} |\tau(\chi)||\psi(N,\chi,\beta_q)| + Q^{3/2}(\log N)^5$$

$$\ll \sum_{1<q\leqslant 2Q} \frac{q^{1/2}}{\phi(q)} \sum_{\chi \bmod q}^{*} |\psi(N,\chi,\gamma q)|(\log Q) + Q^{3/2}(\log N)^5, \qquad (10)$$

since

$$\sum_{m\leqslant Q/q} \frac{1}{\phi(qm)} \ll \frac{\log Q}{\phi(q)},$$

and we have written γ_q for that one of β_m $(m = Q, Q+1, \ldots, 2Q)$ such that $q|m$ and $\sum_{\chi}^{*} |\psi(N,\chi,\beta_m)|$ is maximised. We also put $u(q) = t(m)$, $v(q) = k(m)$. For the values of $q \leqslant (N\beta)^{1/4}$ we take no account of the averaging over q. Since $\beta < N^{-1/2}$ we have (writing u for u(q)),

$$\sum_{\chi}^{*} |\psi(N,\chi,\gamma q)| \leqslant \sum_{\chi}^{*} \max_{y\leqslant N} |\psi(y,\chi,v(q)/u(q))|$$

$$\ll \frac{u^{1/2}}{\phi(u)} \sum_{\chi \bmod uq} \max_{y\leqslant N} |\psi(y,\chi)|$$

since $\chi_1\chi_2$ runs over all characters (mod uq) no more than once as χ_1, χ_2 run over characters mod q and mod u respectively. An appeal to Theorem 2 of [10] then furnishes the bound

$$\frac{u^{1/2}}{\phi(u)} \left(N + N^{3/4}(uq)^{5/8} + N^{1/2}uq\right)(\log N)^3.$$

It quickly follows that

$$\sum_{q\leqslant(N\beta)^{1/4}} \frac{q^{1/2}}{\phi(q)} \sum_{\chi \bmod q}^{*} |\psi(N,\chi,\gamma_q)| \ll N^{7/8}\beta^{-1/8}(\log N)^4.$$

This is a satisfactory esitmate again.

To handle the remaining values of q we need to modify the details of [10]. We must first divide up the range of summation over q, so we now restrict q to lie between Z and 2Z. We write

$$F(s,\chi) = \sum_{n\leqslant u} \chi(n)\Lambda(n)n^{-s}$$

and

$$G(s,\chi) = \sum_{n\leqslant v} \chi(n)\mu(n)n^{-s}$$

where u, v (both not less than 1) will be chosen later in terms of N, Z, and β. We also put $\theta = 1 + (\log N)^{-1}$ and $T = N^2$. We then have (of Lemma 3 of [10]) that

$$\psi(y,\chi) = -\frac{1}{2\pi i}\int_{\theta-iT}^{\theta+iT}\left(\frac{L'}{L}(s,\chi)+F(s,\chi)\right)\frac{y^s}{s}\,ds$$

$$+\ \psi(u,\chi) + O(\log N),$$

for $y \leqslant N$. By partial integration we then obtain

$$\psi(N,\chi,\gamma_q) = -\int_1^N e(\gamma_q y)\,\frac{1}{2\pi i}\int_{\theta-iT}^{\theta+iT}\left(\frac{L'}{L}(s,\chi)+F(s,\chi)\right)y^{s-1}\,ds\,dy$$

$$+\ O\big(N\beta\log N + u\big).$$

The error term above contributes $\ll Z^{3/2}(\log N)^3 u$ to (6) which will be satisfactory providing $u < N\beta^{1/2}$.

Writing

$$h(s) = \int_1^N e(\gamma_q y)y^{s-1}dy$$

(suppressing, in the interests of clarity, the dependence of h on q) we have that h(s) is an entire function of s and for $\sigma \geqslant 1/2$,

$$h(s) \ll N\min\big(1,|t|^{-1/2}\big) \qquad \text{for } t \leqslant 4N\beta,$$

and

$$h(s) \ll N\min\big(1,\ |t|^{-1}\big) \qquad \text{for } t \geqslant 4N\beta,$$

where $s = \sigma + it$. This means that we can follow through all of Vaughan's analysis with the factor h(s) included. Also we have $q^{1/2}/\phi(q)$ in place of his $q/\phi(q)$. This gives , using Vaughan's notation (cf. (20) of [10]),

$$\sum_{Z\leqslant q<2Z}\frac{q^{1/2}}{\phi(q)}\sum_{\chi}{}^{*}\int_{-T'}^{T'}|H(\theta+it,\chi)h(\theta+it)|\,dt$$

$$\ll Z^{-1/2}(\log N)^3\, N\big(1+Z^2u^{-1}\big)^{1/2}\big(1+Z^2v^{-1}T'\big)^{1/2}$$

for $T' \leqslant 4N\beta$, and (cf. (24) of [10])

$$\sum_{Z\leqslant q<2Z}\frac{q^{1/2}}{\phi(q)}\sum_{\chi}{}^{*}\int_{-T'}^{T'}|I(\tfrac{1}{2}+it,\chi)h(\theta+it)|\,dt$$

$$\ll N^{1/2}\big(u^2+Z^2\big)^{1/4}\big(v+T'Z^2\big)^{1/2}(\log N)^4.$$

For integrals with $4N\beta \leqslant t \leqslant T$ the same estimates hold without the factor T' appearing on the right-hand side. The choice $u = 2\beta^{-1/2}Z^{-1}$

(which is less than $N\beta^{1/2}$ as required earlier) and $v = Z^2 N\beta$ then gives an estimate

$$\ll (\log N)^4 \left(NZ^{-1/2} + N^{1/2}Z^{3/2}(N\beta)^{1/2} + N^{3/4}Z^{3/4}(N\beta)^{1/4} \right).$$

Since $(N\beta)^{1/4} \leqslant Z \ll Q$ this completes the proof of Theorem 1.

The proof of Theorem 2 is similar to the above argument with $(N\beta)^{1/4}$ replaced by $P = \exp\left(-(\log N)^{1/2}/2\right)$ and for values of q smaller than this value the required bound quickly follows by partial summation from Gallager's result (the form given in [9] is most convenient).

3. Proof Of Theorem 3.

The following two results are essentially Lemmas 5 and 7 of [7], the only alterations coming from a change in presentation concerning the dependence of the results on the size of "a". We remark that the definition of θ in Lemma 6 of [7] should have been $\theta = \max(T/(Rq), q\delta, 1)$ and not with an "$a\delta$" as stated there, and the "J" occuring in the hypothesis of Lemma 7 should have been an "L".

Lemma 1. *Suppose that* $\varepsilon > 0$, $N \geqslant R, J, M, q \geqslant 1$, $(a,q) = 1$. *Then*

$$\sum_{R \leqslant r < 2R} \sum_{J \leqslant j < 2j} \sum_{M \leqslant m < 2M} \left| \sum_{n \leqslant N/m} a_n b_m e\left(\frac{ajmn}{qr}\right) \right|$$
$$\ll (Na)^{\varepsilon} (JN/q + RJM + qR^2).$$

Lemma 2. *Given the hypotheses of Lemma 1 and two sequences of complex numbers:* $a_n, b_m \ll N^{\varepsilon/3}$. *Then*

$$\sum_{R \leqslant r < 2R} \sum_{J \leqslant j < 2J} \sum_{M \leqslant m < 2M} \left| \sum_{n \leqslant N/m} a_n b_m e\left(\frac{ajmn}{qr}\right) \right|$$

$$\ll N^{\varepsilon} R^{3/2} (R/M + L)^{1/2} |q/a|^{1/2} + (Na)^{\varepsilon} R(J + R/M) N^{1/2} M^{1/2}$$

$$+ (Na)^{\varepsilon} N M^{-1/2} R^{1/2} (J + R/M)^{1/2} (qM/N + 1 + JM/(qR))^{1/2})$$

The proof of Lemma 1 uses the fact that the inner sum is a geometric series, while the proof of Lemma 2 is based on the large sieve and counting the solutions of certain diophantine inequalities.

To prove Theorem 3 we appeal to Heath-Brown's generalized Vaughan identity, whereby a sum of the form $\Sigma\ \Lambda(n)f(n)$ may be decomposed into $O\left((\log N)^{10} \right)$ double sums of the form

$$\sum_{M\leqslant m<2M} \ \sum_{n\leqslant N/m} a_n b_m f(mn)$$

with either

$$\text{(I)} \quad a_n = 1 \text{ or } \log n,\ M \ll N^{2/3}\ (N^{7/10}),\ b_m \ll N^{\varepsilon/6}$$

or

$$\text{(II)} \quad a_n,\ b_m \ll N^{\varepsilon/6},\ N^{1/5} \ll M \ll N^{1/3}\ (N^{1/4};\ N^{2/5})$$

(the values in brackets produce the alternative exponents). The result of Theorem 3 quickly follows.

References.

[1] R. C. Baker and G. Harman, Diophantine approximation by prime numbers, *J. London Math. Soc.*, **(2) 25** (1982), 201-215.

[2] H. Davenport, Multiplicative number theory (ed. revised by Montgomery, H. L.), Springer-Verlag: New York, 1980.

[3] P. X. Gallagher, A large sieve density estimate near $\sigma = 1$, *Invent. Math.* **11** (1970), 329-339.

[4] G. H. Hardy and J. E. Littlewood, Some problems of 'Partitio Numerorum': III On the expression of a number as a sum of primes, *Acta Math.* **44** (1923), 1-70.

[5] ______________________, A further contribution to the study of Goldbach's problem, *Proc. London Math. Soc.* **(2) 22** (1923), 46-56.

[6] G. Harman, On the distribution of α_p modulo one, *J. London Math. Soc.* **(2) 27** (1983), 9-18.

[7] ___________, Diophantine approximation with a prime and an almost-prime", *J. London Math. Soc.* **(2) 29** (1984), 13-22.

[8] D. R. Heath-Brown, Prime numbers in short intervals and a generalized Vaughan identity, *Canad. J. Math.* **34** (1982), 1365-1377.

[9] H. L. Montgomery and R. C. Vaughan, The exceptional set in Goldbach´s problem, *Acta Arith.* **27** (1975), 353-370.

[10] R. C. Vaughan, Mean value theorems in prime number theory, *J. London Math. Soc.* **(2) 10** (1975), 153-62.

[11] ___________, Diophantine approximation by prime numbers III, *Proc. Lond. Math. Soc.* **(3) 33** (1976), 177-192.

[12] ___________, Sommes trigonométriques sur les nombres premiers, *C.R. Acad. Sci. Paris, Ser.* A, **258** (1977), 981-3.

[13] ___________, On the distribution of α_p modulo 1, *Mathematika*, **24** (1977), 135-141.

[14] ___________, The Hardy-Littlewood Method, Cambridge University Press: Cambridge, 1981.

[15] I. M. Vinogradov, The method of trigonometrical sums in the theory of numbers (Translated, revised and annotated by Davenport, A. and Roth, K. F.), Interscience: New York, 1954.

G. Harman
Department of Pure Mathemtics,
University College,
P.O. Box 78,
Cardiff CF1 1XL, Wales, U.K.

THE DISTRIBUTION OF $\Omega(n)$ AMONG NUMBERS WITH NO LARGE PRIME FACTORS

Douglas Hensley

0. Abstract

The main result concerns the distribution of $\Omega(n)$ within

$$S(x,y) = \{ n: 1 \leqslant n \leqslant x \text{ and } p \leqslant y \text{ if } p|n \}.$$

There is an average value k_0 for $\Omega(n)$, and a dispersion parameter V, such that for k not too far from k_0, and for large x, y with

$$2 \log\log x + 1 \leqslant \log y \leqslant (\log x)^{3/4},$$

the number of solutions n of $\Omega(n) = k$ in $S(x,y)$ is roughly $\exp(-V(k - k_0)^2)$ times the number of solutions n of $\Omega(n) = k_0$ in $S(x,y)$.

In the course of the proof, machinery is developed which permits a sharpening in the same range of previous estimates for the local behaviour of $\Psi(x,y)$ as a function of x.

1. Introduction.

The question of the distribution of $\nu(n)$ among natural numbers $n \leqslant x$ with no prime factors $> y$ has received increasing attention in recent years. Alladi's Turan-Kubilius inequality made a good start, and there has been further progress (see [1,2]).

Here it is more natural to deal with $\Omega(n)$, and count prime divisors of n according to their multiplicity. Our methods are best suited to moderately large values of $u := \log x/\log y$, and for most of this work we assume

$$(\log y)^{1/3} \leqslant u \leqslant \sqrt{y} /(2 \log y).$$

This is essentially the same as the region advertised in the abstract, and is technically more convenient.

We adopt most of the standard notation of the subject: The largest prime factor of n is p(n),

$$S(x,y) = \{ n : 1 \leqslant n \leqslant x \text{ and } p(n) \leqslant y \},$$

and

$$\Psi(x,y) = \# S(x,y).$$

Our results have the distinction of giving good estimates for the individual $\Psi_k(x,y)$, where

$$\Psi_k(x,y) := \#\{ n : 1 \leqslant n \leqslant x,\ p(n) \leqslant y \text{ and } \Omega(n) = k\},$$

when k is near the average (over n in S(x,y)) of $\Omega(n)$. This mean is given to a close approximation by

$$k_0 := [\sum_{p \leqslant y} p^{\tau-1}],$$

where $\tau = \tau(x,y)$ is determined by

$$\sum_{p \leqslant y} p^{\tau-1} \log p = \log x.$$

Loosely, $\tau = (\log u + \log\log u)/\log y$, and $k_0 = u + u/\log u$. As k departs from k_0, $\Psi_k(x,y)$ falls off in the typical Gaussian manner, with variance $\approx u/(\log u)^2$, out to $> u^{1/14}$ standard deviations. Very few n in S(x,y) have $\Omega(n)$ farther from k_0.

In the course of the proof we develop considerable machinery which can also be used to study the local behavior of $\Psi(x,y)$ as a function of x.

There are recent and striking results of Hildebrand [5] on this subject. He shows that

$$\Psi(cx,y) = c^{\alpha}\Psi(x,y)\left(1 + O(u^{-1/10})\right)$$

for essentially the entire interesting range of x and y, with α given by

$$\sum_{p \leq y} \frac{\log p}{p^{\alpha}-1} = \log x.$$

This α and our $1 - \tau$ are nearly equal. In fact, $\alpha = 1 - \tau + O(1/u \log^2 y)$ in our range. Later, we will be working with a certain θ defined as τ was except that the primes are "smeared out" a little. The distinction is minor, as $\theta = \tau + O(y^{-5/3+\varepsilon})$ in our range. It will be evident that the error terms in both theorems are large enough that the results hold with τ in place of θ, and without the effect of this smearing on V. For simplicity of exposition though, we do all the mathematics, and state the theorems, in terms of the smeared parameter θ and its associated quantities. In particular the k_0 defined previously, and the subsequent k_0 defined by a smeared analog, are normally equal and at worst differ by 1.

The sharpening promised permits us to replace Hildebrand's $u^{-1/10}$ with $(\log u)^3/\sqrt{u} \log y$ in our narrower range. It may be that the former error term could be improved to like or better sharpness in this narrower range, but this is not obvious.

The starting point for our proofs is the identity

$$\Psi_k(x,y) = \sum_{d=1}^{\infty} Q(d) H_{k-\Omega(d)}(x/d,y) \tag{1.1}$$

where

$$H_m(x,y) = \sum_{n \in S_m(x,y)} \prod_{p^a || n} 1/a!$$

and

$$Q(d) = \prod_{p^a || d} \left(\sum_{j=0}^{a} (-1)^j/j! \right) = \prod_{p^a || d} q_a \text{ , say.}$$

Note that $q_0 = 1$, $q_1 = 0$, and $0 < q_a < 1$ for $a \geq 2$, with $\lim_{a \to \infty} q_a = 1/e$. Thus in (1.1) most $d \leq x$ have $Q(d) = 0$, since most d have a prime divisor of multiplicity 1.

The reason for putting things in terms of the $H_m(x/d,y)$ is that there is a tie to probability. If $Y_1, Y_2, \ldots, Y_j$ are independent, identically distributed random variables on some probability space, with

$$\text{Prob}(Y_i = \log p) = 1/\pi(y)$$

for each $p \leqslant y$, then

$$H_m(t,y) = \frac{\pi(y)^m}{m!} \text{Prob}\left(\sum_1^m Y_i \leqslant \log t\right) . \qquad (1.2)$$

This allows us to transfer the problem of counting $\Psi_k(x,y)$ to the setting of sums of independent random variables. While the concept is then fairly simple, many details must be hammered out.

In Sections 2 and 3 we develop some information about the distribution of the random variables Y_i, and define quantities that later appear in the main results. In Sections 4 and 5 we show that various "exceptional numbers" are rare in $S(x,y)$. In Sec. 6 we return to the main line of argument and obtain sharp estimates of the $H_m(x/d,y)$ for "unexceptional" m and d. In Sec. 7 we prove Theorem 1, the sharper estimate of $\Psi(cx,y)/\Psi(x,y)$ in the range under discussion. In Sec. 8 we prove Theorem 2, showing that the distribution of $\Omega(n)$ in $S(x,y)$ is Gaussian, and that every reasonably central k has as many n in $S(x,y)$ with $\Omega(n) = k$ as expected, to within a factor of $1 + O(u^{-1/4})$.

The origin of the two constraints on u merits some discussion. The lower limit $u = (\log y)^{1/3}$ could easily be relaxed to $u = (\log y)^{\varepsilon}$, and probably to $u = (\log\log y)^{1+\varepsilon}$. But if $u \leqslant \log\log y$, the distribution of mass in

$$\sum_{p \leqslant y} p^{\tau-1}$$

shifts from being packed largely into $(\sqrt{y}\ ,y)$ to being far more spread out. The application of the Berry-Esseen theorem in Sec. 6 breaks down, and all the many calculations along the way are vastly complicated. Happily, there are other ways to study the distribution of $\Omega(n)$ in $S(x,y)$ for smaller u, and Alladi [2] has shown that here too it is Gaussian.

The upper limit seems to be an inherent defect of our method. For $u = y^{1/T}$, the proportion of square-free numbers in $S(x,y)$ is asymptotically $1/\zeta(2(1 - 1/T))$, for $T > 2$. (This follows from Hildebrand´s local behavior result, or from our Theorem 1). As

$T \to 2^+$, $\zeta(2(1 - 1/T)) \to \infty$ and the proportion of square-free numbers drops toward zero.

Since the identity (1.1) is designed to let us recover $S(x,y)$ in full from a weighted version in which only square free numbers receive full weight, it cannot be expected to perform well when the weighted version varies too strongly from the weight-one case.

2. A sense in which the distribution of log p ($p \leq y$) is smooth.

An important distinction in probability is that made between discrete and continuous distributions. Now the distribution of our Y_i, with mass $1/\pi(y)$ at each $\log p$, $p \leq y$ is of course discrete. However, for large y the prime number theorem suggests that this distribution is continuous, with density proportional to $s^{-1}e^s$ on $1 \leq s \leq \log y$. The subsequent analysis would be simpler if it had to do with such a density. This section gives rigorous content to the metaphor above. We show that the distribution of a Y_i is *close to* a continuous distribution with a density that *for most* s is near $Cs^{-1}e^s$.

If we would relax the standards of "close to" we could insist on proportionality to $s^{-1}e^s$ for all large s. But there are stronger results on the local smoothness of primes if a few exceptions are allowed.

Selberg showed that for all $\varepsilon_1 > 0$, all $\varepsilon_2 > 0$, there exists an $x(\varepsilon_1,\varepsilon_2)$ such that if $x > x(\varepsilon_1,\varepsilon_2)$ then [7]

$$\#\Big\{ n \leq x : \Big|\pi(n + n^{\frac{19}{77}+\varepsilon_1}) - \pi(n) - n^{\frac{19}{77}+\varepsilon_1}/\log n\Big| > \varepsilon_2 n^{\frac{19}{77}+\varepsilon_1}/\log n\Big\} < \varepsilon_2 x \ . \tag{2.1}$$

Disallowing exceptions in (2.1) would only permit an exponent of 1/2, even on the Riemann hypothesis.

We now fix an ε_1, $0 < \varepsilon_1 < 1/100$, and let

$$\nu = y^{(\varepsilon_1 - 58/77)} = \nu(y)$$

$$\lambda_p(s) = \nu^{-1}\chi_{[\log p - \nu/2, \log p + \nu/2]}(s)$$

and

$$\lambda(s) = \sum_{p \leq y} \lambda_p(s) .$$

Let m be Lebesque measure.

Lemma 1. *For all* $\varepsilon > 0$ and $\delta > 0$, *if* $U \subset \mathbf{R}$ *is measurable and* $m(U) = \delta$, *then there exist* $k \geq \delta/\varepsilon$ *and* $u_1 < u_2 < \ldots < u_k$ *in* U *such that* $u_{j+1} - u_j \geq \varepsilon$ *for* $1 \leq j \leq k - 1$.

Proof. Clear.

Lemma 2. *For all sufficiently large* y, *and for all* t satisfying $1 \leq t \leq \frac{1}{2}\varepsilon_1 \log y$,

$$m\{ s : \lambda(s) < \frac{99}{100} s^{-1} e^{s} \text{ and } \log y - t \leq s \leq \log y\} \leq \frac{t}{100}.$$

Proof. Fix t and let $U_t = \{ s : \lambda(s) < \frac{99}{100} s^{-1} e^{s}$ and $\log y - t \leq s \leq \log y\}$. Assume $m(U_t) > t/100$. We derive a contradiction.

There must be some interval L of length 1 within $[\log y - t, \log y]$ which intersects U_t in a set of measure greater than 1/100. Let $L' = \{s_1, s_2, \ldots, s_k\}$ be a set of $k \geq \nu/100$ elements of L, with $s_{j+1} - s_j \geq \nu$ for $1 \leq j \leq k - 1$. Such an L' exists by Lemma 1.

For $s_j \in L'$ consider the intervals $[\exp(s_j - \nu/2), \exp(s_j + \nu/2))$ $= [a_j, b_j)$, say. These are disjoint for distinct j, $1 \leq j \leq k - 1$. Fix a particular j and drop the subscripts: $[a,b)$. Temporarily, let $\delta = \frac{19}{77} + \varepsilon_1/2$. For each integer m, $a \leq m \leq a + a^{\delta}$, consider the sequence $\alpha_i^{(m)}$ determined by

$$\begin{aligned} \alpha_1^{(m)} &= m \\ \alpha_{i+1}^{(m)} &= \alpha_i^{(m)} + [(\alpha_i^{(m)})^{\delta}]. \end{aligned} \tag{2.2}$$

These sequences are disjoint for distinct m, collectively they include all but a vanishingly small fraction of the integers in $[a,b)$, and they satisfy

$$\alpha_{i+1}^{(m)} > \alpha_i^{(n)} \quad \text{for } a \leqslant m,\ n \leqslant a + a^{\delta}$$
$$\alpha_i^{(m)} > \alpha_i^{(m-1)} \quad \text{for } a + 1 \leqslant m \leqslant a + a^{\delta}. \tag{2.3}$$

For each m and i, let

$$B(i,m) = \{ n\colon \alpha_i^{(m)} \leqslant n < \alpha_{i+1}^{(m)} \}, \tag{2.4}$$

$$E(i,m) = (\alpha_i^{(m)})^{\delta}/\log \alpha_i^{(m)} ,$$

$$M(m) = \{ i : B(i,m) \subset [a,b)\},$$

and

$$N(m) = \{ i : B(i,m) \subset [a,b) \text{ and } B(i,m) \text{ contains fewer than } 199/200\ E(i,m) \text{ primes}\}.$$

Then the number of primes in $[a_j,b_j)$ is at least

$$\frac{199}{200} \sum_{\substack{i \in M(m) \\ i \notin N(m)}} E(i,m).$$

This lower bound holds for each m, $a \leqslant m \leqslant a + a^{\delta}$. Since $\log \alpha_i^{(m)} \in [s_j - \nu/2,\ s_j + \nu/2]$ and since $[\alpha_i^{(m} + (\alpha_i^{(m)})^{\delta}] - \alpha_i^{(m)} \approx \exp(\delta s_j)$, there are thus at least

$$\frac{994}{1000} \{\#M(m) - \#N(m)\} s_j^{-1} \exp(\delta s_j)$$

primes in $[a_j,b_j)$.

Remark. (This is not to say that we can sum over the various m and get still more primes. But for each fixed m, this is correct).

Now for each m, $a \leqslant m < a + a^{\delta}$,

$$\#M(m) \approx \nu \exp\left((1 - \delta)s_j\right) \gg y^{\varepsilon_1/12} .$$

Thus there are at least

$$\frac{993}{1000}\,\nu\, s_j^{-1} e^{s_j} - \frac{994}{1000}\,\#N(m) s_j^{-1} e^{\delta s_j}$$

primes in $[a_j, b_j)$.

On the other hand since $s_j \in L'$ there are no more than $\frac{99}{100}\,\nu s_j^{-1} e^{s_j}$ primes in $[a_j, b_j)$. Thus

$$\frac{3\nu}{1000}\, s_j^{-1} e^{s_j} \leqslant s_j^{-1} e^{\delta s_j}\, \#N(m),$$

and so

$$\#N(m) \geqslant \frac{3\nu}{1000}\, e^{(1-\delta)s_j} . \tag{2.5}$$

Thus also

$$\sum_{a_j \leqslant m < a_j + a_j^{\delta}} \#N(M) \geqslant \frac{3\nu}{1000}\, e^{(1-\delta)s_j}\left([a_j^{\delta}] - 1\right) \geqslant \frac{\nu}{400}\, e^{s_j} .$$

Now summing over j, we get that there are at least

$$\frac{\nu}{400}\, e^{s_1}\left(\frac{1}{100\nu}\right) = \frac{1}{40000}\, e^{s_1}$$

distinct integers of the form $\alpha_i^{(m)}$, each $\leqslant e^{1+s_1}$, and with fewer than 199/200 E(i,m) primes between $\alpha_i^{(m)}$ and $\alpha_{i+1}^{(m)}$. But according to (2.1), with $\varepsilon_2 = 10^{-6}$, say, there cannot be this high a proportion of integers $n \leqslant e^{1+s_1}$ with so few primes between n and $n + n^{\delta}$. This completes the proof of Lemma 2.

Corollary. *For* y *sufficiently large,* $r \geqslant 0$, *and* $1 \leqslant t \leqslant \frac{1}{2}\varepsilon_1 \log y$, $m\{\, s : \lambda(s)\exp((r-1)s) < s^{-1}(e^{rs} - \frac{1}{100} y^r)$ *and* $\log y - t \leqslant s \leqslant \log y\} \leqslant t/100$.

Proof. The set in question is contained in the set of Lemma 2.

3. Calculus and Statistics.

Here we work out estimates of various quantities related to exponential centering and the Berry-Esseen theorem. Let

$$\hat{G}(r) = \sum_{p \leqslant y} p^{r-1} ,$$

$$\hat{I}(r) = \sum_{p \leqslant y} p^{r-1} \log p , \tag{3.1}$$

$$\hat{J}(r) = \sum_{p \leqslant y} p^{r-1} \log^2 p ,$$

and

$$\hat{K}(r) = \sum_{p \leqslant y} p^{r-1} \log^3 p .$$

Let μ now be the probability measure

$$\frac{1}{\pi(y)} \sum_{p \leqslant y} \delta_{\log p}(t),$$

a sum of equal point masses at each $\log p$, $p \leqslant y$.

Let Y_1, $Y_2 \ldots$ be independent, identically distributed random variables with common measure μ .

Let λ be the probability density function

$$\frac{1}{\pi(y)} \sum_{p \leqslant y} \nu^{-1} \chi_{[\log p - \nu/2,\ \log p + \nu/2]}(s),$$

as in Sec. 2. Let Z_1, $Z_2 \ldots$ be further random variables, independent, and uniformly distributed on $[-\nu/2, \nu/2]$. Then $\lambda(s)$ is the common density of the $(Y_i + Z_i)$'s.

For $r \geqslant 0$ let

$$G(r) = \int_0^\infty e^{(r-1)s} \lambda(s)ds ,$$

$$I(r) = \int_0^\infty se^{(r-1)s} \lambda(s)ds , \tag{3.2}$$

$$J(r) = \int_0^\infty s^2 e^{(r-1)s} \lambda(s)ds ,$$

and

$$K(r) = \int_0^\infty s^3 e^{(r-1)s} \lambda(s)ds .$$

Here λ , and thus G, $\hat{G}$ etc. depends on y implicitly. The defining integrals are convergent for all r since $\lambda(s)$ has bounded support.

The next variation on G, $I, \ldots$ comes from replacing $\pi(t)$ with li(t), the logarithmic integral, and truncating at y. Let

$$\tilde{G}(r) = \int_1^{\log y} s^{-1}e^{rs}\,ds\,,$$

$$\tilde{I}(r) = \int_1^{\log y} e^{rs}\,ds\,,$$

$$\tilde{J}(r) = \int_1^{\log y} se^{rs}\,ds\,, \tag{3.3}$$

and

$$\tilde{K}(r) = \int_1^{\log y} s^2e^{rs}\,ds\,.$$

As the notation is meant to suggest, $\hat{G}$, G and $\tilde{G}$, etc. are nearly equal. We use the prime number theorem:
For fixed C > 0, (see [7])

$$\pi(z) = \mathrm{li}(z) + O_C(ze^{-C\sqrt{\log z}})\,. \tag{3.4}$$

Lemma 3. $\tilde{G}(r) = \hat{G}(r) + O\left(1 + (1 + \frac{1}{r})\exp(r\log y - \sqrt{\log y})\right)$, *uniformly in* r ⩾ 0. *Further,* G *may be replaced with* I, J *or* K.

Proof . $$\hat{G}(r) = \sum_{p\leq y} p^{r-1} = \int_{2^-}^{y} t^{r-1}d\pi(t)$$

$$= t^{r-1}\pi(t)\Big|_{2^-}^{y} + \int_2^y (1-r)t^{r-2}\pi(t)dt$$

$$= y^{r-1}\{\mathrm{li}(y) + O(ye^{-C\sqrt{\log y}})\} +$$

$$+ \int_2^y (1-r)t^{r-2}\left(\mathrm{li}(t) + O(te^{-C\sqrt{\log t}})\right)dt$$

$$= y^{r-1}\mathrm{li}(y) + O(y^re^{-C\sqrt{\log y}})$$

$$+ \int_2^y (1-r)t^{r-2}\mathrm{li}(t)dt + O\left(\int_2^y t^{r-1}e^{-C\sqrt{\log t}}dt\right)$$

$$= \int_2^y t^{r-1}\frac{dt}{\log t} + O(y^re^{-C\sqrt{\log y}}) + O\left(\int_2^y t^{r-1}e^{-C\sqrt{\log t}}dt\right)$$

$$= \int_{\log 2}^{\log y} s^{-1} e^{rs} ds + O(y^r e^{-C\sqrt{\log y}}) + O(\int_{\log 2}^{\log y} e^{rs - C\sqrt{s}} ds),$$

so that

$$\hat{G}(r) = \tilde{G}(r) + O(y^r e^{-C\sqrt{\log y}}) + O(\int_1^{\log y} e^{rs-C\sqrt{s}} ds). \quad (3.5)$$

In estimating with I, J, or K in place of G, the powers of $\log t$ that arise can be subsumed in $\exp(-C\sqrt{\log t})$ by reducing C. Let us take C originally so that after any such reductions, (3.5) holds for G, I, J and K with $C = 2$. It remains to bound

$$\int_1^{\log y} e^{rs-C\sqrt{s}} ds.$$

We need a sublemma.

Lemma 4. *Let* $F(T,r) = \int_1^T e^{rt-\sqrt{t}} dt$. *Then for* $T \geq 1$ *and* $r \geq 0$

(a) $F(T,r) \leq 32 \qquad (r \leq \frac{3}{4} T^{-1/2})$

(b) $F(T,r) \leq \frac{32}{r} e^{rT-\sqrt{T}} \qquad (T^{-1/2} \leq r).$

(The proof presents no special difficulty and is left to the reader).

Now

$$\int_1^{\log y} e^{rs-2\sqrt{s}} ds = \frac{1}{4} \int_4^{4\log y} e^{1/4\, rs-\sqrt{s}} ds \leq 8$$

for $r \leq \frac{3}{\sqrt{\log y}}$, by (a).

For $r > \frac{3}{\sqrt{\log y}}$, though,

$$\int_1^{\log y} e^{rs-2\sqrt{s}} ds \leq \int_1^{\log y} e^{rs-\sqrt{s}} ds \leq \frac{32}{r} e^{r\log y - \sqrt{\log y}},$$

by (b). In both cases,

$$\int_1^{\log y} e^{rs-2\sqrt{s}} ds \ll (1 + \frac{1}{r} e^{r\log y - \sqrt{\log y}}). \quad (3.6)$$

The other error term in (3.5) adds to this to give the claimed error bound of Lemma 3.

Lemma 5. $G = \hat{G}(1 + O(\nu^2))$, *and likewise for* I, J *and* K.

Proof. (For G).

$$|G - \hat{G}| \leq \sum_{p \leq y} \left| p^{r-1} - \nu^{-1} \int_{\log p - \nu/2}^{\log p + \nu/2} e^{(r-1)s}\, ds \right|$$

$$= \sum_{p \leq y} p^{r-1} \left| 1 - \nu^{-1} \int_{-\nu/2}^{\nu/2} e^{(r-1)s} ds \right| \ll \nu^2 \sum_{p \leq y} p^{r-1}.$$

Lemma 6. *For each* $C > 0$, *the following holds uniformly in* $0 < r < C$ *and* $y \geq \exp(1/r)$:

(1) $\tilde{G}(r) = li(y^r) + \log(1/r) + O(1)$,

(2) $\tilde{I}(r) = \frac{1}{r}(y^r - 1) + O(1)$,

(3) $J(r) = \frac{1}{r} y^r(\log y - \frac{1}{r}) + r^{-2} + O(1)$,

(4) $\tilde{K}(r) = \frac{1}{r} y^r(\log^2 y - \frac{2\log y}{r} + 2r^{-2}) + 2r^{-3} + O(1)$

(5) $$(\tilde{G}\tilde{J} - \tilde{I}^2)(r) = \frac{y^{2r}}{r^4 \log^2 y}\left(1 + \frac{4}{r\log y} + O\left(\frac{1}{r^2 \log^2 y}\right)\right)$$

$$+ r^{-2}\log\left(\frac{1}{r}\right)\left(y^r(r\log y - 1) + 1\right) + O(\log r)$$

Proof. We have $\tilde{G}(r) = \int_1^{\log y} \frac{1}{rs} e^{rs}\, r ds = li(y^r) + \int_r^1 t^{-1} e^t dt.$ The last integral is $\log r + O(1)$, uniformly in $0 < r \leq C$. The rest is also a simple calculus exercise.

From Lemmas 5 and 6 we have the

Corollary. Lemma 3 *holds with* G, I, J *and* K *in place of* $\hat{G}$, $\hat{I}$, $\hat{J}$, *and* $\hat{K}$ *respectively, when* $\frac{1}{\log y} \leq r \leq C$.

Now from Lemmas 3, 5 and 6, we have uniformly in $\frac{1}{\log y} \leqslant r \leqslant C$ that

$$GJ - I^2 = \tilde{G}\tilde{J} - \tilde{I}^2 + O\left(\frac{y^r}{r\log y} + \frac{y^r e^{-\sqrt{\log y}}}{r^2}\right). \tag{3.7}$$

Let

$$\sigma^2 = (GJ - I^2)/G^2$$

$$\alpha = I/G \tag{3.8}$$

and

$$\beta = \frac{1}{G}\int_0^\infty |s - \alpha|^3 \, e^{(r-1)s}\lambda(s)ds.$$

We plan later to modify the density function λ (s) of the $Y_i + Z_i$´s to $G^{-1}e^{(r-1)s}\lambda(s)$, which is also a probability density function, and with mean α, standard deviation σ and absolute third moment β. These three statistical parameters are needed to apply the Berry-Esseen theorem. (The central limit theorem with explicit error estimates).

Until now we have left r in a wide range.

In Lemma 6 (1), this splits naturally into two regions: $li(y^r)$ predominant, and log (1/r) predominant. In the latter case, the calculus becomes very involved. This case is also the one associated with small u = log x/loy y where traditional methods have worked so well. Accordingly, we shall here treat only the case of large u from now on. We assume

$$(\log y)^{1/3} \leqslant u \leqslant y^C. \tag{3.9}$$

Defining ξ as usual be $\xi > 0$, $e^\xi - 1 = u\xi$, we now restrict attention to r satisfying

$$|r\log y - \xi| \leqslant 2. \tag{3.10}$$

Then (3.10) is contained in $(\frac{1}{\log y} \leqslant r \leqslant C + 1)$, and for any r satisfying (3.10), all the previous results of this section are valid. From now on, we assume (3.9) and (3.10).

Lemma 7. *Given* (3.9) *and* (3.10),

(1) $r = (\log u + \log\log u + O(1))/\log y$,

(2) $\tilde{G}(r) = \mathrm{li}(y^r) + \log\log y + O(\log\log u)$,

(3) $\tilde{I}(r) = \frac{1}{r}(y^r - 1) + O(1)$,

(4) $\tilde{J}(r) = \frac{1}{r}\, y^r(\log y - \frac{1}{r}) + r^{-2} + O(1)$,

(5) $\tilde{K}(r) = \frac{1}{r}\, y^r\left(\log^2 y - \frac{2\log y}{r} + 2r^{-2}\right) + 2r^{-3} + O(1)$,

(6) $G(r) = y^r\left((r\log y)^{-1} + (r\log y)^{-2} + O((r\log y)^{-3})\right)$,

(7) $(\tilde{G}\tilde{J} - \tilde{I}^2)(r)$, *and* $(GJ - I^2)(r)$, *both equal*

$$\frac{y^{2r}}{r^4\log^2 y}\left\{1 + \frac{4}{r\log y} + O\left(\frac{1}{r^2\log^2 y}\right)\right\}.$$

(8) $\sigma^2 = r^{-2}\left(1 + \frac{4}{r\log y} + O\left(\frac{1}{r^2\log^2 y}\right)\right)$

$= \left(\frac{\log^2 y}{\log^2 u}\right)\left(1 + O\left(\frac{\log\log u}{\log u}\right)\right)$, *and*

(9) $\log y - \alpha = \frac{1}{r}\left(1 + O\left(\frac{1}{r\log y}\right)\right) = \frac{\log y}{\log u}\left(1 + O\left(\frac{\log\log u}{\log u}\right)\right)$.

Proof. A routine, if lengthy, calculation.

Now let $h(r,x,y) = G(r)(\log x)/I(r)$. Frequently we will abbreviate this to $h(r)$, or just h. Then

$$\frac{dh}{dr} = -(GJ - I^2)\log x/I^2 = -\sigma^2 G^2\log x/I^2 = -\frac{\log x}{\log^2 u}\left(1+O\left(\frac{\log\log u}{\log u}\right)\right). \tag{3.11}$$

Proof. Immediate from the definition and from Lemma 7.

Let θ be the (unique from (3.11)) r such that $I(r) = \log x$. Then $|\theta\log y - \xi| \leq 1$. (See (3.13) below). Let $h_0 = G(\theta)$, and $n_0 = [h_0]$.

Remark. In $S(x,y)$, the mean and median value of $\Omega(n)$ is quite close to h_0, as we shall see.

Now, more calculus :

$$h(\theta - \frac{1}{\log y}) = h_0 + \frac{u}{\log^2 u}\left(1 + O(\frac{\log\log u}{\log u})\right), \qquad (3.12)$$

$$h(\theta + \frac{1}{\log y}) = h_0 - \frac{u}{\log^2 u}\left(1 + O(\frac{\log\log u}{\log u})\right),$$

(*Proof.* Immediate from (3.11).) ;

$$\theta \log y = \xi + O(e^{-\sqrt{\log y}}). \qquad (3.13)$$

(*Proof.* Here $I(r) = \tilde{I}(r)(1 + O(e^{-\sqrt{\log y}}))$ from Lemma 3. Now $\tilde{I}(\xi/\log y) = \log x$, so $I(\xi/\log y) = \log x(1 + O(e^{-\sqrt{\log y}}))$. Now $dI/dr = J \asymp \log x \log y$ for $r = \xi/\log y + O(e^{-\sqrt{\log y}})$, so a change of $O(\frac{1}{\log y} e^{-\sqrt{\log y}})$ in r will bring $I(r)$ to $\log x$.) ;

$$\beta\sigma^{-3} = O(1), \qquad (3.14)$$

uniformly in (r,y) satisfying (3.9) and (3.10).

Proof. Lemma 7 has an estimate of σ. To estimate β, we cut the defining integral at $\frac{1}{2}\log y$ and at α. For $s < \frac{1}{2}\log y$, the integrand is $\ll \log^3 y\, e^{(r-1)s}\lambda(s)$. Using the definition of $\lambda(s)$ and the prime number theorem,

$$\int_1^{\frac{1}{2}\log y} e^{(r-1)s}\lambda(s)ds < \int_1^{\frac{1}{2}\log y} s^{-1}e^{rs}ds < \frac{1}{r}y^{r/2}.$$

For $\frac{1}{2}\log y < s < \alpha$, the integral in (3.8) is

$$< \int_{\frac{1}{2}\log y}^{\alpha} \frac{1}{\log y}|s-\alpha|^3 e^{rs}ds \ll \frac{1}{\log y}e^{r\alpha}r^{-4} \ll \frac{y^r}{r^4\log y},$$

from Lemma 7, (9), and

$$\int_{\alpha}^{\log y} (s-\alpha)^3 e^{(r-1)s}\lambda(s)ds \ll r^{-3} \int_{\frac{1}{2}\log y}^{\log y} s^{-1}e^{rs}ds \ll \frac{y}{r^4 \log y},$$

again by the prime number theorem. Together with $G \approx y^r/r\log y$ and $\sigma^2 \approx 1/r^2$ from Lemma 7, this gives (3.14).

Now let $r = r(h)$ be the inverse function of $h(r)$, and $\mathcal{G}(h) = (h \log G(r(h)) - h \log h + h + (1 - r(h) \log x)$. Then

$$d\mathcal{G}/dh = \log G - \log h \tag{3.15}$$

$$d^2\mathcal{G}/dh^2 = -\frac{1}{\log x}\left(\frac{IJ}{GJ-I^2}\right)$$
$$= \frac{\log^2 u}{u}\left(1 + O\left(\frac{\log\log u}{\log u}\right)\right),$$

uniformly in $h_0 - \frac{u}{\log^2 u} \leqslant h \leqslant h_0 + \frac{u}{\log^2 u}$.

Further, if $V = -\, d^2\mathcal{G}/dh^2 \,\big|_{h_0}$, and $\Delta h = h - h_0$, then

$$d^2\mathcal{G}/dh^2 = -V(1+O(\Delta h(\log u)/u)) \text{ for } |h - h_0| \leqslant \frac{u}{\log^2 u}.$$

Proof. Only the last claim is at all difficult. We expand

$$G = y^r\left(\frac{1}{r\log y} + \frac{1}{r^2\log^2 y} + \frac{2}{r^3\log^3 y} + \frac{6}{r^4\log^4 y} + O\left(\frac{1}{r^5\log^5 y}\right)\right),$$

and I, J and K to like accuracy. Then

$$\frac{d}{dh}\log\left(\frac{GJ-I^2}{IJ}\right) = \frac{-I^2}{(GJ-I^2)\log x}\left(\frac{I^3K - J^3G}{(GJ-I^2)IJ}\right),$$

and $I^3K - J^3G \ll y^{4r}/r^7\log y$. On the other hand, $(GJ - I^2)IJ \approx y^{4r}/r^6\log y$, so

$$\left|\frac{d}{dh}\log\left(\frac{GJ-I^2}{IJ}\right)\right| \ll \frac{1}{r}\left|\frac{dr}{dh}\right| \ll \frac{\log u}{u}.$$

Now let $\sigma_0 = \sigma(\theta), = \sigma(r(h_0))$. Then again uniformly in $h_0 - \frac{u}{\log^2 u} \leqslant h \leqslant h_0 + \frac{u}{\log^2 u}$,

$$\left|\frac{d}{dh}\log(\sigma^2)\right| < \frac{\log u}{u}, \tag{3.16}$$

and

$$\sigma^2 = \sigma_0^2 \left(1 + O\left(\frac{\Delta h \log u}{u}\right)\right).$$

Proof. $\sigma^2 = \frac{GJ-I^2}{G^2}$. Now

$$\frac{d}{dh}\log\left(\frac{GJ-I^2}{G^2}\right) = \frac{d}{dh}\log\left(\frac{GJ-I^2}{IJ}\right) + \frac{d}{dh}\log\left(\frac{IJ}{G^2}\right).$$

But $\log (IJ/G^2) = 2 \log (I/G) + \log (J/I)$. Expanding as before and simplifying now gives (3.16).

We make one last observation.

Given (3.9) and (3.10), and moreover $|r - \theta| \leqslant (\log y)^{-1}$, for $r \leqslant 2/3$

$$\frac{1}{r-1} = \frac{1}{1-\theta}\left(1 + O\left(\frac{\Delta h \log^2 u}{u \log y}\right)\right). \tag{3.17}$$

Proof. Plug in Lemma 7 (1) and the given conditions.

4. Exclusion of numbers with many prime powers.

Here we show that in the identity

$$\Psi(x,y) = \sum_d Q(d)\Psi'(x/d,y) \tag{4.1}$$

the contribution due to terms with $d \geqslant K$ is small for large K under the hypothesis

$$(\log y)^{1/3} \leqslant u \leqslant \frac{\sqrt{y}}{2 \log y}. \tag{4.2}$$

Remark. It is roughly at $u = \sqrt{y}$ that we turn a kind of corner. For smaller u, the proportion of square-free numbers in S(x,y) is positive, while for larger u, it is asymptotically zero. Thus for larger u it is increasingly difficult to recover $\Psi(x,y)$ from the weighted sum $\Psi'(x,y)$ which counts only square-free numbers with full weight. In all our theorems, we assume (4.2).

We first skip ahead to (6.8) and borrow a result:

$$\Psi'(x,y) > \frac{\log u}{\log x} \exp\left(h_0 + (1 - \theta)\log x\right). \tag{4.3}$$

Lemma 8. *Uniformly in* x *and* y *satisfying* (4.2), *and in* d, $1 \leqslant d \leqslant x$,

$$\Psi'(x/d,y) \ll (d^{\theta-1} \log x)\Psi'(x,y).$$

Proof. $\Psi'(x/d,y) = \sum_{m=0}^{\infty} \frac{\pi(y)^m}{m!} \mathrm{Prob}\left(\sum_1^m Y_i \leqslant \log x - \log d\right)$. Now

$$\mathrm{Prob}\ (\sum_1^m\ Y_i \leqslant \log x/d) \leqslant \mathrm{Prob}\ (\sum_1^m\ Y_i + Z_i \leqslant \log x - \log d + \frac{m\nu}{2})$$

$$\leqslant \frac{G(\theta)^m}{\pi(y)^m} \int_0^{(\log x/d) + m\nu} e^{(1-\theta)s} f(s)^{(m)}\, ds\ .$$

Thus

$$\Psi'(x/d,y) \leqslant \sum_0^{\infty} \frac{G(\theta)^m}{m!} e^{(1-\theta)(\log x - \log d + \frac{1}{2}m\nu)} \tag{4.4}$$

$$\leqslant (x/d)^{1-\theta} \sum_0^{\infty} \frac{1}{m!} G(\theta)^m e^{m\nu/2}$$

$$= (x/d)^{1-\theta} \exp\{e^{\nu/2} G(\theta)\}.$$

But $e^{\nu/2} = 1 + O(y^{-5/6 + \varepsilon_1})$ and $G(\theta) \approx u$ so $e^{\nu/2}G(\theta) = G(\theta) + O(uy^{-4/5}) = G(\theta) + O(1)$ so that $\Psi'(x/d,y) \ll (x/d)^{1-\theta} e^{G(\theta)}$. But $G(\theta) = h_0$, so from (4.3) we get $\Psi'(x/d,y) \ll d^{\theta-1}\Psi'(x,y)\log x$.

From Lemma 8, we have

$$\sum_{d>K} Q(d)\Psi'(x/d,y) \ll \left(\Psi'(x,y)\log x\right) \sum_{d>K} Q(d)d^{\theta-1}. \tag{4.5}$$

Lemma 9.

$$\sum_{d>K} Q(d)d^{\theta-1} \ll K^{(\theta-1/2)}\log y.$$

Remark. This lemma is of course useless if $\theta > 1/2$. That is why we had to assume (4.2), which ensures $\theta < 1/2$, and a bit more: $\theta \geqslant \frac{1}{2} - \frac{1.2}{\log y}$ for large y.

Proof of lemma 9. We have

$$\sum_{d=1}^{\infty} Q(d)d^{\theta-1} = \prod_{p\leqslant y}\Big(1+\sum_{j=2}^{\infty} q_j p^{j(\theta-1)}\Big). \tag{4.6}$$

Let $M_p = 1+\sum_{j=2}^{\infty} q_j p^{j(\theta-1)}$, and let (J_p), $p \leqslant j$ be independent random variables with mass at j of $\frac{1}{M_p} q_j p^{j(\theta-1)}$. Then

$$\sum_{d>K} Q(d)d^{\theta-1} = \Big(\prod_{p\leqslant y} M_p\Big)\,\mathrm{Prob}\Big(\sum_{p\leqslant y} J_p \log p > \log K\Big). \tag{4.7}$$

Let $N_p(s) = \sum_{j=0}^{\infty} \frac{1}{M_p} q_j p^{j(\theta-1)}\delta_j(s)$. Then

$\mathrm{Prob}\Big(\sum_{p\leqslant y} J_p \log p > \log K\Big) =$

$$\int_{\log K}^{\infty} \prod_{p\leqslant y}{}^{*} N_p(s/\log p)\,ds \leqslant \int_{0}^{\infty} e^{\gamma(s-\log K)} \prod_{p\leqslant y}{}^{*} N_p(s/\log p)\,ds$$

where Π^* denotes convolution, and $\gamma > 0$.

This last integral is

$$\leqslant e^{-\gamma \log K} \prod_{p\leqslant y} M_p^{-1}\Big(1+\sum_{j=2}^{\infty} q_j p^{(\theta-1)j+\gamma j}\Big).$$

With $\gamma = \frac{1}{2} - \theta$ this last product is

$$\ll \prod_{p\leqslant y}\Big(1-\frac{1}{p}\Big)^{-1} \ll \log y.$$

From Lemma 9 and (4.5), we get

$$\sum_{d>K} Q(d)\Psi'(x/d,y) \ll \log x \log y\; K^{\theta-1/2}\Psi'(x,y),$$

uniformly in $d \leqslant x$, and (x,y) satisfying (4.2).

5. Exclusion of atypical Ω(k).

Here we show that in $S(x,y)$, $\Omega(k)$ is close to h_0 most of the time. For small u, we could simply refer to Alladi´s Turan-Kubilius inequality, but its range does not extend to u as large as those included in (4.2), which we assume.

Lemma 10. (a) *For* $1 \leq B \leq \frac{1}{2}\sqrt{u}/\log u$,

$$\sum_{\substack{k \in S(x,y) \\ \Omega(k) \leq h_0 - B\sqrt{u}/\log u}} 1 \ll (e^{-B^2/3}\log^2 x)\Psi(x,y)$$

(b) For $B \geq 1$,

$$\sum_{\substack{k \in S(x,y) \\ \Omega(k) \geq h_0 + B\sqrt{u}/\log u}} 1 \ll \left(e^{-B^2/12} + e^{-\frac{1}{4}(\frac{1}{2} - \theta)B\sqrt{u}/\log u}\right) \times \Psi(x,y)\log^2 x.$$

Proof (a). The sum on the left is equal to

$$\sum_d Q(d) \sum_{m \leq h_0 - B\sqrt{u}/\log u - \Omega(d)} H_m(x/d,y)$$

$$\leq \sum_d Q(d) \sum_{m \leq h_0 - B\sqrt{u}/\log u} H_m(x/d,y)$$

Let $M = [h_0 - B\sqrt{u}/\log u]$, and $r = r(M)$, so $I(r)/G(r) = \log x/M$. Then since $\theta < \frac{1}{2} - \frac{1.2}{\log y}$, and in view of (3.12), $r < \frac{1}{2} - \frac{0.1}{\log y} < \frac{1}{2}$.

Now for $m \leq M$,

$$H_m(x/d,y) = \frac{\hat{G}^m}{m!} \int_0^{\log(x/d)} e^{(1-r)s} d\left(\text{Prob}\left(\sum_1^m Y_i = s\right)\right) \leq \frac{\hat{G}^m}{m!} (x/d)^{1-r}. \tag{5.1}$$

Since $\hat{G} = G(1 + O(y^{-3/2}))$ from Sec.3 and the definition of ν, and since $m \leq u(1 + o(1))$ here, and $u < \sqrt{y}$,

$$\hat{G}^m = G^m(1 + O(1/y)).$$

Thus

$$H_m(x/d,y) \ll \frac{G^m}{m!} (x/d)^{1-r}$$

and

$$\sum_d Q(d) \sum_{m\leq M} H_m(x/d,y) \ll x^{1-r} \sum_d Q(d)d^{r-1} \sum_{m\leq M} G^m/m! \; . \tag{5.2}$$

Since $M < h_0$, $r > \theta$ so $G > M$. Thus

$$\sum_{m\leq M} G^m/m! < MG^M/M! \; . \tag{5.3}$$

Now

$$\sum_d Q(d)d^{r-1} = \prod_{p\leq y} \Big(1 + \sum_{j=2}^{\infty} q_j p^{(r-1)j}\Big)$$

$$< \prod_{p\leq y} \Big(1 + \sum_{2}^{\infty} q_j p^{-j/2}\Big) \leq \exp\big(\tfrac{1}{2}\mathrm{loglog}\; y + O(1)\big).$$

Hence

$$\sum_d Q(d)d^{r-1} \ll \sqrt{\log y}. \tag{5.4}$$

Now recalling that $M = M(B)$, we have for $B \leq \frac{1}{2}\sqrt{u}/\log u$,

$$Mx^{1-r}G^M/M! \ll x\sqrt{u}\,\exp(M\log G - M\log M + M - r\log x). \tag{5.5}$$

The quantity exponentiated in (5.5) simplifies to

$$h_0 - \theta\log x - \tfrac{1}{2}\big(1 + o(1)\big)(h_0 - M)^2 u^{-1}\log^2 u.$$

Thus

$$\sum_{\substack{k\in S(x,y)\\ \Omega(k)\leq M}} 1 \ll x\sqrt{u}\; e^{(h_0-\theta\log x)}\exp\big(-\tfrac{1}{3}(h_0-M)^2u^{-1}\log^2 u\big). \tag{5.6}$$

From Sec.4 we have $\Psi'(x,y) \gg \frac{x}{\log x}\exp\,(h_0 - \theta\log x)$, and (a) follows.

Proof (b). Let $K = \exp(\frac{1}{4}\frac{B\sqrt{u}}{\log u})$. The quantity on the left of (b) is

$$\sum_{n\geq h_0+B\sqrt{u}/\log u} \Psi_n(x,y) = \tag{5.7}$$

$$\sum_{d\leq K} Q(d) \sum_{n\geq h_0+B\sqrt{u}/\log u} H_{n-\Omega(d)}(x/d,y) + O\big(K^{\theta-1/2}\log y\,\log x\,\Psi(x,y)\big),$$

from (4.8).

For $d \leqslant K$, $\Omega(d) \leqslant \log K/\log 2$, and for $n \geqslant h_0 + B\sqrt{u}/\log u$, $n - \Omega(d) \geqslant h_0 + \frac{1}{2}B\sqrt{u}/\log u$. Now consider

$$\sum_{d \leqslant K} Q(d) \sum_{n \geqslant h_0 + \frac{1}{2}B\sqrt{u}/\log u} H_n(x/d,y).$$

This is larger than the double sum on the right of (5.7). For each $n \geqslant h_0 + \frac{1}{2}B\sqrt{u}/\log u$, the corresponding $r = r(n)$ is less than θ. As in the proof of (a),

$$H_n(x/d,y) < d^{r-1} \exp(\tfrac{1}{2} ny^{-3/2})x \exp\big(\mathcal{G}(r)\big). \tag{5.8}$$

But $\mathcal{G}(r)$ is concave and decreasing in h for $h > h_0$. From (3.15), with M now denoting $[h_0 + \frac{1}{2}B\sqrt{u}/\log u]$,

$$\mathcal{G}(r(M)) \leqslant \mathcal{G}(\theta) - \frac{1}{12} B^2.$$

Further,

$$d\mathcal{G}/dh \leqslant -\frac{1}{3} \frac{B \log u}{\sqrt{u}}$$

for $h \geqslant M$. Thus

$$\sum_{n \geqslant M} \exp\big(\mathcal{G}(n) + \tfrac{1}{2}ny^{-3/2}\big) < \sqrt{u}\, e^{(-B^2/12)+h_0-\theta\log x}, \tag{5.9}$$

and so

$$\sum_{d \leqslant K} Q(d) \sum_{n \geqslant M} H_n(x/d,y) \ll xe^{h_0-\theta\log x} \sqrt{u}\, e^{-B^2/12} \sum_{d} Q(d)d^{\theta-1} \tag{5.10}$$

$$\ll \log^2 x\, \Psi(\chi,y)e^{-B^2/12}.$$

6. Application of the Berry-Esseen theorem.

We now confine our attention to $n \in [h_0 - u/\log^2 u,\ h_0 +$

$u/\log^2 u]$, and $1 \leq d \leq \exp(\frac{1}{2}\sqrt{u}\log y/\log u)$, and estimate $H_n(x/d,y)$. Under these circumstances, we have

Lemma 11.

$$\frac{H_n(x/d,y)}{H_n(x,y)} = d^{r-1}\left(1 + O\left(\frac{(1+\log d)\log u}{\sqrt{u}\log y}\right)\right).$$

(*Here* d *need not be an integer*).

Proof. Let $f_n(s) = \frac{\pi(y)}{G} e^{(r-1)s}\lambda(s)$, where $r = r(n)$ and $G = G(r)$. Let $X_1, X_2 \ldots$ be independent random variables with density $f_n(s)$. Then

$$E(X_i) = \frac{1}{n}\log x \tag{6.1}$$

$$\mathrm{Var}(X_i) = \sigma^2(r) \tag{6.2}$$

$$E(|X_i - \frac{1}{n}\log x|^3) = \beta(r). \tag{6.3}$$

From Sec.3, $\beta\sigma^{-3} \ll 1$, so by the Berry-Esseen theorem, for any $a < b$,

$$\mathrm{Prob}\left(\sum_1^n X_i \in [\log x - b, \log x - a]\right) = \Phi\left(\frac{b}{\sigma\sqrt{n}}\right) - \Phi\left(\frac{a}{\sigma\sqrt{n}}\right) + O\left(\frac{1}{\sqrt{n}}\right).$$

We take $b = u^{-1/4}\sigma\sqrt{n}$ and $a = -1$. Then

$$\Phi\left(\frac{b}{\sigma\sqrt{n}}\right) - \Phi\left(\frac{a}{\sigma\sqrt{n}}\right) + O\left(\frac{1}{\sqrt{n}}\right) = \frac{1}{\sqrt{2\pi}}u^{-1/4} + O(u^{-1/2}),$$

so

$$\mathrm{Prob}\left(\sum_1^n X_i \in [\log x - b, \log x + 1]\right) = \frac{1}{\sqrt{2\pi}}u^{-1/4} + O(u^{-1/2}).$$

Now from Sec.2 and 3, $f_n(s)$ is $\geq \frac{1}{3G}\left(\frac{y^r}{\log y}\right)$ throughout $(\alpha, \log y)$ with the possible exception of a set of measure $o(\log y/\log u)$. Thus there is a "rectangular block", of width $(\log y - \alpha)$ and mass asymptotically equal to 1/3, and solid except for a possible missing mass of $o(1)$. Under these circumstances, we may apply the results of Sec.6 of [4] to $f_n(s) * f_n(s)$, which has a "block" with no exceptions. We conclude that $f_n^{(n)}(s)$ can be written as $q_1(s) + q_2(s)$, such that

$$q_1(s) \geqslant 0, \; q_2(s) \geqslant 0,$$

$$\int_{-\infty}^{\infty} q_2(s) \leqslant (.98)^n,$$

and

$$\left|\frac{d}{ds} q_1(s)\right| < \frac{\log^2 u}{u\log^2 y}.$$

Thus from (6.2), we get

$$q_1(\log x) = \frac{1}{\sqrt{2\pi n}\sigma}\left(1 + O(u^{-1/4})\right) = \frac{\log u}{\sqrt{2\pi u}\,\log y}\left(1 + O\left(\frac{\log\log u}{\log u}\right)\right). \tag{6.4}$$

Since the $q_i(s)$ depend on r, and since $q_1(\log x)$ will appear several times, we introduce the notation $Q(r) = q_1(\log x)$.

Now consider

$$\int_{\log x-b}^{\log x-c} e^{(1-r)s} f_n^{(n)}(s)ds,$$

for $-1 < c < b$. From (6.3) and (6.4),

$$q_1(s) = Q(r) + O\left(\frac{(\log x-s)\log^2 u}{u\log^2 y}\right) \tag{6.5}$$

uniformly in the range of n and r under consideration and in s, $\log x - b \leqslant s \leqslant \log x + 1$. Thus

$$\int_{\log x-b}^{\log x-c} e^{(1-r)s} f_n^{(n)}(s)\,ds = Q(r)\int_{\log x-b}^{\log x-c} e^{(1-r)s}\,ds \tag{6.6}$$

$$+ O\left((.98)^n e^{(1-r)(\log x-c)}\right) + O\left(x^{1-r} e^{(r-1)c} \frac{\log^2 u}{u\log^2 y}(1 + |c|)\right),$$

$$= \frac{Q(r)x^{1-r}}{1-r} e^{(r-1)c}\left\{1 + O\left(\frac{\sqrt{u}\,\log y}{\log u} e^{(r-1)(b-c)}\right) + \right.$$

$$0\left(\frac{\sqrt{u}\log y}{\log u}(.98)^n\right) + 0\left(\frac{\log u}{\sqrt{u}\log y}(1+|c|)\right)\}.$$

If we now restrict c to $-\frac{1}{2} \leqslant c \leqslant \frac{1}{2}\frac{\sqrt{u}\log y}{\log u}$, these error terms reduce to $0\left(\frac{(1+c)\log u}{\sqrt{u}\log y}\right)$. Thus uniformly in that range of c, and in $|n - h_0| \leqslant u/\log^2 u$,

$$\int_{\log x-b}^{\log x-c} e^{(1-r)s} f_n^{(n)}(s)\, ds = \frac{Q(r)x^{1-r}e^{(r-1)c}}{1-r}\left\{1 + 0\left(\frac{(1+c)\log u}{\sqrt{u}\log y}\right)\right\}. \tag{6.7}$$

In particular, with $n = n_0$, $r = \theta + 0\left(\frac{\log^2 u}{\log x}\right)$ from (3.11) so with $c = 1$,

$$\int_{\log x-b}^{\log x-1} e^{(1-r)s} f_n^{(n)}(s)\, ds \gg \frac{\log u}{\sqrt{u}\log y} x^{1-r}.$$

But

$$H_n(x,y) = \frac{\pi(y)^n}{n!}\mathrm{Prob}\left(\sum_1^n Y_i \leqslant \log x\right) \geqslant \frac{\pi(y)^n}{n!}\mathrm{Prob}\left(\sum_1^n Y_i + Z_i \leqslant \log x-1\right)$$

$$\geqslant \frac{G(r)^n}{n!}\int_{\log x-b}^{\log x-1} e^{(1-r)s} f_n^{(n)}(s)\, ds$$

$$\gg \frac{G(r)^n e^n}{n^n}\frac{x^{1-r}\log u}{u\log y},$$

by Stirling's formula. Now $\frac{G(r)^n e^n x^{-r}}{n^n} = \exp(\mathcal{G}(n))$, and from (3.15), since $n = h_0 + O(1)$, $\mathcal{G}(n) = \mathcal{G}(h_0) + 0\left(\frac{\log^2 u}{u}\right)$. Thus $H_{n_0}(x,y) \gg \frac{x\log u}{u\log y}\exp(\mathcal{G}(h_0))$. But $\mathcal{G}(h_0) = x^{-\theta}e^{h_0}$, and $u\log y = \log x$, so

$$H_{n_0}(x,y) \gg x^{(1-\theta)}e^{h_0}\log u/\log x. \tag{6.8}$$

Since $\Psi'(x,y) > H_{n_0}(x,y)$ this proves (4.3).

We now return to a consideration of general n and c. Clearly

$$\int_0^{\log x-b} e^{(1-r)s} f_n^{(n)}(s)\, ds \leqslant x^{(1-r)}e^{(r-1)b} \tag{6.9}$$

$$\ll Q(r)x^{1-r}e^{(r-1)c}\left(\frac{(1+c)\log u}{\sqrt{u}\log y}\right),$$

so that in (6.7) the lower limit of integration could just as well be zero. Now $\sum_1^n Y_i = \sum_1^n (Y_i + Z_i) + O(uv)$, and a change in c of $O(uv)$ changes $e^{(r-1)c}$ by a factor of (easily) $1 + o(\frac{\log u}{\sqrt{u}\log y})$. Thus

$$\mathrm{Prob}\left(\sum_1^n Y_i \leqslant \log x - c\right) = \frac{G(r)^n}{\pi(y)^n}\frac{Q(r)x^{1-r}e^{(r-1)c}}{1-r}\left(1 + O\left(\frac{(1+c)\log u}{\sqrt{u}\log y}\right)\right), \tag{6.10}$$

and so

$$H_n(xe^{-c},y) = \frac{G(r)^n}{n!}x^{1-r}\frac{e^{(r-1)c}Q(r)}{1-r}\left(1 + O\left(\frac{(1+c)\log u}{\sqrt{u}\log y}\right)\right), \tag{6.11}$$

Now with c = log d, we get Lemma 11.

7. $\Psi(cx,y)$.

Now we narrow the range of c a bit, and assume

$$\exp(-\tfrac{1}{6}\sqrt{u}\log y/\log u) \leqslant c \leqslant 1,$$
$$(\log y)^{1/3} \leqslant u \leqslant \tfrac{1}{2}\sqrt{y}/\log y. \tag{7.1}$$

Given (7.1), we have uniformly in that rane of x, y, and c,

Theorem 1.

$$\Psi(cx,y) = c^{1-\theta}\Psi(x,y)\left(1 + O\left(\frac{\log^3 u(1-\log c)}{\sqrt{u}\log y}\right)\right).$$

Remark. This improves on both the range and accuracy of (11.5) of [4] (which had a slightly different definition of θ), where the error factor was $1 + O(u^{-1/7})$. It is also stronger in its range of validity than [5], which had $1 + O(u^{-1/10})$ over a wider range, extending essentially to u = y. The present approach, dependent as it is on the weighted sum $\Psi'(x,y)$, presents stubborn difficulties when $u > y^{1/2}$, as then the proportion of square-free numbers in $S(x,y)$ tends to zero. This makes it hard to recover $\Psi(x,y)$ from $\Psi'(x,y)$.

To prove Theorem 1 we first exclude atypical cases. From Lemma 10 of Sec.5, we have

$$\sum_{|n-h_0| > \frac{u}{2\log^2 u}} \Psi_n(x,y) \ll_\varepsilon \Psi(x,y)e^{-(u^{1-\varepsilon})}. \tag{7.2}$$

Now let $K = \min\{e^{u/\log^3 u}, e^{\sqrt{u}\log y/3\log u}\}$. We have

$$\Psi(cx,y) = \sum_{|n-h_0|<\frac{u}{\log^2 u}} \sum_{d\leqslant K} Q(d)H_{n-\Omega(d)}(cx/d,y) \tag{7.3}$$

$$+ O_\varepsilon(\Psi(x,y)\exp(-u^{1-\varepsilon})) + O(\sum_{d>K} Q(d)\Psi'(x/d,y)).$$

But

$$\sum_{d\leqslant K} Q(d) \sum_{|n-h_0| \leqslant u/\log^2 u} H_{n-\Omega(d)}(cx/d,y) \tag{7.4}$$

$$= \sum_{d\leqslant K} Q(d) \sum_{|n-h_0| \leqslant u\log^2 u} H_n(cx/d,y)$$

$$+ O(\sum_{d\leqslant K} Q(d) \sum_{|n-h_0| > 2u/3\log^2 u} H_{n-\Omega(d)}(cx/d,y)).$$

This last error term is $\ll_\varepsilon \Psi(x,y)\exp(-u^{1-\varepsilon})$, from Lemma 10. Thus

$$\Psi(cx,y) = \sum_{d\leqslant K} Q(d) \sum_{|n-h_0| \leqslant u/\log^2 u} H_n(cx/d,y) \tag{7.5}$$

$$+ O(\sum_{d>K} Q(d)\Psi'(x/d,y)) + O_\varepsilon(\Psi(x,y)\exp(-u^{1-\varepsilon})).$$

The error terms simplify to

$$\Psi(x,y) \times O\{\log x \log y \exp(-(\tfrac{1}{2} - \theta)u/\log^3 u) +$$

$$\log x \log y \exp(-(\tfrac{1}{2} - \theta)\sqrt{u} \log y/\log u) + \exp(-u^{1-\varepsilon}) \}$$

and finally to

$$O\left(\Psi(x,y)e^{-\sqrt{u}/3\log^2 u}\right).$$

That is,

$$\Psi(cx,y) = \sum_{d\leqslant K} Q(d) \sum_{|n-h_0| \leqslant u/\log^2 u} H_n(cx/d,y) + O(\Psi(x,y)e^{-\sqrt{u}/3\log^2 u}). \tag{7.6}$$

Recall $\Delta h = h - h_0$, or here, $n - h_0$. For $|\Delta h| \leqslant u/\log^2 u$, $\frac{dr}{dh} \approx -\frac{\log^2 u}{\log x}$ from (3.11). Thus in this range,

$$r = \theta + O(\Delta h \log^2 u/\log x). \tag{7.7}$$

From Lemma 11 then, uniformly over the range of (7.1),

$$\frac{H_n(cx/d,y)}{H_n(x,y)} = (c/d)^{1-\theta}\{1 + O(\frac{\log u(1+\log (d/c))}{\sqrt{u}\log y}) + O(\frac{\Delta h\log^2 u(1+\log (d/c))}{\log x})\}. \tag{7.8}$$

Now the error term of (7.8) that involves Δh is smallest precisely when $H_n(x,y)$ is largest. So it will pay to consider carefully how $H_n(x,y)$ varies with n. From (6.11), we have

$$H_n(x,y) = \frac{G(r)^n}{n!} x^{1-r} \frac{Q(r)}{1-r} \{ 1 + O(\frac{\log u}{\sqrt{u}\log y})\}. \tag{7.9}$$

Now $Q(r) = \frac{1}{\sqrt{2\pi n}\,\sigma(r)}(1 + O(u^{-1/4}))$, and from (3.16), this is constant in $|n - h_0| \leqslant u/\log^2 u$ to within a factor of $1 + O(u^{-1/4})$. Thus

$$H_n(x,y) = \frac{x^{1-r}}{1-\theta} \frac{G(r)^n e^n}{n^n} \frac{1+O(u^{-1/4})}{2\pi\sigma_0 h_0} \tag{7.10}$$

$$= \frac{x(1+O(u^{-1/4}))}{(1-\theta)2\pi\sigma_0 h_0} \exp(\mathcal{G}(n))$$

$$= \frac{x(1+O(u^{-1/4}))}{(1-\theta)2\pi\sigma_0 h_0} x^{-\theta} e^{h_0}\left(1 + O(\frac{(\Delta h)^2\log^4 u}{u^3})\right)e^{-\frac{1}{2}V(\Delta h)^2}$$

$$= \frac{x^{1-\theta} e^{h_0}}{(1-\theta)2\pi\sigma_0 h_0} \left(1 + O(u^{-1/4})\right) e^{-\frac{1}{2}V(\Delta h)^2}.$$

Remark. (Foreshadowing the Erdos-Kac type result of the next section. This is the corresponding result for $\Psi'(x,y)$.)

If we now sum the error due to the second "O" of (7.8) in estimating $\sum_{d\leq K} \sum_{|n-h_0| \leq u/\log^2 u} Q(d)H_n(cx/d,y)$, it comes to, say, Error_2, with

$$\mathrm{Error}_2 \ll \frac{c^{1-\theta} x^{1-\theta} \log^3 u}{\log^2 x} e^{h_0} \times \tag{7.11}$$

$$\times \sum_{|n-h_0| \leq u/\log^2 u} e^{-\frac{1}{2} V(\Delta h)^2} |\Delta h| \sum_{d\leq K} Q(d) d^{\theta-1}\left(1 + \log\left(\frac{d}{c}\right)\right).$$

To estimate the inner sum, we go to a lemma.

Lemma 12

(a) $\sum_d Q(d)d^{\theta-1}\left(1 + \log(d/c)\right) \ll (\log y)^{3/2}(1 + \log(1/c))$

(b) *If further* $u \leq y^{1/3}$, *then* $\theta \leq 2/5$ *and the sum of* (a) *is* $\ll (1 + \log(1/c))$.

Proof (a) The sum to be estimated is

$$\ll \left(1 + \log(1/c)\right) \frac{d}{dr}\Big(\prod_{p\leq y}\Big(1 + \sum_{j=2}^{\infty} q_j p^{j(r-1)}\Big)\Big)\Big|_{r=\theta}.$$

This derivative is

$$\ll \prod_{p\leq y}\left(1 + \frac{1}{2} p^{2(\theta-1)}\right) \sum_{p\leq y} p^{2(\theta-1)} \log p.$$

Since $\theta < 1/2$, this is $\ll \log^{3/2} y$. And if $u \leq y^{1/3}$, then $\theta \approx \log u/\log y \leq 2/5$ so the derivative in question is $O(1)$. Thus

$$\mathrm{Error}_2 \ll \frac{Ac^{1-\theta} x^{1-\theta} e^{h_0} \log^3 u}{\log^2 x} \sum_{|n-h_0| \leq u/\log^2 u} |\Delta h| e^{-\frac{1}{2} V(\Delta h)^2},$$

where

$$A = \begin{cases} 1 + \log(1/c) & (u \leqslant y^{1/3}) \\ (1 + \log(1/c))\log^{3/2} y & (u > y^{1/3}). \end{cases}$$

The sum above is $\ll \sqrt{u}/\log u$ from (3.15). Thus the quantity in (7.11) is

$$\ll \frac{Ac^{1-\theta}x^{1-\theta}e^{h_0}\sqrt{u}\log^2 u}{\log^2 x}.$$

This, however, is small compared to $\Psi'(x,y)$. If fact,

$$\Psi'(x,y) \gg \frac{\sqrt{u}}{\log u}\,\frac{x^{1-\theta}e^{h_0}\log u}{\log x}$$

from (6.8). So the ratio of Error_2 to $\Psi'(x,y)$ is $\ll Ac^{1-\theta}\log u/\sqrt{u}\log y$. In either case $(u \gtrless y^{1/3})$, this is

$$\ll c^{1-\theta}\log^3 u(1 + \log(1/c))/\sqrt{u}\log y.$$

Therefore this error term is within the error allowed for in Theorem 1.

We now consider the other error term in (7.8). Summed over d in (7.5), for any fixed eligible n, this comes to, say, Error_1,

$$\mathrm{Error}_1 \ll \left\{\frac{c^{1-\theta}(1 + \log(1/c))\log u}{\sqrt{u}\log y}\sum_d Q(d)d^{\theta-1}\right\} H_n(x,y). \qquad (7.12)$$

But $\sum_d Q(d)d^{\theta-1} \ll 1$ $(u \leqslant y^{1/3})$, and $\ll \sqrt{\log u}$ $\left(y^{1/3} < u \leqslant \frac{\sqrt{y}}{2\log y}\right)$ as in the proof of Lemma 12. Thus (7.12) simplifies to

$$\mathrm{Error}_1 \ll (1 + \log(1/c))\log^{3/2} u\; c^{1-\theta}/(\sqrt{u}\log y). \qquad (7.13)$$

This is smaller than the other error, which proves Theorem 1.

8. The distribution of $\Omega(k)$ in $S(x,y)$.

Theorem 2. *Uniformly in* $(\log y)^{1/3} \leqslant u \leqslant \frac{1}{2}\sqrt{y}/\log y$, *as* $y \to \infty$,

(a) $\sum_{|n-h_0|>u^{4/7}} \Psi_n(x,y) \ll \Psi(x,y)\exp(-u^{1/8})$;

(b) *For* $|n - h_0| \leqslant u^{4/7}$,

$$\Psi_n(x,y) = e^{-\frac{1}{2}V(n-h_0)^2}\Psi_{n_0}(x,y)\left(1 + O(u^{-1/4})\right).$$

Proof. Part (a) is simple. In Lemma 10 put $B = u^{1/14}\log u$. For part (b), we put $K = \exp(u^{5/12})$, and have

$$\Psi_n(x,y) = \sum_{d\leqslant K} Q(d)H_{n-\Omega(d)}(x/d,y) + O\left(\exp(-u^{2/5})\right)\Psi'(x,y) \quad (8.1)$$

from (4.8).

Let $\overline{\Psi}_n(x,y) = \sum_{d\leqslant K} Q(d)H_{n-\Omega(d)}(x/d,y)$. From (7.10),

$$\Psi'(x,y) \ll \frac{\sqrt{u}}{\log u} H_{n_0}(x,y). \quad (8.2)$$

On the other hand, $\overline{\Psi}_{n_0}(x,y) > H_{n_0}(x,y)$. Thus

$$\overline{\Psi}_{n_0}(x,y) \gg \frac{\log u}{\sqrt{u}} \Psi'(x,y). \quad (8.3)$$

Thus to prove (b) of Theorem 2, in view of (8.1) and (8.3) we need only show that for $|n - h_0| \leqslant u^{4/7}$,

$$\overline{\Psi}_n(x,y) = e^{-\frac{1}{2}V(n-h_0)^2}\overline{\Psi}_{n_0}(x,y)\left(1 + O(u^{-1/4})\right). \quad (8.4)$$

Now consider the component terms of $\overline{\Psi}_n(x,y)$. For integer d, $1 < d \leqslant K$, from Lemma 11 we have

$$H_{n-\Omega(d)}(x/d,y) = d^{r(n-\Omega(d))-1}H_{n-\Omega(d)}(x,y). \quad (8.5)$$

From (3.11),

$$r(n - \Omega(d)) = r(n) + O(\log^2 u \log d/\log x).$$

For $d \leqslant K$, $\log d \leqslant \sqrt{\log x}/\log u$, so

$$d^{r(n-\Omega(d))-1} = d^{r(n)-1}\left(1 + O\left(\frac{\log^2 u \log^2 d}{\log x}\right)\right).$$

Thus

$$H_{n-\Omega(d)}(x/d,y) = d^{r(n)-1} H_{n-\Omega(d)}(x,y) \times$$

$$\times\left(1 + O\left(\frac{\log u \log d}{\sqrt{u}\log y}\right) + O\left(\frac{\log^2 u \log^2 d}{\log x}\right)\right).$$

Now we estimate $H_{n-\Omega(d)}(x,y)/H_n(x,y)$. We have $\Omega(d) \leqslant \sqrt{\log x}/\log u$ and $\leqslant 2u^{5/12}$, and $r(n-\Omega(d)) = r(n) + O(\log^2 u \log d /\log x)$. Now from (6.11), with $r' = r(n-\Omega(d))$ and $r = r(n)$ for the moment,

$$H_n(x,y) = \frac{x^{1-r}}{\sqrt{2\pi n}} \frac{G(r)^n e^n}{n^n} \frac{Q(r)}{1-r} \left(1 + O\left(\frac{\log u}{\sqrt{u}\log y}\right)\right) \tag{8.6}$$

while

$$H_{n-\Omega(d)}(x,y) = \frac{x^{1-r'}}{\sqrt{2\pi(n-\Omega(d))}} \frac{G(r')^{n-\Omega(d)} e^{n-\Omega(d)}}{(n-\Omega(d))^{n-\Omega(d)}} \frac{Q(r')}{1-r'} \left(1+O\left(\frac{\log u}{\sqrt{u}\log y}\right)\right).$$

Thus

$$\frac{H_{n-\Omega(d)}(x,y)}{H_n(x,y)} = \frac{\sqrt{n}}{\sqrt{n-\Omega(d)}} \frac{1-r}{1-r'} \frac{Q(r')}{Q(r)} \left(1 + O\left(\frac{\log u}{\sqrt{u}\log y}\right)\right) \tag{8.7}$$

$$\times \exp\{G(n - \Omega(d)) - G(n)\}.$$

The product of all but the last factor here is $1 + O(u^{-1/4})$, from (6.4) and (3.11). As for $\exp\{G(n - \Omega(d)) - G(n)\}$,

$$|dG/dh| \ll u^{-3/7}\log^2 u \text{ for } n - \Omega(d) \leqslant h \leqslant n,$$

since $|\Delta h| \leqslant u^{4/7}$ and $|dG/dh| \ll |\Delta h| \log^2 u/u$ from (3.15). Thus

$$|G(n - \Omega(d)) - G(n)| \ll \Omega(d)u^{-3/7}\log^2 u.$$

Together with (8.7), this gives for $d \leqslant K$ that

$$\frac{H_{n-\Omega(d)}(x,y)}{H_n(x,y)} = 1 + O(u^{-1/4}) + O(u^{-3/7}\log^2 u \log d). \tag{8.8}$$

We now show

$$\overline{\Psi}_n(x,y) = \left(1 + O(u^{-1/4})\right) \sum_{d \leq K} Q(d) d^{r(n)-1} H_n(x,y). \tag{8.9}$$

Proof. We have

$$H_{n-\Omega(d)}(x/d,y) = H_n(x,y) d^{r(n)-1}\{1 + O(u^{-1/4}) + O(\frac{\log u \log d}{\sqrt{u} \log y}) +$$

$$O(\frac{\log^2 u \log^2 d}{\log x}) + O(\frac{\log^2 u \log d}{u^{3/7}})\}. \tag{8.10}$$

If we sum the errors in (8.10) over $d \leq K$, then, we get, aside from the acceptable error due to the $O(u^{-1/4})$,

$$\text{Error} \ll \sum_{d \leq K} Q(d) d^{r(n)-1} \{\frac{\log^2 u \log^2 d}{\log x} + \frac{\log^2 u \log d}{u^{3/7}}\} H_n(x,y). \tag{8.11}$$

As in Lemma 12,

$$\sum_{d=1}^{\infty} Q(d) d^{r(n)-1} \log^i d \ll \log^i u \sum_{d=1}^{\infty} Q(d) d^{r(n)-1} \tag{8.12}$$

for $i = 1$ or 2. Thus the error bounded in (8.11) is

$$\ll \left(\frac{\log^4 u}{\log x} + \frac{\log^3 u}{u^{3/7}}\right) \overline{\Psi}_n(x,y),$$

which gives (8.9). In view of (7.10), it remains only to show that

$$\sum_d Q(d) d^{r(n)-1} = \left(1 + O(u^{-1/4})\right) \sum_d Q(d) d^{\theta-1}.$$

(The sums will be more nearly equal if truncated so we are just taking the worst case).

More precisely, a simple induction argument shows that

$$\sum_{d \leq K} Q(d) d^{r-1} / \sum_{d \leq K} Q(d) d^{\theta-1}$$

is monotone in K, for fixed r and θ.

Now

$$\sum_{d=1}^{\infty} Q(d) d^{r-1} = \prod_{p \leq y} \left(1 + \sum_{j=2}^{\infty} q_j p^{(r-1)j} \right) \tag{8.13}$$

so

$$\left\{ \sum_{d=1}^{\infty} Q(d)d^{r(n)-1} / \sum_{d=1}^{\infty} Q(d)d^{\theta-1} \right\} \tag{8.14}$$

$$= \prod_{p \leqslant y} \left(1 + O\left\{ \sum_{j=2}^{\infty} q_j p^{(\theta-1)j} (p^{(r-\theta)j} - 1 \right\} \right)$$

$$= \prod_{p \leqslant y} \left(1 + O\left(\frac{u^{-3/7} \log^2 u \log p \quad p^{2(\theta-1)}}{\log y} \right) \right)$$

Now we already observed that $\sum_{p \leqslant y} p^{2(\theta-1)} \log p \ll \log y$, so this equals

$$1 + O(u^{-3/7} \log^2 u) = 1 + O(u^{-1/4}).$$

To summarize, there is a $n_0 \approx \sum_{p \leqslant y} p^{\theta-1}$, with θ determined essentially by the condition $\sum_{p \leqslant y} p^{\theta-1} \log p = \log x$. To a looser approximation, $n_0 = u(1 + \frac{1}{\log u})$.

In $S(x,y)$, the distributionn of $\Omega(k)$ is roughly normal, with mean n_0 and standard deviation $\approx 1/\sqrt{V}$, where V is defined by (3.15), so that the standard deviation is loosely $\sqrt{u}/\log u$. Out to a distance of at least $u^{4/7}$ from n_0, that is, $\approx u^{1/14} \log u$ standard deviations, the number of k in $S(x,y)$ with $\Omega(k) = n$ is, to within an error factor of $1 + O(u^{-1/4})$, given by

$$e^{-\frac{1}{2}V(n-h_0)^2} \Psi_{n_0}(x,y).$$

References.

1. K. Alladi, The Turan-Kubilius inequality for integers without large prime factors, *J. Fur die reine u. angew. Math.* **335** (1982) 180-196.

2. ________________, An Erdös-Kac theorem for integers without large prime factors, *Acta Arith.* (to appear).

3. P. D. T. A. Elliot, Probabilistic Number Theory I, Grundlehren der mathematischen Wissenchaften **239**, Springer Verlag, NY 1979 (p. 74).

4. D. Hensley, A property of the counting function of integers with no large prime factors, *J. of Number Th.* **22** (1986), 46-74.

5. A. Hildebrand, On the local behavior of $\Psi(x,y)$, *Trans. Am. Math. Soc.* (1986), to appear.

6. K. Prachar, Primzahlverteilung, Grundlehren der mathematischen Wissenschaften **41**, Berlin 1957.

7. A. Selberg, On the normal density of primes in small intervals and the difference between consecutive primes. *Arch. Math. Naturvid* **47**, No. 6 (1943) 87-105.

D. Hensley
Texas A&M University,
College Station,
Texas 77843, U.S.A.

ON THE SIZE OF $\sum_{n<x} d(n)e(nx)$

Takeshi Kano

1. In his famous Habilitationsschrift of 1854 on trigonometric series and integration theory, Riemann gave the following interesting example which shows his high ingenuity of analysis and arithmetic as well.

Let us define first

$$D(x) = \begin{cases} x - [x] - 1/2 & \ldots\ x \notin \mathbf{Z} \\ 0 & \ldots\ x \in \mathbf{Z} \end{cases}$$

and

$$c_n = \sum_{k|n} (-1)^k.$$

Now we consider the two series

$$\sum_{n=1}^{\infty} D(nx + 1/2)/n, \tag{1}$$

and

$$-\frac{1}{\pi} \sum_{n=1}^{\infty} \frac{c_n}{n} \sin(2\pi nx).$$

Then Riemann states that the function which is defined by (1) for all rational values of x can be expressed by the trigonometric series (2), and it is unbounded in every fixed interval, hence it follows that it is by no means integrable in his sense. This was finally established by Chowla and Walfisz [3], and later Wintner [12] made additional remarks. We combine their results in the following

Theorem 1. *Both of (1) and (2) converge to the same value for almost all x including all algebraic numbers, while they diverge on a dense set of transcendental numbers. The function thus defined by (1) and (2) belongs to L^p for any $p > 0$, but it is discontinuous*

almost everywhere.

Here we mention that such exceptional set of transcendental numbers x is defined by certain relations between convergents of the continued fraction expansion of x.

Now one sees that (1) and (2) are linked with

$$\sum_{n=1}^{\infty} D(nx)/n \tag{3}$$

and

$$-\frac{1}{\pi}\sum_{n=1}^{\infty} \frac{d(n)}{n} \sin(2\pi nx), \tag{4}$$

respectively. Formally, (1) = (2) if and only if (3) = (4), and we have the same assertion for (3) = (4) as Theorem 1.

Also it is known [2] that the complex series

$$\sum_{n=1}^{\infty} \frac{d(n)}{n} e(nx) \tag{5}$$

converges for all algebraic irrational values of x, while it diverges on a dense set of transcendental numbers.

Next we shall show

Theorem 2. *The series*

$$\sum_{n=1}^{\infty} \frac{d(n)}{n^{1/2+\varepsilon}} e(nx), \qquad \varepsilon > 0, \tag{6}$$

converges for almost all x *including all algebraic irrational numbers, while it diverges on a dense set of transcendental numbers.*

Proof. The last statement follows (trivially) from the corresponding fact in (5). The second part is obvious from

$$\sum_{n \leq N} d(n)\, e(nx) = o(N^{1/2+\varepsilon}), \qquad \varepsilon > 0, \tag{7}$$

which holds for all algebraic irrational numbers. This can be achieved if we employ Roth's theorem instead of Liouvilles' in the proof of Hilfssatz 32 of Walfisz [10].

The first assertion can be proved trivially if we appeal to the deep L^2-theorem of L. Carleson [cf. 1] because

$$\sum_{n=1}^{\infty} \left(\frac{d(n)}{n^{1/2+\varepsilon}} \right)^2 < \infty ,$$

which shows that (7) holds for almost all x. It is still possible to deduce the first assertion from the following estimate due to Erdös [5]:

$$\sum_{n \leqslant N} d(n)\, e(nx) = O(\sqrt{N} \log N), \text{ for almost all } x. \tag{8}$$

We remark, at first, that Theorem 2 seems sharp in the sense that it will likely be impossible to make $\varepsilon = 0$ in (6). As a matter of fact, Walfisz [11] made a conjecture that

$$\sum_{n \leqslant N} d(n)\, e(nx) = \Omega(\sqrt{N}), \tag{9}$$

would hold for all irrational values of x. Obviously (9) implies that

$$\sum_{n=2}^{\infty} \frac{d(n)}{\sqrt{n}}\, e(nx)$$

diverges for all irrational x.

Next we shall show that (6) is not summable by Abel´s method. In fact we can prove

Theorem 3. *The series*

$$\sum_{n=2}^{\infty} \frac{d(n)}{n \log n}\, e(nx) \tag{10}$$

is not summable for any x *, by Abel's method, on a dense set of transcendental numbers.*

For the proof we apply the following known Tauberian theorem of mean type.

Theorem 4. *If the series*

$$\sum_{n=1}^{\infty} c_n \tag{11}$$

is summable to s *by Abel's method and satisfies the condition*

$$\sum_{n \leq N} nc_n = o(N),$$

then (11) *is necessarily convergent to* s.

Proof of Theorem 3. Chowla [2: Theorem 5] proved that

$$\sum_{n \leq N} d(n)\, e(nx) = o(N \log N)$$

holds for all irrational x, which implies

$$\sum_{n=2}^{N} n \frac{d(n)}{n \log n} e(nx) = o(N).$$

Thus if (10) be Abel summable, then Theorem 4 shows that (10) is necessarily convergent. But this is not always the case since Chowla [2: Theorem 7] proved that

$$\sum_{n=2}^{\infty} \frac{d(n)}{n \log n} \cos(2n\pi x)$$

diverges on a dense set of transcendental x.

In view of this theorem and the following lemma, it is clear that (6) is also non-summable by Abel's method on a dense set of transcendental x.

Lemma. *If* (11) *is Abel summable, then for any monotonically decreasing sequence* d_n *of real numbers, the series*

$$\sum_{n=1}^{\infty} c_n d_n$$

is also Abel summable.

Proof. Apply partial summaiton to $\sum_{n=1}^{N} d_n c_n x^n$.

Now we shall return to Theorem 2. Walfisz [11] showed that for almost all x,

$$\sum_{n \leq N} d(n)\, e(nx) = \Omega(\sqrt{N \log N}\, (\log\log N)^{3/2}), \tag{12}$$

which implies the following

Theorem 5. *The series*

$$\sum_{n=3}^{\infty} \frac{d(n)}{\sqrt{n \log n}\,(\log\log n)^{3/2}}\, e(nx)$$

diverges almost everywhere.

Thus, in view of Theorems 2 and 5, we may naturally ask the following question:

Does $\sum_{n=2}^{\infty} \frac{d(n)}{\sqrt{n}\log n} e(nx)$ converge almost everywhere?

If the answer is "Yes", then we replace the O in (8) by o, and if the answer is "No", then we improve (12) up to

$$\sum_{n\leqslant x} d(n)\, e(nx) = \Omega(\sqrt{N} \log N) \tag{13}$$

for almost all x, which shows that (8) is a correct estimate.
A. Oppenhiem [8] pointed out that by the method of Hardy and Littlewood he could show for all irrational x

$$\sum_{n\leqslant N} r(n)\, e(nx) = \Omega(\sqrt{N}),$$

where as usual r(n) stands for the number of representations of n as the sum of two integral squares. Also we remark that Erdös [5] observes that for almost all x,

$$\sum_{n\leqslant N} r(n)\, e(nx) = O(\sqrt{N} \log N).$$

2. In this section we shall consider a certain generalization of the equation (3) = (4). If we put

$$A_n = \sum_{k|n} a_k,$$

the we have the formal identity

$$\sum_{n=1} \frac{a_n}{n} D(nx) = -\frac{1}{\pi} \sum_{n=1}^{\infty} \frac{A_n}{n} \sin(2n\pi x), \tag{14}$$

which is shown to be true for all real x, by Davenport [4], for special a_n such that $a_n = \mu(n)$, $\lambda(n)$ (Liouville), $\Lambda(n)$ (von Mangoldt). Actually he proved that for all irrational x

$$\sum_{n=1}^{\infty} \frac{\mu(n)}{n} D(nx) = -\frac{1}{\pi} \sin(2\pi x), \tag{15}$$

$$\sum_{n=1}^{\infty} \frac{\lambda(n)}{n} D(nx) = -\frac{1}{\pi} \sum_{n=1}^{\infty} \frac{\sin(2\pi n^2 x)}{n^2}, \tag{16}$$

$$\sum_{n=1}^{\infty} \frac{\Lambda(n)}{n} D(nx) = -\frac{1}{\pi} \sum_{n=1}^{\infty} \frac{\log n}{n} \sin(2\pi nx). \tag{17}$$

His method of proof depends on the deep estimate such as

$$\sum_{n\leq N} \mu(n)\, e(nx) = O\big(N(\log N)^{-K}\big), \qquad K > 1,$$

by virtue of Vinogradov's method. Segal [9] reinvestigated the identity (14) through a different approach by using complex analysis. He obtained

Theorem 6. *If the Dirichlet series* $\sum_{n=1}^{\infty} a_n n^{-s}$ *converges absolutely and uniformly for* Re $s \geq 1 + \varepsilon$ $(\varepsilon > 0)$, *then* (14) *holds in the sense that for given* x *either both sides converge to the same value, or both diverge.*

However, unfortunately, this theorem dos not tell us for what values of x do both sides converge or diverge. In spite of this fact, we can somewhat simplify the proof of Theorem 1 by virtue of it.

It will be worth observing that the series on the r.h.s. of (16) is actually the one that Riemann is reputed to have given in his lecture as an example of "almost" everywhere non-differentiable continuous functions. Later Hardy [7] proved that it is non-differentiable for all irrational values of x. On the one hand Gerver found that it is in fact *differentiable* at only particular rational points [6]. Now we shall show

Theorem 6. *The function defined by (17) is discontinuous only at integral points, and can be differentiated at non-integral points.*

Proof. This is immediate from the following closed expression for the r.h.s. of (17) which is valid for $0 < x < 1$:

$$-\frac{1}{\pi}\sum_{n=1}^{\infty}\frac{\log n}{n}\sin(2\pi nx) = -\left\{\log\Gamma(x) + \frac{1}{2}\log(\sin\pi x) + (\gamma + \log 2\pi)x\right\} + \left(\log\sqrt{2}\,\pi + \gamma/2\right), \qquad (18)$$

where γ is Euler's constant. (18) is a consequence of the Fourier series expansion of $\log\Gamma(x)$, which was obtained by Kummer.

References.

[1] L. Carleson, On convergence and growth of partial sums of Fourier series, *Acta Math.* **116** (1966), 135-157.

[2] S. Chowla, Some problems of diophantine approximation (I), *Math. Z.* **33** (1931), 544-563.

[3] S. Chowla and A. Walfisz, Ueber eine Riemannsche Identität, *Acta Arith.* **1** (1936), 87-112.

[4] H. Davenport, On some infinite series involving arithmetical functions, *Quart. J. Math.*, (2), **8** (1937), 8-13.

[5] P. Erdös, *J. Indian Math. Soc.*, **12** (1948), 67-74.

[6] J. Gerver, The differentiability of the Riemann function at certain rational multiples of π, *Amer. J. Math.*, **92** (1970), 33-55.

[7] G. H. Hardy, Weierstrass's non-differentiable function, *Trans. Amer. Math. Soc.*, **17** (1916), 301-325.

[8] A. Oppenheim, The approximate functional equation for the multiple theta-function and the trignometric sums associated therewith, *Proc. London Math. Soc.*, **28** (1928), 476-483.

[9] S.L. Segal, On an identity between infinite series of arithmetic functions, *Acta Arith.*, **28** (1976), 345-348.

[10] A. Wafisz, Ueber einige trigonometrische Summen , *Math. Z.* **33** (1931), 564-601.

[11] A. Walfisz, Ueber einige trigonometrische Summen II, *Math. Z.* **35** (1932), 774-788.

[12] A. Wintner, On a trigonometrical series of Riemann, *Amer. J. Math.*, **59** (1937), 629-634.

T. Kano
Okayama University,
Okayama, Japan.

ANOTHER NOTE ON BAKER´S THEOREM

D. W. Masser and G. Wüstholz

1. Introduction.

Recently G. Wüstholz [5], [6] proved a theorem in transcendence which includes and greatly extends many classical results. In particular it generalizes Baker´s famous theorem [2] on linear forms in logarithms, and places it within the context of arbitrary commutative group varieties.

Now although Wüstholz´s Theorem has a rather general setting, the main innovations in his proof are primarily analytic and not related specifically to the theory of group varieties. So they may be well illustrated with particular examples. When the underlying group variety is a product of multiplicative groups, the result reduces simply to Baker´s Theorem. Thus the aim of the present article is to give a proof of Baker´s Theorem using the methods of Wüstholz, but without reference to group varieties. Our exposition follows to a large part a course of lectures given by Wüstholz himself at Ann Arbor in May 1984; as noted there, many of the technical complications of [5] and [6] disappear altogether.

We shall prove the following version of Baker´s Theorem.

Theorem. *For* $n \geqslant 2$ *let* $\beta_1,\ldots,\beta_{n-1}$ *be algebraic numbers with* $1,\beta_1,\ldots,\beta_{n-1}$ *linearly independent over the rational field* $\mathbb{Q}$, *and let* $\alpha_1,\ldots,\alpha_{n-1}$ *be non-zero algebraic numbers with logarithms* $\ell_1,\ldots,\ell_{n-1}$ *not all zero. Then the number*

$$\alpha_1^{\beta_1} \cdots \alpha_{n-1}^{\beta_{n-1}} = \exp(\beta_1\ell_1 + \cdots + \beta_{n-1}\ell_{n-1})$$

is transcendental.

As usual in transcendence, the proof proceeds by contradiction.

If $\alpha_1^{\beta_1}\cdots\alpha_{n-1}^{\beta_{n-1}}$ is algebraic, we construct from Siegel's Lemma an auxiliary function $\phi(z_1,\ldots,z_{n-1})$, analytic in $z_1,\ldots,z_{n-1}$, which has many zeroes. We then use the Schwarz Lemma to deduce that $\phi(z_1,\ldots,z_{n-1})$ has many more zeroes. Up to here Wüstholz's proof follows exactly the classical lines, so we omit the details (see for example [2]). The conclusion is as follows.

Lemma. *For any* $C \geq 1$ *the following holds for all sufficiently large integers* D. *There exists a non-zero polynomial* P *in* $\mathbf{Z}[x_1,\ldots,x_n]$, *of total degree at most* D, *such that the function*

$$\phi(z_1,\ldots,z_{n-1}) = P\left(e^{z_1},\ldots,e^{z_{n-1}},\ e^{\beta_1 z_1 + \cdots + \beta_{n-1}z_{n-1}} \right)$$

satisfies

$$(\partial/\partial z_1)^{\tau_1} \cdots (\partial/\partial z_{n-1})^{\tau_{n-1}} \phi(s\ell_1,\ldots,s\ell_{n-1}) = 0$$

for all non-negative integers $\tau_1,\ldots,\tau_{n-1}$, s *with*

$$\tau_1 + \cdots + \tau_{n-1} \leq D^{1+1/(2n-2)}, \quad s \leq CD^{1/2}.$$

The last step is to prove that $\phi(z_1,\ldots,z_{n-1})$ has too many zeroes. For example in [2] this is done by means of generalized Vandermonde determinants, and Kummer theory is used in some of the later quantitative refinements. Wüstholz proceeds by proving a zero estimate that is essentially algebraic in nature. To emphasize this we formulate it over the polynomial ring

$$R = K[x_1,\ldots,x_n],$$

where K is any algebraically closed field of zero characteristic. We identify $\mathbf{Q}$ with the prime field of K, and we write $K^\times$ for the set of non-zero elements of K. For elements $\beta_1,\ldots,\beta_{n-1}$ of K we introduce the fundamental operators

$$\Delta_i = x_i(\partial/\partial x_i) + \beta_i x_n(\partial/\partial x_n) \quad (1 \leq i \leq n-1)$$

acting on R. It is easy to verify that these are commuting derivations on R (and a better reason for this will be given shortly). We then have

Proposition. (Wüstholz). *Suppose* $1, \beta_1, \ldots, \beta_{n-1}$ *are linearly independent over* $\mathbf{Q}$. *For an integer* $D \geq 1$ *and real* $S \geq 1$, $T \geq 1$ *suppose* P *is a polynomial in* R *of total degree at most* D *and* $(\xi_1^{(s)}, \ldots, \xi_n^{(s)})$ $(0 \leq s \leq S)$ *are distinct points of* $(K^{\times})^n$ *such that*

$$\Delta_1^{\tau_1} \ldots \Delta_{n-1}^{\tau_{n-1}} P(\xi_1^{(s)}, \ldots, \xi_n^{(s)}) = 0$$

for all non-negative integers $\tau_1, \ldots, \tau_{n-1}$, s *with*

$$\tau_1 + \ldots + \tau_{n-1} \leq T, \quad s \leq S .$$

Then if

$$T^{n-1} S \geq n^{2n} D^n, \qquad T \geq n^2 D$$

the polynomial P *is identically zero.*

The rest of this article is devoted to a proof of the Proposition. We see here how it supplies the required contradiction to the Lemma.

For this we note first the basic relation

$$(\partial/\partial z_i) P(e^{z_1}, \ldots, e^{z_{n-1}}, X) = (\Delta_i P)(e^{z_1}, \ldots, e^{z_{n-1}}, X) \; (1 \leq i \leq n-1)$$

for $X = \exp(\beta_1 z_1 + \ldots + \beta_{n-1} z_{n-1})$ and any polynomial P; this is the real reson why $\Delta_1, \ldots, \Delta_{n-1}$ are commuting derivations. By iteration we obtain

$$\left(\frac{\partial}{\partial z_1}\right)^{\tau_1} \ldots \left(\frac{\partial}{\partial z_{n-1}}\right)^{\tau_{n-1}} \phi(z_1, \ldots, z_{n-1}) = \Delta_1^{\tau_1} \ldots \Delta_{n-1}^{\tau_{n-1}} P(e^{z_1}, \ldots, e^{z_{n-1}}, X)$$

Hence the polynomial P of the Lemma satisfies the vanishing conditions of the Proposition at the points

$$(\xi_1^{(s)},\ldots,\xi_n^{(s)}) = \left(e^{s\ell_1},\ldots,e^{s\ell_{n-1}},e^{s(\beta_1\ell_1+\ldots+\beta_{n-1}\ell_{n-1})}\right) \quad (0 \leqslant s \leqslant S) \qquad (1)$$

with

$$T = D^{1+1/(2n-2)}, \quad S = CD^{1/2}.$$

It therefore suffices to take

$$C = n^{2n}, \; D \geqslant n^{4(n-1)}$$

in the Lemma to obtain a contradiction. Note that the distinctness of the points (1) is an immediate consequence of the linear independence of $1,\beta_1,\ldots,\beta_{n-1}$ and the fact that $\ell_1,\ldots,\ell_{n-1}$ are not all zero.

2. Jacobians.

Let P be a prime ideal of R, and regard R as embedded in the corresponding local ring R_P. We shall be considering matrices M with entries in R_P, and we write $\mathrm{rank}_P M$ for the rank of M taken modulo P.

Let $\mathbf{D}_1,\ldots,\mathbf{D}_k$ be commuting derivations on R. For an ideal I of R we define the Jacobian $J_{\mathbf{D}}(I)$ of I with respect to the system $\mathbf{D} = (\mathbf{D}_1,\ldots,\mathbf{D}_k)$ as follows. It is the infinite matrix with k columns whose rows are indexed by elements of I; for P in I and an integer j with $1 \leqslant j \leqslant k$ the entry corresponding to P and j is $\mathbf{D}_j P$. In practice no ambiguity will arise from not specifying the order of the rows.

We consider first the system $\Delta = (\Delta_1,\ldots,\Delta_{n-1})$ defined in Section 1 for $1,\beta_1,\ldots,\beta_{n-1}$ linearly independent over $\mathbf{Q}$. We say that a prime ideal P of R is general if $x_1\ldots x_n$ is not in P; equivalently, if the variety of P in K^n contains a point in $(K^\times)^n$.

Jacobian Lemma. *Suppose* $1 \leqslant r \leqslant n$ *and* P *is a general prime ideal of* R *of rank* r. *Then*

$$\mathrm{rank}_P\, J_\Delta(P) = \min(r,n-1)\;.$$

Proof. If $r = n$ then P contains $x_1 - \xi_1, \ldots, x_n - \xi_n$ for $\xi_1, \ldots, \xi_n$ in $K^\times$. The corresponding finite submatrix **B** of $J_\Delta(P)$ has a square minor of order n-1 whose determinant is $x_1 \ldots x_{n-1}$; and since this is not in P we deduce that **B**, and hence also $J_\Delta(P)$, has rank n-1 modulo P as desired.

Henceforth we assume $1 \leqslant r < n$. Let $D = (D_1, \ldots, D_n)$ be the system formed from

$$D_i = \partial/\partial x_i \quad (1 \leqslant i \leqslant n) .$$

Then $J_D(P)$ is the usual Jacobian associated with P and it is well-known that

$$\operatorname{rank}_P J_D(P) = r . \tag{2}$$

Consider the formal expression

$$L = \beta_1 \log x_1 + \ldots + \beta_{n-1} \log x_{n-1} - \log x_n . \tag{3}$$

Since the derivatives

$$D_1 L = \beta_1/x_1, \; \ldots, \; D_{n-1} L = \beta_{n-1}/x_{n-1}, \; D_n L = -1/x_n$$

are in R_P, we can consider the matrix $J_D(L,P)$ obtained by adjoining an initial row to $J_D(P)$. An easy (but crucial) calculation now gives

$$J_D(L,P)\mathbf{B} = J_\Delta(0,P), \quad J_D(P)\mathbf{B} = J_\Delta(P), \tag{4}$$

where $J_\Delta(0,P)$ is obtained from $J_\Delta(P)$ by adding an initial row of zeroes.

Now in general we have

$$\operatorname{rank} \mathbf{C} + \operatorname{rank} \mathbf{B} - n \leqslant \operatorname{rank} \mathbf{CB} \leqslant \operatorname{rank} \mathbf{C}$$

if **B** has n rows and **C** has n columns. Applying this to (4), we find that

$$\mathrm{rank}_{P}\, J_D(L,P) - 1 \leqslant \mathrm{rank}_{P}\, J_\Delta(0,P) = \mathrm{rank}_{P}\, J_\Delta(P) \leqslant r. \qquad (5)$$

Assume the lemma is false. Then (5) implies

$$\mathrm{rank}_{P}\, J_D(L,P) \leqslant r .$$

Comparing this with (2), we conclude that there exist finitely many elements P of P and elements A, A_P of R, with A not in P, such that

$$AD_iL \equiv \sum_P A_P D_i P \pmod{P} \quad (1 \leqslant i \leqslant n). \qquad (6)$$

We now interpret the expression (3) and the relations (6) locally on the variety V of P. We can find a smooth point $\pi = (\xi_1,\ldots,\xi_n)$ on V at which none of the polynomials $x_1,\ldots,x_{n-1},A$ vanish. Then V can be parametrized near π by means of equations

$$x_i = \xi_i F_i(t_1,\ldots,t_r) \quad (1 \leqslant i \leqslant n), \qquad (7)$$

where $F_1,\ldots,F_n$ are power series in the variables $t_1,\ldots,t_r$ which converge for $t_1,\ldots,t_r$ sufficiently small. The Jacobian matrix with entries $\partial F_i/\partial t_s$ $(1 \leqslant i \leqslant n,\ 1 \leqslant s \leqslant r)$ therefore has rank r.

Since the constant terms of $F_1,\ldots,F_n$ are 1, we can define convergent power series

$$Y_i(t_1,\ldots,t_r) = \log F_i(t_1,\ldots,t_r) \quad (1 \leqslant i \leqslant n)$$

with zero constant terms. Then

$$F_i \partial Y_i/\partial t_s = \partial F_i/\partial t_s \quad (1 \leqslant i \leqslant n,\ 1 \leqslant s \leqslant r) ,$$

and we deduce easily that the Jacobian matrix with entries $\partial Y_i/\partial t_s$ $(1 \leqslant i \leqslant n,\ 1 \leqslant s \leqslant r)$ also has rank r. In particular $Y_1,\ldots,Y_n$ are not all zero, so the vector space they generate over **Q** has dimension m satisfying $1 \leqslant m \leqslant n$. Let $y_1,\ldots,y_m$ be a basis consisting of a subset of $Y_1,\ldots,Y_n$. Then the Jacobian matrix with entries $\partial y_j/\partial t_s (1 \leqslant j \leqslant m,\ 1 \leqslant s \leqslant r)$ also has rank r.

We can now apply Corollary 1 (p.253) of Ax´s well-known paper

[1] on Schanuel's Conjecture for power series. We conclude that the functions

$$y_1,\dots,y_m,\ e^{y_1},\dots,e^{y_m}$$

generate a field of transcendence degree at least m+r over K. Since $e^{y_1},\dots,e^{y_m}$ are among $F_1,\dots,F_n$ which generate a field of transcendence degree r over K, we deduce that $y_1,\dots,y_m$ are algebraically independent over K.

But now the relations (6) lead to a final contradiction as follows. Write $\beta_n = -1$ and consider the function $\lambda = \sum_{i=1}^{n} \beta_i Y_i$. Then

$$\partial\lambda/\partial t_s = \sum_{i=1}^{n} (\beta_i/F_i)(\partial F_i/\partial t_s) \quad (1 \leqslant s \leqslant r).$$

For P in $\mathcal{R}$ let $\overline{P}$ be the function of $t_1,\dots,t_r$ obtained from the substitution (7); clearly

$$\partial\overline{P}/\partial t_s = \sum_{i=1}^{n} \xi_i \overline{D_i P}(\partial F_i/\partial t_s) \quad (1 \leqslant s \leqslant r).$$

Making the substitution (7) in (6) gives

$$\beta_i \overline{A}/(\xi_i F_i) = \sum_P \overline{A}_P \overline{D_i P} \qquad (1 \leqslant i \leqslant n),$$

and on multiplying by $\xi_i \partial F_i/\partial t_s$ and summing over i we obtain

$$\overline{A}\partial\lambda/\partial t_s = \sum_P \overline{A}_P \partial\overline{P}/\partial t_s \qquad (1 \leqslant s \leqslant r).$$

Since each P is now in $\mathcal{P}$ we have $\overline{P} = 0$ identically; consequently, since

$$\overline{A}(0,\dots,0) = A(\xi_1,\dots,\xi_n) \neq 0$$

we deduce $\partial\lambda/\partial t_s = 0$ for all s. Thus λ is the constant $\lambda(0,\dots,0) = 0$.

Finally there are rationals q_{ij} such that

$$Y_i = \sum_{j=1}^{m} q_{ij} y_j \qquad (1 \leqslant i \leqslant n)$$

and therefore

$$0 = \lambda = \sum_{i=1}^{n} \sum_{j=1}^{m} \beta_i q_{ij} y_j .$$

Since $y_1,\ldots,y_m$ are algebraically independent over K, we deduce

$$\sum_{i=1}^{n} q_{ij}\beta_i = 0 \quad (1 \leqslant j \leqslant m) .$$

Then since $\beta_1,\ldots,\beta_n$ are linearly independent over $\mathbf{Q}$, we conclude that $q_{ij} = 0$ for all i,j, leading to $Y_1 = \ldots = Y_n = 0$, the desired contradiction. This completes the proof of the Jacobian Lemma.

3. Integration.

Let $\mathbf{D} = (\mathbf{D}_1,\ldots,\mathbf{D}_k)$ be a system of commuting derivations on $\mathcal{R}$, and let T be a non-negative integer. For an ideal I of $\mathcal{R}$ we define $\int I d^T \mathbf{D}$ (the notation was suggested by a remark of D.J. Lewis) as the ideal generated by the polynomials P for which all the derivatives

$$\mathbf{D}_1^{e_1} \ldots \mathbf{D}_k^{e_k} P \quad (e_1 \geqslant 0,\ldots,e_k \geqslant 0,\ e_1+\ldots+e_k \leqslant T)$$

lie in I. Clearly

$$\int I d^{T+1}\mathbf{D} = \int(\int I d^T \mathbf{D}) d\mathbf{D} = \int(\int I d\mathbf{D}) d^T \mathbf{D} \tag{8}$$

and it is easy to verify the inclusions

$$I^{T+1} \subseteq \int I d^T \mathbf{D} \subseteq I . \tag{9}$$

We shall also need the remark that with $D = (D_1,\ldots,D_n)$ as in Section 2 the equality

$$\int I d^T \mathbf{D} = \int I d^T D \tag{10}$$

for T = 1 implies the same equality for all $T \geqslant 0$. This is proved by induction on T. For suppose $T \geqslant 1$ and (10) holds with T replaced by each t with $0 \leqslant t \leqslant T$. In this case write I_t for either side of (10). Then a polynomial P lies in $\int Id^{T+1}\mathbf{D} = \int I_T d\mathbf{D}$ if and only if P, $\mathbf{D}_j P$ are in $I_T = \int I_{T-1} dD$ for all j. This in turn holds if and only if P, $D_i P$, $\mathbf{D}_j P$, $D_i \mathbf{D}_j P$ are in I_{T-1} for all i,j. Now the commutators $D_i \mathbf{D}_j - \mathbf{D}_j D_i$ are themselves derivations and therefore linear combinations of $D_1,\ldots,D_n$ with coefficients in R. Thus P lies in $\int Id^{T+1}\mathbf{D}$ if and only if P, $\mathbf{D}_j P$, $D_i P$, $\mathbf{D}_j D_i P$ are in I_{T-1} for all i,j; and on retracing steps we see that this is equivalent to P lying in

$$\int(\int I_{T-1} d\mathbf{D}) dD = \int I_T dD = \int Id^{T+1} D \ .$$

This completes the proof of the remark.

The main lemma of this section concerns the system $\Delta = (\Delta_1,\ldots,\Delta_{n-1})$ defined in Section 1 for $1,\beta_1,\ldots,\beta_{n-1}$ linearly independent over $\mathbf{Q}$.

Integration Lemma. *Let* $1 \leqslant r \leqslant n$, *and suppose* P *is a general prime ideal of* R *of rank* r. *Then* $\int Pd^T\Delta$ *is primary with radical* P, *and its length is at least* $\binom{T+\rho}{\rho}$, *where* $\rho = \min(r,n-1)$.

Proof. Assume $r \neq n$ to begin with. We start by showing that

$$\int Pd\Delta = \int PdD \ . \tag{11}$$

In one direction, since $\Delta_1,\ldots,\Delta_{n-1}$ are linear combinations of $D_1,\ldots,D_n$ with coefficients in R, it is clear that

$$\int Pd\Delta \supseteq \int PdD. \tag{12}$$

For the opposite inclusion we shall express $D_1,\ldots,D_n$ back as linear combinations of $\Delta_1,\ldots,\Delta_{n-1}$ in a restricted sense. By the Jacobian Lemma, the matrices $J_\Delta(P)$, $J_D(P)$ have equal ranks modulo P. It follows that the relation $J_D(P)\mathbf{B} = J_\Delta(P)$ of Section 2 can be inverted in the form $AJ_D(P) \equiv J_\Delta(P)\mathbf{A} \pmod{P}$, where $\mathbf{A}$ is a matrix with entries in R and A is in R but not in P. Hence there exist elements A_{ij} of R such that

$$AD_i P \equiv \sum_{j=1}^{n-1} A_{ij} \Delta_j P \pmod{\mathcal{P}} \quad (1 \leqslant i \leqslant n) \tag{13}$$

for all P in $\mathcal{P}$. It follows from this easily that

$$\int \mathcal{P} d\Delta \subseteq \int \mathcal{P} dD \ . \tag{14}$$

Now (12) and (14) together give (11). From our opening remark we deduce that in fact

$$\int \mathcal{P} d^T \Delta = \int \mathcal{P} d^T D \ .$$

By the Corollary (p.164) in a recent paper of Seibt [4], the integral $\int \mathcal{P} d^T D$ is simply the (T+1)-th symbolic power $\mathcal{P}^{(T+1)}$ of $\mathcal{P}$; that is, the unique isolated primary component of the ordinary power $\mathcal{P}^{T+1}$. And this is known to have length $\binom{T+r}{r}$. For, passing to the localization $R_{\mathcal{P}}$, the length of $\mathcal{P}^{(T+1)}$ is the dimension of $R_{\mathcal{P}} / \mathcal{P}^{(T+1)} R_{\mathcal{P}}$ as a vector space over $F = R_{\mathcal{P}} / \mathcal{P} R_{\mathcal{P}}$. But the former quotient is the same as $R_{\mathcal{P}} / \mathcal{P}^{T+1} R_{\mathcal{P}}$, which by standard results (see, e.g., [7], Theorem 25 (p.301) and the Remark (p.310)) is isomorphic to the vector space over F, of dimension $\binom{T+r}{r}$, of all polynomials in r variables of total degree at most T (these remarks are due to M. Hochster). This completes the case $r \neq n$.

Finally suppose r = n. Then $\mathcal{P}$ is maximal, and we see at once from (9) that $\mathcal{J} = \int \mathcal{P} d^T \Delta$ is primary with radical $\mathcal{P}$. We now descend to the ring $R' = K[x_1,\ldots,x_{n-1}]$, and we put $\mathcal{P}' = R' \cap \mathcal{P}$, $\mathcal{J}' = R' \cap \mathcal{J}$. Since the derivations

$$\Delta_i' = x_i D_i \quad (1 \leqslant i \leqslant n-1)$$

act like $\Delta_1,\ldots,\Delta_{n-1}$ on R', it is clear that

$$\mathcal{J}' = \int \mathcal{P}' d^T \Delta'$$

for the system $\Delta' = (\Delta_1',\ldots,\Delta_{n-1}')$. But $\Delta_1',\ldots,\Delta_{n-1}'$ are linear combinations of $D_1,\ldots,D_{n-1}$ and moreover there are converse relations of the form (13) (with, e.g. $A = x_1\ldots x_{n-1}$). It follows that we can use the preceding arguments to prove that

$$\int P' d^T \Delta' = \int P' d^T D'$$

for $D' = (D_1,\ldots D_{n-1})$ in R'. The right-hand side is just the symbolic power $P'^{(T+1)}$, whose length is $\binom{T+n-1}{n-1}$. Also there is a natural injection from $R'_{P'}/J'R'_{P'}$ to R_P/JR_P as vector spaces over

$$R'_{P'}/P'R'_{P'} = R_P/PR_P = K .$$

It follows that the length of J is at least the length of J', which is $\binom{T+n-1}{n-1}$; this completes the proof in the case $r = n$.

4. Proof of Proposition.

This is by contradiction. We suppose there exists a non-zero polynomial P of total degree $D \geqslant 1$, and distinct points $\pi_s = (\xi_1^{(s)},\ldots,\xi_n^{(s)})$ $(0 \leqslant s \leqslant S)$ of $(K^\times)^n$, such that

$$\Delta_1^{\tau_1}\ldots\Delta_{n-1}^{\tau_{n-1}} P(\xi_1^{(s)},\ldots,\xi_n^{(s)}) = 0$$

for all non-negative integers $\tau_1,\ldots,\tau_{n-1},s$ with

$$\tau_1 + \ldots + \tau_{n-1} \leqslant T, \quad s \leqslant S.$$

It suffices to assume S is an integer but that the weaker inequalities

$$T^{n-1}(S + 1) \geqslant n^{2n}D^n, \quad T \geqslant n^2 D \tag{15}$$

hold; from these we shall deduce our contradiction.

For any ideal I of R we define I^* as the contracted extension simultaneously with respect to the maximal ideals $P_0,\ldots,P_S$ corresponding to the points $\pi_0,\ldots,\pi_S$. We put

$$T' = [T/n]$$

and we let I_r be the ideal generated by the polynomials

$$\Delta_1^{\tau_1}\cdots\Delta_{n-1}^{\tau_{n-1}} P \quad (\tau_1 \geqslant 0,\ldots,\tau_{n-1} \geqslant 0,\ \tau_1+\ldots+\tau_{n-1} \leqslant (r-1)T')$$

of total degrees at most D.

We start by observing that since $nT' \leqslant T$ the generators of I_{n+1} all vanish at the points $\pi_0,\ldots,\pi_S$. Consequently all the ideals $I^*_1,\ldots,I^*_{n+1}$ are proper and non-zero.

Next, we prove that if $1 \leqslant r < n$ and I^*_r has rank $m < n$, then I^*_{r+1} has rank strictly larger than m. For this it will suffice to deduce a contradiction if I^*_{r+1} has rank m. But in this case let P be a prime component of I^*_{r+1} of rank m. Evidently P is general, and, since $I^*_r \subseteq I^*_{r+1} \subseteq P$, it follows that P is also a prime component of I^*_r; let Q be the corresponding primary component. It is clear from the definitions that $I_r \subseteq \int I_{r+1} d^{T'}\Delta$, and since $I_{r+1} \subseteq I^*_{r+1} \subseteq P$, we get

$$I_r \subseteq \int P d^{T'}\Delta. \tag{16}$$

By the Integration Lemma the right-hand side is primary with radical P; hence localizing (16) at P yields

$$Q \subseteq \int P d^{T'}\Delta.$$

Comparing lengths and using once more the Integration Lemma, we find that the length $\ell(Q)$ of Q satisfies

$$\ell(Q) \geqslant \binom{T'+m}{m} \geqslant ((T'+1)/m)^m > (T/mn)^m > (T/n^2)^m . \tag{17}$$

On the other hand, Q is an isolated primary component of rank m of the ideal I_r generated by polynomials of total degrees at most D; so by the Corollary (p.419) of [3] we have the estimate

$$\ell(Q) \leqslant D^m.$$

By (15) this contradicts (17).

So the assertion about ranks is established; in other words, the ranks of $I^*_1,\dots,I^*_{n+1}$ strictly increase until they reach n, and then remain stationary. In particular I^*_n and I^*_{n+1} must have rank n. We have already noted that I^*_{n+1} has general prime components $P_0,\dots,P_S$; hence these are all prime components of I^*_n as well; let $\mathcal{Q}_0,\dots,\mathcal{Q}_S$ be the corresponding primary components. As above we find that

$$\mathcal{Q}_s \subseteq \int P_s d^{T'} \Delta \quad (0 \leqslant s \leqslant S),$$

and now the Integration Lemma yields

$$\ell(\mathcal{Q}_s) \geqslant \binom{T'+n-1}{n-1} > (T/n^2)^{n-1} .$$

Thus

$$\sum_{s=0}^{S} \ell(\mathcal{Q}_s) > (T/n^2)^{n-1}(S+1) .$$

But once again the Corollary (p.419) of [3] gives

$$\sum_{s=0}^{S} \ell(\mathcal{Q}_s) \leqslant D^n$$

which by (15) is another contradiction. This completes the proof of the Proposition.

References.

[1] J. Ax, On Schanuel's conjectures, *Annals of Math.* **93** (1971), 252-268.

[2] A. Baker, Transcendental Number Theory, Cambridge 1975.

[3] D.W. Masser and G. Wüstholz, Fields of large transcendence

degree generated by values of elliptic functions, *Invent. Math.* **72** (1983), 407-464.

[4] P. Seibt, Differential filtrations and symbolic powers of regular primes, *Math. Z.* **166** (1979), 159-164.

[5] G. Wüstholz, Multiplicity estimates on group varieties, to appear.

[6] G. Wüstholz, The analytic subgroup theorem, to appear.

[7] O. Zariski and P. Samuel, Commutative algebra Vol. II, Springer, New York 1968.

D.W. Masser
Dept. of Mathematics,
University of Michigan,
Ann Arbor, MI 48109,U.S.A.

G. Wüstholz
Max-Planck-Institut für Mathematik,
Gottfried-Claren-Strasse 26,
5300 Bonn 3, Fed. Rep.of Germany.

SUMS OF POLYGONAL NUMBERS

Melvyn B. Nathanson

Let $m \geqslant 1$. The k-th polygonal number of order m+2 is the sum of the first k terms of the arithmetic progression 1, 1+m, 1+2m, 1+3m,... The polygonal numbers of orders 3 and 4 are the triangular numbers and squares, respectively.

In his note to Book IV, Article 29, of Diophantus's *Arithmetica*, Fermat [2] wrote, "Every number is either a triangular number or the sum of two or three triangular numbers; every number is a square or the sum of two, three, or four squares; every number is a pentagonal number or the sum of two, three, four or five pentagonal numbers; and so on *ad infinitum*".

Lagrange [4] proved that every number is the sum of four squares. Gauss [3] showed that every number is the sum of three triangular numbers, or, equivalently, that every non-negative integer $n \equiv 3 \pmod 8$ is the sum of three odd squares. Weil [8] presented proofs of these theorems that use only techniques available to Fermat.

Gauss [3] also proved that a positive integer n is the sum of three squares if and only if n is not of the form $4^a(8k + 7)$.

For $m \geqslant 5$, Cauchy [1] proved that every number is the sum of m polygonal numbers of order m, and that at most four of the polygonal summands are different from 0 or 1. Legendre [5] proved that, for $m \geqslant 1, 2, 3 \pmod 4$, every sufficiently large integer is the sum of four polygonal numbers of order m, and , for $m \equiv 0 \pmod 4$, every sufficiently large integer is the sum of five polygonal numbers of order m, at least one of which is 0 or 1.

Uspensky and Heaslet [7, p.380] stated that "Cauchy showed that other parts of the Fermat theorem can be derived in a comparatively elementary but rather long way" from the triangular number theorem. Recently, Weil [9, p.102] wrote that from the triangular number theorem "one can derive (not quite easily, but at any rate elementarily) all of Fermat's further assertions." The purpose of this paper is to give short and easy proofs of the Fermat-Cauchy theorem (Theorem 1), of Legendre's results (Theorems 2-5), and of some further refinements of these results on sums of polygonal numbers (Theorems 6-8).

Pepin [6] published tables of representations of all integers $n \leqslant 120m$ as sums of m polygonal numbers of order m, at most four of which are different from 0 or 1. (There are mistakes in these tables, but they are easily corrected.) It suffices, therefore, to prove Cauchy's theorem only for $n \geqslant 120m$.

Denote the k-th polygonal number of order m + 2 by

$$p_m(k) = \frac{m}{2}(k^2 - k) + k.$$

Then $p_m(0) = 0$, $p_m(1) = 1$, $p_m(2) = m + 2, \ldots.$

Lemma 1. *Let* L *denote the length of the interval defined by the inequalities*

$$\frac{1}{2} + \sqrt{6\left(\frac{n}{m}\right) - 3} < b \leqslant \frac{2}{3} + \sqrt{8\left(\frac{n}{m}\right) - 8}. \tag{1}$$

Then

(i) $L \geqslant 4$ *if* $n \geqslant 108m$,

(ii) $L \geqslant hm$ *if* $n \geqslant 7h^2m^3$.

Proof. A simple computation shows that

$$L = \sqrt{8\left(\frac{n}{m}\right) - 8} - \sqrt{6\left(\frac{n}{m}\right) - 3} + \frac{1}{6} \geqslant g$$

if

$$\frac{n}{m} \geqslant 7\left(g - \frac{1}{6}\right)^2 + 5 \tag{2}$$

The right side of (2) is 107.86 for g = 4. This proves (i).

Let $g \geqslant 3$. Then $7g^2 \geqslant 7\left(g - (\frac{1}{6})\right)^2 + 5$ and so $L \geqslant g$ for $n \geqslant 7g^2 m$. This yields (ii) for $g = hm$.

Lemma 2. *Let* $m \geqslant 3$ *and* $n \geqslant 2m$. *Let* a, b, r *be non-negative integers such that* $0 \leqslant r < m$ *and*

$$n = \frac{m}{2}(a - b) + b + r. \tag{3}$$

If

$$\frac{1}{2} + \sqrt{6(\frac{n}{m}) - 3} < b \leqslant \frac{2}{3} + \sqrt{8(\frac{n}{m}) - 8}$$

then

$$\text{(i)} \quad b^2 \leqslant 4a$$

$$\text{(ii)} \quad 3a < b^2 + 2b + 4$$

Proof. Equation (3) implies that

$$a = (1 - \frac{2}{m})b + 2(\frac{n - r}{m}). \tag{4}$$

Therefore,

$$b^2 - 4a = b^2 - 4(1 - \frac{2}{m})b - 8(\frac{n - r}{m}) \leqslant 0$$

if

$$0 \leqslant b \leqslant 2(1 - \frac{2}{m}) + \quad 4(1 - \frac{2}{m})^2 + 8(\frac{n-r}{m}) \ .$$

Since $m \geqslant 3$ and $0 \leqslant r/m < 1$, it follows that $b^2 \leqslant 4a$ if

$$0 \leqslant b \leqslant \frac{2}{3} + \sqrt{8(\frac{n}{m}) - 8} \ .$$

Similarly, using (4), we obtain

$$b^2 + 2b + 4 - 3a = b^2 - (1 - \frac{6}{m})b - \left(6(\frac{n-r}{m}) - 4\right) > 0$$

if

$$b > (\frac{1}{2} - \frac{3}{m}) + \sqrt{(\frac{1}{2} - \frac{3}{m})^2 + 6(\frac{n-r}{m}) - 4} \ .$$

Therefore, $3a < b^2 + 2b + 4$ if

$$b > \frac{1}{2} + \sqrt{6(\frac{n}{m}) - 3}\ .$$

Lemma 3. *Let* a *and* b *be non-negative integers. If*

$$\text{(i)}\quad b^2 \leqslant 4a$$

$$\text{(ii)}\quad 3a < b^2 + 2b + 4$$

and if for some $d \geqslant 1$ *either*

$$\text{(iii)}\quad a/d^2 \equiv b/d \equiv 1 \pmod 2$$

or

$$\text{(iv)}\quad a/d^2 \equiv 2 \pmod 4 \text{ and } b/d \equiv 0 \pmod 2,$$

then there exist non-negative integers s, t, u, v *such that*

$$a = s^2 + t^2 + u^2 + v^2$$

$$b = s + t + u + v.$$

Proof. Suppose that (iii) holds with $d = 1$. Then a and b are odd, hence $4a - b^2 \equiv 3 \pmod 8$. Since $4a - b^2 \geqslant 0$ by (i), Gauss's theorem implies that there exist odd integers $x \geqslant y \geqslant z > 0$ such that

$$4a - b^2 = x^2 + y^2 + z^2. \tag{5}$$

The integer $b + x + y + z$ is even. Choose $\pm z$ so that $b + x + y \pm z \equiv 0 \pmod 4$. Define integers s, t, u, v, as follows:

$$s = \frac{b + x + y \pm z}{4}$$

$$t = \frac{b + x}{2} - s = \frac{b + x - y \mp z}{4}$$

$$u = \frac{b + y}{2} - s = \frac{b - x + y \mp z}{4}$$

$$v = \frac{b \pm z}{2} - s = \frac{b - x - y \pm z}{4}.$$

Then

$$a = s^2 + t^2 + u^2 + v^2$$

$$b = s + t + u + v$$

$$s \geqslant t \geqslant u \geqslant v.$$

To prove that s, t, u, v are non-negative, it is enough to show that the integer $v = (b - x - y \pm z)/4 \geq 0$, or, equivalently, $(b - x - y \pm z)/4 > -1$. The worst case is $(b - x - y - z)/4 > -1$, or $x + y + z < b + 4$. The maximum value of $x + y + z$ subject to the constraint (5) is $\sqrt{12a - 3b^2}$, and so it suffices to prove that $\sqrt{12a - 3b^2} < b + 4$, or $3a < b^2 + 2b + 4$. This is precisely (ii), and so s, t, u, v are non-negative integers.

Suppose (iv) holds with $d = 1$. Then $a - (b/2)^2 \equiv 1$ or $2 \pmod 4$. It follows from (i) that $a - (b/2)^2 = (4a - b^2)/4 \geqslant 0$. By Gauss's theorem, there exist non-negative integers $X \geqslant Y \geqslant Z$ such that $a - (b/2)^2 = X^2 + Y^2 + Z^2$. Let $x = 2X$, $y = 2Y$, $z = 2Z$. Then $4a - b^2 = x^2 + y^2 + z^2$. If k is an even integer, then $k^2 \equiv 2k \pmod 8$. Since a, b, x, y and z are even, it follows that

$$0 \equiv 4a = b^2 + x^2 + y^2 + z^2 \equiv 2(b + x + y + z) \pmod 8$$

and so $b + x + y + z \equiv 0 \pmod 4$. Then $s = (b + x + y + z)/4$ is an integer. Define t, u, v as above. The proof continues as in case (iii) with $d = 1$.

Suppose that (iii) or (iv) holds with $d \geqslant 2$. Let $A = a/d^2$ and $B = b/d$. Then

$$B^2 = b^2/d^2 \leqslant 4a/d^2 = 4A$$

and

$$\begin{aligned} B^2 + 2B + 4 &= (b^2 + 2db + 4d^2)/d^2 \\ &\geqslant (b^2 + 2b + 4)/d^2 \\ &> 3a/d^2 = 3A. \end{aligned}$$

It follows that there are non-negative integers S, T, U and V such that

$$A = S^2 + T^2 + U^2 + V^2$$

$$B = S + T + U + V.$$

Let $s = dS$, $t = dT$, $u = dU$, $v = dV$. Then s, t, u, v are non-negative integers satisfying $a = s^2 + t^2 + u^2 + v^2$ and $b = s + t + u + v$. This concludes the proof.

Lemma 4. *Let* $m \geq 1$. *Then* n *is the sum of four polygonal numbers of order* $m+2$ *if and only if* $n = (m(a - b)/2) + b$, *where* $a = s^2 + t^2 + u^2 + v^2$ *and* $b = s + t + u + v$ *for non-negative integers* s, t, u, v.

Proof. This follows directly from the representation $p_m(k) = (m(k^2 - k)/2) + k$.

Theorem 1. *Let* $m \geq 3$ *and* $n \geq 108m$. *Then* n *is the sum of* $m+2$ *polygonal numbers of order* $m+2$, *of which at most four are different from* 0 *or* 1.

Proof. By Lemma 1, the interval (1) has length at least 4, and so it contains at least two consecutive odd positive integers. Therefore, the set $S = \{b + r\}$, where b is an odd positive integer in the interval and $r = 0, 1, 2, \ldots, m-2$, contains a complete set of residues modulo m. Choose $b + r$ in the set S so that $n \equiv b + r \pmod m$. Define a by equation (4) of Lemma 2. Then a and b are odd positive integers that satisfy the hypotheses of Lemma 3 with $d = 1$ in (iii). Apply Lemma 4 with $n-r$ in place of n. Then $n-r$ is a sum of four polygonal numbers of order $m+2$. Since $0 \leq r \leq m-2$, it follows that n is a sum of $r+4 \leq m+2$ polygonal numbers of order $m+2$.

Theorem 2. *Let* $m \geq 3$, m *odd, and* $n \geq 28m^3$. *Then* n *is the sum of four polygonal numbers of order* $m+2$.

Proof. By Lemma 1, the interval (1) contains at least $2m$ consecutive integers. Since m is odd, there is an odd integer b in this interval such that $n \equiv b \pmod m$. Let $r = 0$. Define a by equation (4). Then $a \equiv b \equiv 1 \pmod 2$ and the Theorem follows from Lemmas 3(iii) and 4.

Theorem 3. *Let* $m \geqslant 3$, m *even, and* $n \geqslant 7m^3$. *If* n *is odd, then* n *is the sum of four polygonal numbers of order* $m+2$. *If* n *is even, then* n *is the sum of five polygonal numbers of order* $m+2$, *at least one of which is* 1.

Proof. By Lemma 1, the interval (1) contains at least m consecutive integers. If n is odd, choose b in this interval so that $n \equiv b \pmod{m}$. Then b is odd since m is even. Let $r = 0$ and define a by equation (4).

If n is even, choose b in the interval so that $n \equiv b + 1 \pmod{m}$. Then b is odd. Let $r = 1$ and define a by equation (1).

In both cases, $a \equiv b \equiv 1 \pmod{2}$ and the Theorem follows from Lemmas 3(iii) and 4.

Theorem 4. *Let* $m \equiv 0 \pmod{4}$, $m \geqslant 4$, *and* $n \geqslant 28m^3$. *If* n *is even, then* n *is the sum of four polygonal numbers of order* $m+2$.

Proof. By Lemma 1, the interval (1) contains at least $2m$ consecutive numbers. Choose b_1 and b_2 in this interval such that $b_2 - b_1 = m$ and $n \equiv b_1 \equiv b_2 \pmod{m}$. Define $a_i = (2(n-b_i)/m) + b_i$ for $i = 1, 2$. Then a_1, a_2, b_1, b_2 are positive even integers, and

$$a_2 - a_1 = \frac{2}{m}(b_1 - b_2) + b_2 - b_1 = m - 2 \equiv 2 \pmod{4}.$$

It follows that $a_i \equiv 2 \pmod{4}$ for $i = 1$ or 2. Choose i so that $a_i \equiv 2 \pmod{4}$. Let $a = a_i$ and $b = b_i$. Let $r = 0$. Then $a \equiv 2 \pmod{4}$ and $b \equiv 0 \pmod{2}$, and the Theorem follows from Lemmas 3(iv) and 4.

Theorem 5. *Let* $m \equiv 2 \pmod{4}$, $m \geqslant 6$, *and* $n \geqslant 7m^3$. *If* $n \equiv 2 \pmod{4}$, *then* n *is the sum of four polygonal numbers of order* $m+2$.

Proof. By Lemma 1, the interval (1) contains at least m consecutive integers. Choose b in this interval such that $b \equiv n \pmod{m}$. Then $b \equiv 0 \pmod{2}$. Let $x = (n-b)/m$ and $r = 0$. Define a by equation (4). Then

$$a = b + 2x = n - (m-2)x \equiv 2 \pmod{4}.$$

The Theorem follows from Lemmas 3(iv) and 4.

Legendre's results (Theorems 2-5) show that every sufficiently large integer n is a sum of four polygonal numbers of order m+2 *unless* $m+2 \equiv n \equiv 0 \pmod 4$. The following propositions refine this exceptional case. Corollary 1 of Theorem 6 is due to Legendre [6].

Theorem 6. *Let* $m \geq 3$, $k \geq 1$, *and* $n \geq 74^k m^3$. *If* $m \equiv 2 \pmod{2^{k+2}}$ *and* $n \equiv 2^{2k} \pmod{2^{2k+1}}$, *or if* $m \equiv 2 + 2^{k+1} \pmod{2^{k+2}}$ *and* $n \equiv 0 \pmod{2^{2k+1}}$, *then* n *is a sum of four polygonal numbers of order* m+2.

Proof. Let $m = 2^{k+1}m' + 2$ and $n = 2^{2k}n'$. Theorem 6 is equivalent to the statement that n is the sum of four polygonal numbers of order m+2 if $m' \not\equiv n' \pmod 2$.

By Lemma 1, the interval (1) contains at least $2^k m$ consecutive integers. Chose b in this interval such that $n \equiv b \pmod m$ and $x = (n-b)/m \equiv 2^{k-1} \pmod{2^k}$. Let $x = 2^k x' + 2^{k-1}$. Apply Lemma 3 with $d = 2^k$ in (iii). Then

$$\frac{b}{d} = \frac{n - mx}{d}$$

$$= 2^k n' - (2^k m' + 1)(2x' + 1)$$

$$\equiv 1 \pmod 2.$$

Let $a = n-(m-2)x$. If $m' \not\equiv n' \pmod 2$, then

$$\frac{a}{d^2} = \frac{2^{2k}n' - 2^{k+1}m'(2^k x' + 2^{k-1})}{2^{2k}}$$

$$= n' - m'(2x' + 1)$$

$$\equiv n' - m' \equiv 1 \pmod 2.$$

Let $r = 0$. Then a, b, r satisfy (3) of Lemma 2, and the Theorem follows from Lemmas 3 and 4.

Corollary 6.1 *Let* $m \geq 10$ *and* $n \geq 28m^3$. *If* $m \equiv 2 \pmod 8$ *and* $n \equiv 4 \pmod 8$, *or if* $m \equiv 6 \pmod 8$ *and* $n \equiv 0 \pmod 8$, *then* n *is a* *sum of four polygonal numbers of order* $m+2$.

Corollary 6.2 *Let* $m \geqslant 18$ *and* $n \geqslant 112m^3$. *If* $m \equiv 2 \pmod{16}$ *and* $n \equiv 16 \pmod{32}$, *or if* $m \equiv 10 \pmod{16}$ *and* $n \equiv 0 \pmod{32}$, *then* n *is a sum of four polygonal numbers of order* $m+2$.

Theorem 7. *Let* $m \geqslant 3$, $k \geqslant 1$, $1 \leqslant j \leqslant k+1$, *and* $n \geqslant 74^{2k+1}m^3$. *If* $m \equiv 2 \pmod 4$ *and* $n \equiv 2^{2k+1} \pmod{2^{2k+2}}$, *or if* $m \equiv 2 + 2^j \pmod{2^{j+1}}$ *and* $n \equiv 0 \pmod{2^{2k+2}}$, *then* n *is a sum of four polygonal numbers of order* $m+2$.

Proof. Let $m = 2^j m' + 2$ and $n = 2^{2k+1}n'$. Theorem 7 is equivalent to the statement that n is a sum of four polygonal numbers of order m+2 if $m' \not\equiv n' \pmod 2$.

By Lemma 1, the interval (1) contains at least 2^{2k+1} consecutive integers. Choose b in this interval such that $n \equiv b \pmod m$ and $x = (n - b)/m \equiv 2^{2k+1-j} \pmod{2^{2k+2-j}}$. Let $x = 2^{2k+2-j}x' + 2^{2k+1-j}$. Let $a = n - (m - 2)x$. Apply Lemma 3 (iv) with $d = 2^k$. Then

$$b = n - mx = 2^{2k+1}n' - 2^{2k+1-j}(2^j m' + 2)(2x' + 1)$$

and so $b/d \equiv 0 \pmod 2$. If $m' \not\equiv n' \pmod 2$, then

$$a = n - (m - 2)x = 2^{2k+1}(n' - m'(2x' + 1))$$

and so $a/d^2 \equiv 2 \pmod 4$. The Theorem follows from Lemmas 3 and 4.

Corollary 7.1 *Let* $m \geqslant 3$ *and* $n \geqslant 448m^3$. *If* $m \equiv 2 \pmod 4$ *and* $n \equiv 8 \pmod{16}$, *or if* $m \equiv 6 \pmod 8$ *and* $n \equiv 0 \pmod{16}$, *then* n *is the sum of four polygonal numbers of order* $m+2$.

Corollary 7.2 *Let* $m \geqslant 3$ *and* $n \geqslant 7168m^3$. *If* $m \equiv 2 \pmod 4$ *and* $n \equiv 32 \pmod{64}$, *or if* $m \equiv 6, 10,$ *or* $14 \pmod{16}$ *and* $n \equiv 0 \pmod{64}$, *then* n *is a sum of four polygonal numbers of order* $m+2$.

The Fermat-Cauchy theorem that every non-negative integer is the sum of m+2 polygonal numbers of order m+2 is best possible in the sense that there exist integers (for example, 2m+3 and 5m+6) that cannot be represented as the sum of m+1 polygonal numbers of order m+2. It is natural to define F(m+2) as the smallest number f such that every sufficiently large integer is the sum of f polygonal numbers of order m + 2. Clearly F(3) = 3 and F(4) = 4.

Theorem 8. *Let* $m \geq 3$. *If* $m+2 \equiv 1, 2, \text{ or } 3 \pmod 4$, *then* $F(m+2) = 3$ *or* 4. *If* $m+2 \equiv 0 \pmod 4$, *then* $F(m+2) = 4$ *or* 5.

Proof. Let $P_m(x)$ denote the number of polygonal numbers of order m+2 that do not exceed x. Since $p_m(k) = (m(k^2 - k)/2)+k$, it follows that $P_m(x) = \sqrt{(2/m)x + O(1)}$. Let $Q_m(x)$ denote the number of integers n not exceeding x such that n can be written as the sum of two polygonal numbers of order m+2. If $m \geq 3$, then

$$Q_m(x) \leq P_m(x)^2 = (2/m)x + O(\sqrt{x}) \leq (2/3)x + O(\sqrt{x})$$

and so there are infinitely many positive integers that are not sums of two polygonal numbers of order m+2. Therefore, $F(m+2) \geq 3$.

By Theorems 2, 3, and 4, if m is odd or $m \equiv 0 \pmod 4$, then $F(m+2) \leq 4$. Therefore, F(m+2) = 3 or 4 for $m+2 \equiv 1, 2, \text{ or } 3$ (mod 4).

If n is the sum of three polygonal numbers of order m+2, then there exist nonnegative integers t, u, v such that

$$n = p_m(t) + p_m(u) + p_m(v).$$

This is equivalent to

$$8mn + 3(m - 2)^2 = (2mt - m + 2)^2 + (2mu - m + 2)^2 + (2mv - m + 2)^2.$$

Let $m+2 \equiv 0 \pmod 4$. Then $m = 4m'+2$ and

$$\begin{aligned} N &= (2m' + 1)n + 3(m')^2 \\ &= (2m't + t - m')^2 + (2m'u + u - m')^2 + (2m'v + v - m')^2. \end{aligned}$$

Since $2m' + 1$ is odd, hence relatively prime to 8, there exists an entire congruence class r(mod 8) such that

$$N = (2m' + 1)n + 3(m')^2 \equiv 7 \pmod 8$$

for $n \equiv r \pmod 8$. Then N is not a sum of three squares, and so n is not a sum of three polygonal numbers of order m+2 if $n \geqslant 0$ and $n \equiv r \pmod 8$. Therefore, $F(m+2) \geqslant 4$ if $m+2 \equiv 0 \pmod 4$.

By Theorem 3, $F(m+2) \leqslant 5$ for m even, and so $F(m+2) = 4$ or 5 if $m+2 \equiv 2 \pmod 4$. This concludes the proof.

The exact value of F(m+2) is not known for any $m \geqslant 3$.

References.

[1] A. Cauchy, Démonstration du théorème général de Fermat sur les nombres polygones, *Mém. Sc. Math. et Phys. de l'Institut de France*, (1) **14** (1813-15), 177-220 = *Oeuvres*, (2) vol.6, 320-353.

[2] P. Fermat, quoted in T. L. Heath, Diophantus of Alexandria, Dover: New York, 1964, p.188.

[3] C. F. Gauss, Disquisitiones Arithmeticae, Yale University Press: New Haven and London, 1966.

[4] J. L. Lagrange, Démonstration d'un théorème d'arithmetique, *Nouveaux Mémoires de l'Acad. royale des Sc. et Belles-L. de Berlin*, 1770, pp.123-133 = *Oeuvres*, vol.3, pp.189-201.

[5] A.- M. Legendre, Théorie des nombres, 3rd ed., vol.2, 1830, pp.331-356.

[6] T. Pepin, Démonstration du théorème de Fermat sur les nombres polygones, *Atti Accad. Pont. Nuovi Lincei* **46** (1892-3), 119-131.

[7] J. V. Uspensky and M. A. Heaslet, Elementary Number Theory, McGraw-Hill: New York and London, 1939.

[8] A. Weil, Sur les sommes de trois et quatres carrés, *L'Enseignement Mathématique* **20** (1974), 215-222.

[9] A. Weil, Number Theory, An Approach through History from Hammurabi to Legendre, Birkhauser: Boston, 1983.

Note (added November, 1985). The following two articles are related to the subject of this paper :

L. E. Dickson, All positive integers are sums of values of a quadratic function of x, *Bull. Amer. Math. Soc.* **33** (1927), 713-720.

G. Pall, Large positive integers are sums of four or five values of a quadratic function, *Amer. J. Math.* **54** (1932), 66-78.

M. B. Nathanson
Rutgers University,
Newark, New Jersey 07102

Office of the Provost and Vice President
for Academic Affairs,
Lehman College (CUNY),
Bronx, New York 10468, U.S.A.

ON THE DENSITY OF B_2-BASES

Andrew D. Pollington

A sequence A of positive integers is called a Sidon sequence or a B_2-sequence if the pairwise sums are all distinct. If, in addition every non-zero integer appears in the set of differences we call A a B_2-basis.

Let $A(n)$ denote the number of elements of A not exceeding n. Erdös, see [2], has shown that lim inf $n^{-1/2} A(n) = 0$ for every B_2-sequence A, and that there is a B_2-sequence A satisfying

$$\limsup n^{-1/2} A(n) \geqslant 1/2.$$

In 1981, Ajtai, Kolmós and Szemeredi [1] gave a random construction of a Sidon sequence for which

$$A(n) > \frac{1}{1000} (n \log n)^{1/3} \quad \text{for} \quad n > n_0.$$

It is the purpose of this note to show that the same results can be obtained for B_2-bases.

Theorem 1. *There is a B_2-basis A for which*

$$\limsup n^{-1/2} A(n) \geqslant 1/2 .$$

Proof. Following Erdös, [2], p. 90, let A_p, p a prime, denote the set of numbers

$$a_k = 2p(k + p) + (k^2)_p \qquad k = 1, 2, \dots, p - 1 \tag{1}$$

where $(k^2)_p$ is the least positive residue of k^2 mod p. Then A_p is a B_2-sequence. If $a, a' \in A_p$, $a \neq a'$, then

$$p < |a - a'| < 2p^2 - p. \tag{2}$$

Let $\mathcal{P}$ denote a sequence of primes $p_1 < p_2 < \dots$, for which

$$p_{r+1} \geq 2p_r^6. \tag{3}$$

Put

$$\mathcal{D}_1' = A_{p_1}, \quad \mathcal{D}_n' = \mathcal{D}_{n-1} \cup A_{p_n} \tag{4}$$

where

$$\mathcal{D}_n = \mathcal{D}_n' \cup \{b_n, b_n + m\},$$

m is the least positive integer which is not in $\mathcal{D}_n' - \mathcal{D}_n'$ and b_n is the least positive integer for which neither b_n or $b_n + m$ are of the form $a_i + a_j - a_k$, $a_i, a_j, a_k \in \mathcal{D}_n'$. Then $b_n + m < 2|\mathcal{D}_n'|^3$.
$|\mathcal{D}_n| = \sum_{i=1}^{n} (p_i + 1) < p_n^2$. So $b_n + m < 2p_n^6$. Clearly if $\mathcal{D}_n'$ is a B_2-sequence then so is $\mathcal{D}_n$.

Lemma. *$\mathcal{D}_n'$ is a B_2-sequence.*

Proof. We use induction an n. Since $\mathcal{D}_1' = A_{p_1}$, $\mathcal{D}_1'$ is B_2. Now suppose that $\mathcal{D}_{n-1}$ is B_2. It suffices to show that

$$a_1 - a_2 = a_3 - a_4, \quad a_i \in \mathcal{D}_n' \tag{5}$$

with $a_1, a_2 \geq a_3 > a_4$ cannot hold.

If (5) holds then $a_1 \in A_{p_n}$ and $a_4 \in \mathcal{D}_{n-1}$.
If $a_3 \in A_{p_n}$, then

$$a_1 - a_2 \leq 2p_n^2 - p_n \quad \text{by (2)}$$

and

$$a_3 - a_4 > 2p_n^2 - 2p_{n-1}^6 \geq a_1 - a_2 \quad \text{violating (5).}$$

If $a_3 \in \mathcal{D}_{n-1}'$ then

$$a_1 - a_2 > p_n > 6p_{n-1}^2 > a_3 - a_4 \quad \text{by (2) and (3)}$$

which again violates (5).

Put $A = \bigcup_{n=1}^{\infty} \mathcal{D}_n$. Then A is B_2, since $\mathcal{D}_1 \subset \mathcal{D}_2 \subset \ldots$. A is clearly a B_2-basis. For each $p_n \in P$ there are at least $p_n - 1$ elements of A less than $4p_n^2 - p_n$. Hence

$$\limsup A(n)\, n^{-1/2} \geqslant 1/2 \; .$$

Theorem 2. *There exist* B_2*-bases* A*, for which*

$$A(n) > \frac{1}{10^3} (n \log n)^{1/3} \qquad \text{for all } n > n_0 .$$

Note. The greedy algorithm gives a B_2-basis A, with $A(n) > cn^{1/3}$. Pollington and Vanden Eynden [3] have constructed a B_2-basis $a_1 < a_2 < \ldots$ with $a_k \in [c(k-1)^3, ck^3]$ where c is a fixed constant.

Theorem 2 follows immediately from a slight adaptation of the random construction of a B_2-sequence given by Ajtai, Kolmos and Szemeredi, [1]. If $x \leqslant y$ are positive integers, then the triple $(x, y, x + y)$ is called a general triangle. To obtain their B_2-sequence, Ajtai, et. al construct a sequence B_i of sets of positive integers with the following properties:

i) B_i is a subset of the interval $[2.10^i, 3.10^i]$

ii) $|B_i| = [\frac{1}{100}\, i^{1/3}\, 10^{i/3}]$

iii) B_i is B_2

iv) the set $A_i = \bigcup_{j \leqslant i} B_j$ generates less than $10^{1.26i}$ general triangles

v) for no pair $b, b' \in B_i$, $b > b'$, is the difference $b - b'$ in $A_{i-1} - A_{i-1}$.

We can use the same construction, except, infinitely often we choose to replace B_i by a pair $\{b_i, b_i + m_i\}$, where as in Theorem 1, m_i is the least positive integer not in $A_{i-1} - A_{i-1}$ and b_i is chosen so that $b_i \in [2.10^i, 3.10^i]$. If this change is made sufficiently infrequently we still have

$$A(n) > \frac{1}{10^3} (n \log n)^{1/3} \quad \text{for all } n > n_0,$$

but now $\bigcup_{i=1}^{\infty} A_i$ is a B_2-basis.

References.

[1] Ajtai, Kolmós, Szemeredi, A dense infinite Sidon sequence, *Europ. J. Combinatorics* (1981) 2, 1-11.

[2] Halberstam and Roth, Sequences. Oxford University Press, 1966.

[3] Pollington and Vanden Eynden. The integers as differences of a sequence, *Canad. Bull. Math.* Vol. 24 (4), 1981, 497-499.

A. D. Pollington
386 TMCB
Brigham Young University
Provo, Utah 84601
USA

STATISTICAL PROPERTIES OF EIGENVALUES OF THE HECKE OPERATORS

Peter Sarnak

0. Introduction.

Two basic questions concerning the Ramanujan τ-function concern the size and variation of these numbers :

(i) Ramanujan conjecture: $|\tau(p)| \leq 2p^{11/2}$ for all primes p.

(ii) "Sato-Tate" conjecture: $a_p = \frac{\tau(p)}{p^{11/2}}$ is equidistributed with respect to

$$d\mu(x) = \begin{cases} \frac{1}{2\pi}\sqrt{4-x^2}\,dx & \text{if } |x| \leqslant 2 \\ 0 & \text{otherwise} \end{cases}$$

as $p \to \infty$. We refer to the last as the semicircle distribution.

Concerning the above the following is known: (i) has been proved by Deligne [1]. However its generalization to a general GL(2) cusp form, as well as to more general groups is far from being solved. (ii) This conjecture is motivated by related questions for L-functions of elliptic curves [8]. It is conjectured to be true for $\tau(p)$ as well as for "typical" cusp forms in GL(2). It certainly does not hold for all cusp forms and we will consider this again later. Our aim here is to outline results which prove averaged versions of (i) and (ii) in general.

I have benefited immeasurably from discussions with R. Phillips and I. Piatetski-Shapiro and some of the results quoted here are from joint work with them.

1. Classical Hecke Operators.

We begin by considering the simplest example of Hecke

operators. Let $\Gamma = SL(2, \mathbf{Z})$ and $h = \{ z \mid \text{Im } z > 0\}$. Let H be the Hilbert space $L^2(\Gamma/h)$, that is of all Γ invariant functions on h which are square summable over a fundamental domain F for Γ with respect to $d\omega(z) = \frac{dxdy}{y^2}$. The operators in question are then defined by

$$\begin{cases} T_n f(z) = \frac{1}{\sqrt{n}} \sum\limits_{\substack{ad=n \\ b \bmod d}} f\left(\frac{az+b}{d}\right) \\ T_\infty f(z) = -\Delta f(z) = -y^2\left(\frac{\partial^2}{\partial x^2} + \frac{\partial^2}{\partial y^2}\right) f(z) \end{cases} \tag{1.1}$$

for $n = 1, 2, \ldots$.

It is well known that $\{T_n\}$ forms a commutative family of self-adjoint operators. Furthermore H decomposes into Hecke invariant subspaces

$$H = \{1\} \oplus E \oplus \text{Cusp}$$

where $\{1\}$ spans the constant functions, E is spanned by Eisenstein series [3] and Cusp is orthogonal to these and consists of cuspidal functions. On Cusp we have a simultaneous orthonormal basis of $\{T_n\}$ which we denote by $u_j(z)$:

$$\begin{cases} T_p u_j = \rho_j(p) u_j \\ T_\infty u_j = (\frac{1}{4} + r_j^2) u_j = \lambda_j u_j \end{cases} \tag{1.2}$$

where $\lambda_1 \leq \lambda_2 \leq \lambda_3 \ldots$. Thus we use the λ's to order the u_j's.

For these cusp forms u_j, very little is known about $\rho_j(p)$ or r_j. Very interesting computations of $\rho_1(p)$ for $p < 1000$ and r_j for small j appear in Stark [10] and Hejhal [3]. For these, the Ramanujan conjecture takes the form

$$|\rho_j(p)| \leq 2 \tag{1.3}$$

for all j and primes p.

We note that since the Ramanujan conjecture holds for the Eisenstein series $E(z, \frac{1}{2} + it)$, as one checks easily by a calculation, we can restate the Ramanujan conjecture purely in terms of the spectrum of T_p. Thus the following is equivalent to (1.3). For p a prime,

$$\left\{ |\langle T_p f,f\rangle| \le 2\langle f,f\rangle \quad \text{for all } f \in L^2(\Gamma/h) \text{ for which } \langle f,1\rangle = 0. \right. \tag{1.3'}$$

Put another way $\sigma(T_p|_{\{1\}^\perp}) \subset [-2,2]$. Here $\sigma(T)$ is the spectrum of T. On the other hand $T_p 1 = (p^{1/2} + p^{-1/2})1$ and indeed

$$n(p) := \|T_p\| = p^{1/2} + p^{-1/2} > 2. \tag{1.4}$$

It is known that

$$|\rho_j(p)| \leqslant 2\,(p^{1/5} + p^{-1/5}). \tag{1.5}$$

(This was communicated to the author in a letter from S.J. Patterson 1981).

Definition 1.6. Let X be a topological space. We say that a sequence x_j in X is μ-equidistributed where μ is a Radon measure on X, if for all $f \in C_c(X)$,

$$\lim_{N\to\infty} \frac{1}{N} \sum_{j\leqslant N} f(x_j) \to \int_X f(x)\, d\mu(x)\ . \tag{1.6}$$

The Sato-Tate conjecture for the numbers $\rho_j(p)$, states that for fixed j, $\rho_j(p)$ is μ-equidistributed, where μ is the semicircle distribution.

Our approach here is to study these questions concerning $\rho_j(p)$ in both variables j and p. Thus we consider seriously the operator $T_p|_{Cusp}$ i.e. the variation in j for fixed p. Our first result is a density result concerning the number of exceptions T_p may have to the Ramanujan conjecture. We recall Weyl's law, see Selberg [9]

$$N(K) = \#\,\{r_j \leqslant K\} \sim \frac{1}{12}\,K^2. \tag{1.7}$$

For $\alpha \geqslant 2$ (and p fixed) we set

$$N(\alpha,K) = \#\{j \mid r_j \leqslant K,\ |\rho_j(p)| \geqslant \alpha\}.$$

Theorem 1.1.

$$N(\alpha,K) \ll K^{2-\frac{\log \alpha/2}{\log p}} .$$

In particular almost all $\rho_j(p)$ *(in the sense of density in* j*) lie in* [-2,2].

Concerning the variation of the $\rho_j(p)$ in j and p, let

$$x_j = (\rho_j(2),\ \rho_j(3),\ \rho_j(5),\ldots)$$

so that $x_j \in X = \prod_p [-n(p),\ n(p)]$.

Theorem 1.2. $\{x_j\}$, $j = 1,2,\ldots$ *is* μ *equidistributed in* X *where* $\mu = \prod_p \mu_p$ *and*

$$d_{\mu_p}(x) = \begin{cases} \dfrac{(1+p)\sqrt{4-x^2}}{2\pi(n(p)^2 - x^2)} & \textit{if } |x| < 2 \\ 0 & \textit{otherwise} . \end{cases}$$

The following Corollary was first proved by Phillips and Sarnak [7] by completely different methods. In that paper approximate eigenfunctions for T_p were constructed directly.

Corollary 1.3. *Let* α_m, β_m, $m = 1,2,\ldots,k$ *be numbers satisfying* $-2 \leqslant \alpha_m < \beta_m \leqslant 2$ *and let* $p_1,p_2,\ldots,p_k$ *be* k *primes. Then*

$$\lim_{K\to\infty} \frac{1}{K^2} \#\{r_j \leqslant K \mid \rho_j(p_m) \in [\alpha_m,\ \beta_m],\ m = 1,\ldots,k\} > 0 .$$

It follows that any given finite sequence of numbers, satisfying the Ramanujan bound may be approximated by the eigenvalues of a cusp form.

In the above we study the behavior of $\rho_j(p)$ as a vector in p as $j \to \infty$. If, as expected, the Sato-Tate holds for each j, we might hope that the interchange of the two limits would agree. It is clear that

$$\lim_{p\to\infty} \mu_p = \mu \quad \text{the semicircle distribution!}$$

What this shows is that in this way of averaging the numbers $\rho_j(p)$, we do have equidistribution with respect to the semicircle. There are obvious advantages in averaging over j, since if for example we consider cusp forms for $\Gamma_0(N)$, $N > 1$, then there is a subset of the j's (the number of which whose $r_j \leq K$, is of order K) for which the Sato-Tate conjecture is false. These are cusp forms coming from the Maass-Hecke construction [4]. Of course these disappear in our averaging and indeed we still find that the generic cusp form has the semicircle behaviour. These Maass-Hecke cusp forms have their eigenvalues equidistributed with respect to μ_p above, with $p = 1$! The measures μ_p therefore interpolate between this distribution at $p = 1$, and the semicircle at $p = \infty$.

A final comment concerning the semicircle. As $p \to \infty$ the operators T_p are presumably becoming random, at least that is what we are showing. For it is known that the eigenvalues of a random Hermitian matrix, whose size tends to ∞, become distributed according to the semicircle distribution. This is due to Wigner (see [6]) and is known as the Wigner semicircle law.

We will discuss the general case in Section 4. We first turn to a general phenomenon which is at the heart of the above considerations.

2. A Weyl Law.

In this section we describe an extension of the classical Weyl theorem on eigenvalues of the Laplacian to the case where we have a family of operators commuting with the Laplacian. Let M be a compact Riemannian manifold and $\tilde{M} \stackrel{\Delta}{=} S$ its universal cover. Let G be the isometry group of S and so $\Gamma = \Pi_1(M)$ is a discrete subgroup of G. Δ will denote the Laplacian on M or S. Now suppose we are given a family of operators $T_1, T_2, \ldots$ on $L^2(M)$ for which the family $\Delta, T_1, T_2, \ldots$ is commutative. We take the T_j to be bounded, with say $\|T_k\| = n_k$. We may then simultaneously diagonalize the family:

$$\begin{cases} T_k u_j = \rho_j(k)\, u_j \\ T_\infty u_j = -\Delta\, u_j = \lambda_j u_j \end{cases} \tag{2.1}$$

where $\{u_j\}_{j=1,2,\ldots}$ is an orthonomal basis for $L^2(M)$, and are ordered by increasing λ_j. The asymptotics of λ_j is well known, this being Weyl's law

$$N(\lambda) = \#\{\lambda_j \leq \lambda\} \sim C\lambda^{n/2} \tag{2.2}$$

where C is an appropriate non-zero constant and n = dim M. Let $B_k = \{z \in \mathbf{C} \mid |z| \leq n_k\}$ and

$$X = \prod_k B_k. \tag{2.3}$$

For j = 1,2,... we obtain a point x_j in X where

$$x_j = (\rho_j(1), \rho_j(2), \rho_j(3) \ldots).$$

The question is how do these x_j's distribute themselves in X as $j \to \infty$? To obtain an answer we assume further the T_k's are "Hecke like" operators. So we assume T_k to be selfadjoint (normal would suffice) and is of the form

$$T_k f(x) = \sum_{\ell=0}^{n_k} f(S_\ell^{(k)} x) \tag{2.4}$$

where $S_\ell^{(k)} \in G$. The important assumption is that T_k: $L^2(\Gamma/S) \to L^2(\Gamma/S)$, which can be arranged with appropriate $S_\ell^{(k)}$ if the commensurator of Γ in G is non-trivial [11].

For $\nu_1, \nu_2, \ldots, \nu_r \in \mathbf{N}$ let

$$M(\nu_1, \nu_2, \ldots, \nu_r) = \text{the number of words of the type} \begin{cases} \omega_1\omega_2 \ldots\omega_r \equiv I \pmod{\Gamma} \text{ where } \omega_k \text{ is a} \\ \text{word in } S_1^{(k)}, S_2^{(k)} \ldots S_{n(k)}^{(k)} \text{ of length } \nu_k. \end{cases} \tag{2.5}$$

In this case, since we are assuming that the T_k's are self-adjoint, our space X in (2.3) is a product of intervals.

Theorem 2.1 *Let T_k be as above, then the sequence $\{x_j\}_{j=1,2,\ldots} \in X$ is μ equidistributed, where μ is the measure given by the moments*

$$\int_X t_1^{\nu_1} \ldots t_k^{\nu_k} \, d\mu(t_1, \ldots) = M(\nu_1, \ldots, \nu_k) .$$

Notice that since X is compact, one sees easily that μ exists and is unique. We now examine some simple instances of the above theorem.

Example 2.2. Suppose that the original manifold M admits a non-trivial isometry $S : M \to M$ of order k (k may be infinite). Let $T : L^2 \to L^2$ be the unitary operator given by

$$Tf(x) = f(Sx).$$

T commutes with Δ and let u_j be as above with

$$T u_j = \omega_j u_j \qquad j = 1,2,\ldots \quad .$$

Clearly $|\omega_j| = 1$. The theorem then asserts that ω_j is μ-equidistributed on the circle where

(i) μ puts mass $1/k$ at the k-th roots of 1 if $k < \infty$.

(ii) μ is $d\Theta/2\pi$ on the circle, if $k = \infty$.

Example 2.3. $M = S' = R/\mathbf{Z}$, $\Delta = \frac{d^2}{dx^2}$, $u_j(x) = e^{2\pi ijx}$. Let $\alpha_1,\ldots,\alpha_k \in \mathbf{R}$ and $T_k(x) = x + \alpha_k$. In this case $\rho_j(k) = e^{2\pi ij\alpha_k}$. The theorem thus asserts that the sequence $j(\alpha_1,\alpha_2,\ldots,\alpha_k)$, $j = 1,2,\ldots$ is μ-equidistributed in the k-torus. Clearly $M(\nu_1,\ldots,\nu_k) = 0$ if $1,\alpha_1,\alpha_2,\ldots,\alpha_k$ are linearly independent over $\mathbf{Q}$, so that in this case the sequence is equidistributed with respect to Lebesgue measure. This is the well known result of Weyl [12].

The main application of the theorem is however to the Hecke operators in symmetric spaces. In the case of $\Gamma = SL(2,\mathbf{Z})$ as in Section 1, there are added complications in the proof of the above type of theorem due to the noncompactness. We will outline the proof in that case in the next section. The proof of Theorem 2.1 in the general case combines the ideas outlined in the next section, with the standard derivation of Weyl's law via differential equation methods - e.g. small time behavior of the fundamental solution to the wave equation on $\tilde{M}$.

In the $\Gamma = SL(2,\mathbf{Z})$ case of Section 1, if we ignore the difficulties coming from the Eisenstein series (which in this case are not difficult to overcome) we can compute the number $M(\nu)$ for T_p

quite easily from the well known identity

$$T_p^n T_p = T_{p^{n+1}} + T_{p^{n-1}} .$$

We find

$$N(\nu) = \int_{-\infty}^{\infty} t^\nu \, d\mu_p(t) = \begin{cases} 0 , & \text{if } \nu \text{ is odd} \\ \sum_{j=0}^{n} \left(\binom{2n}{n-j} - \binom{2n}{n-j-1}\right) p^{-j}, & \text{if } \nu = 2n. \end{cases}$$

The inverse moment problem is easily solved giving the μ_p´s in Theorem 1.2. The fact that μ is a product of the μ_p´s follows from the multiplicative property of the Hecke operators.

3. Outline of Proofs.

We now outline proofs of the results in Section 1, details will appear elsewhere. The basic ingredient is the Selberg trace formula but it is not the full formula that is needed. Indeed such a formula cannot be used to prove Theorem 2.1. Basically what we need is the "singularity at 0" in the trace formula.

Consider the case of $\Gamma = SL(2,\mathbf{Z})$. Let $k(z,\zeta)$ be a point pair invariant [3], which we assume to have very small support. That is $k(z,\zeta) = 0$, if $d(z,\zeta) > \varepsilon$, where $d(z,\zeta)$ is the non-Euclidian distance from z to ζ. Let

$$K(z,\zeta) = \sum_{\gamma \in \Gamma} k(z,\gamma\zeta) . \tag{3.1}$$

We have the spectral expansion [3]

$$K(z,\zeta) = \sum_j h(r_j)u_j(z)\overline{u_j(\zeta)} + \frac{1}{4\pi}\int_{-\infty}^{\infty} h(t)E(z, {1/2} + it)\overline{E(\zeta, {1/2} + it)} \, dt. \tag{3.2}$$

For what follows we ignore the contribution from the Eisenstein series since in this case as was mentioned before they are known explicitly, and may be dealt with easily. It follows that

$$T_p^\nu K(z,\zeta) = \sum_j h(r_j)(\rho_j(p))^\nu u_j(z)\, \overline{u_j(\zeta)} + \dots \tag{3.3}$$

and hence

$$[T_p^{\nu}\, K(z,\zeta)]_{z=\zeta} = \sum_j h(r_j)(\rho_j(p))^{\nu} |u_j(\zeta)|^2 + \dots \quad (3.4)$$

However one can calculate $[T_p^{\nu}\, K(z,\zeta)]_{z=\zeta}$ asymptotically as $\varepsilon \to 0$:

$$[T_p^{\nu} K(z,\zeta)]_{z=\zeta} = \frac{1}{p^{\nu/2}} \sum_{\gamma \in \Gamma} k(S_{i_1}\, S_{i_2} \dots S_{i_\nu}\, z,\ \gamma\zeta)|_{z=\zeta}$$

so that unless ζ is the fixed point of some $\gamma^{-1} S_{i_1} S_{i_2} \dots S_{i_\nu}$, the above is zero for ε small enough.

On integrating with respect to ζ one finds the main contribution comes from exactly those $S_{i_1} S_{i_2} \dots S_{i_\nu} \equiv I \pmod{\Gamma}$. This, combined with (3.4) leads naturally to the asymptotics

$$\frac{1}{N(K)} \sum_{|r_j| \leq K} (\rho_j(p))^{\nu} \sim M(\nu) \, . \quad (3.5)$$

Theorems 2.1 and 1.2 follow from this type of argument. If one is more careful in the analysis in the case $\Gamma = SL(2,\mathbf{Z})$, and keeps track of all contributions above, one finds: (i) that the contribution from the continuous spectrum is controlled by the constant term of the Eisenstein series which is essentially the zeta function.
(ii) the number of terms $\gamma\, S_{i_1} S_{i_2} \dots S_{i_\nu}$ with fixed points in F is easily majorized by elementary bounds for class numbers of binary quadratic forms. This leads to the inequality:

$$K > p^k \Rightarrow \sum_{|r_j| \leq K} |\rho_j(p)|^{2k} \leq 2^k K^2 + p^{2k} 2^k \quad (3.6)$$

Theorem 1.1 is an immediate consequence.

4. General Case.

The results in this section are joint with I. Piatetski-Shapiro. The first thing to observe is that the measures μ_p are none other than the spherical Plancherel measures for $SL_2(\mathbf{Q}_p)$, see for example MacDonald [5]. He uses the variable Θ where $x = 2\cos\Theta$. One may also see that this is so by carrying out the above proof using the adelic trace formula for $GL_2(\mathbf{Q})/GL_2(\mathbf{A_Q})$ [2].

The case of a compact quotient such as that coming from a quaternion algebra and its generalizations, is particularly simple and an analogue of Theorem 1.2 may be proved in complete generality, i.e. for a reductive algebraic group defined over a number field. In this case the existence of a limiting distribution follows from Theorem 2.1 but the point is that one can avoid solving the inverse moment problem, since these limiting distributions are spherical Plancherel measures, which have been computed in complete generality - see MacDonald [5]. In the general noncompact case such as $G = SL(n,\mathbf{R})$, $\Gamma = SL(n,\mathbf{Z})$ there are technical problems coming from the continuous spectrum. We expect the same answer for the limiting distribution, but so far have not been able to verify it in general.

For $GL(n,\mathbf{Z})$ the eigenvalues of the p-th Hecke operators on u_j (cusp forms) may be parametrized by $\alpha_j^{(1)}(p), \ldots, \alpha_j^{(n)}(p)$ where $\alpha_j^{(1)} \cdots \alpha_j^{(n)} = 1$. The corresponding limiting distribution for these is the spherical Plancherel measure for $SL(n,\mathbf{Q}_p)$, and lives on the n-1 torus. As in Section 1, one takes the limit $p \to \infty$ of these measure and this turns out to be the measure

$$d\mu(\Theta_1,\ldots,\Theta_{n-1}) = C_n \prod_{k<j} |e^{i\Theta_k} - e^{i\Theta_j}|^2 \, d\Theta_1 \ldots d\Theta_{n-1} \qquad (4.1)$$

where

$$k,j = 1,2,\ldots,n \text{ and } \Theta_1 + \Theta_2 \ldots + \Theta_n = 0.$$

This gives a natural generalization of the semicircle or Sato-Tate distribution. Indeed the above results prove this conjecture in the average over the cusp forms (in the sense of Section 1). There are other theoretical ways of arriving at the measure in (4.1), we note in particular that it is the measure obtained by projecting Haar measure on $SU(n)$ to its maximal torus. If $n = 2$ then the measure (4.1) is $C_2 \sin^2\Theta \, d\Theta$ which is of course the semicircle distribution for the variable $\rho = 2 \cos \Theta$.

References.

[1]. Deligne, P., La conjecture de Weil I. *Publ. Math.* IHES, **43** (1974) 273-307.

[2]. Gelbart, S., Automorphic forms on Adele group, Anal of Math. Studies, **83**, 1975.

[3]. Hejhal, D., The Selberg trace formula for PSL(2,ℝ), Vol.2, S.L.N. 1001, 1983.

[4]. Maass, H., Über eine neue Art von Nichtanalytischen Automorphismen ..., *Math. Ann.* 121, 1949 pp. 141-183.

[5]. MacDonald, I.G., Spherical functions on groups of P-adic type, TATA Inst. Series, 1971.

[6]. Mehta, M.L., Random matrices, Academic Press, 1967.

[7]. Phillips, R. and Sarnak, P., Preprint.

[8]. Serre, J.P., Abelian ℓ-adic representations, Benjamin, 1968.

[9]. Selberg, A., Göttingen lectures, 1954.

[10]. Stark, H., Fourier coefficients of Maass wave forms, in Modular forms ed. Rankin, Ellis Horwood, 1985.

[11]. Venkov, A.B., Spectral theory of automorphic forms, *Proc. Steklov. Inst.* 1982, No. 4 (English translation).

[12]. Weyl, H., Über die Gleichverteilung von Zahlen Mod. eins, *Math. Ann.* **77** , 1914, 313-352.

P.Sarnak,
Stanford University,
Stanford, CA, 94305 U.S.A.

TRANSCENDENCE THEORY OVER NON-LOCAL FIELDS

Hans-Bernd Sieburg

1. Summary.

For any commutative ring R let Val(R) denote the set of all multiplicative real valuations. Let $\delta : \mathrm{Val}(R) \to \mathbf{R}$ denote the map given by $\phi \to \delta(\phi) := \inf \{\phi(a) : a \in R, a \neq 0\}$. Here **R** is the field of real numbers. In the first part of the present paper we show that for $\delta(\phi) > 0$ the quotient field of R "is" either an algebraic extension of the field **Q** of rational numbers, if and only if ϕ is Archimedian , or an algebraic extension of a rational function field in arbitrarily many variables, if and only if ϕ is non-Archimedian. Local fields are contained in the class of rings (R,ϕ) with $\delta(\phi) = 0$.

The second part of the paper is devoted to transcendence questions over groundfields k which are quotient fields of non-Archimedian valued rings (R,ϕ) with $\delta(\phi) > 0$. Our results include axiomatic formulations of the methods of Schneider, Gelfond and Baker. We also derive transcendence measures for certain elements of the completion of k.

2. Classification of groundfields.

Let the notation be as above. The trivial valuation 1 given by $1(a) = 1$ for $a \neq 0$ and $1(0) = 0$ has $\delta(1) = 1$. To provide less trivial examples we consider $R = \mathbf{Z}$, the ring of rational integers. For a fixed prime number p let $| \ |_p$ denote the p-adic valuation. Then $\delta(| \ |_p) = 0$. If $| \ |$ denotes the ordinary absolute value on **Z** then $\delta(| \ |) = 1$. Furthermore, let $d \in \mathbf{Z}$, $d \neq 0$ and not a square. Let $| \ |_1$, $| \ |_2$ denote the extensions of $| \ |$ to $\mathbf{Z}[\sqrt{d}]$. Then, for $i = 1, 2$, $\delta(| \ |_i) = \sqrt{d} - 1$ or 1 depending on $d > 0$ or

$d < 0$. Finally, consider $R = A[X_1,\ldots,X_m]$, where $m \geq 1$ is an integer and A denotes an arbitrary integral domain. Let $|\ |_\infty$ be the discrete non-Archimedean valuation with $|P|_\infty := e^{\deg(P)}$, where $\deg(P)$, for $P \neq 0$, is the total degree of P and $\deg(0) := -\infty$. Then $\delta(|\ |_\infty) = 1$.

Definition. Let **D** denote the class of all commutative rings R having a real multiplicative valuation ϕ such that $\delta(\phi) > 0$ and $\phi(b) > \frac{1}{\delta(\phi)}$ for at least one $b \in R$.

These rings have the following properties

Lemma 1. *Let* $(R,\phi) \in$ **D**. *Then*

(1) R *is an infinite, non-trivially valued integral domain which is not a field,*

(2) R *is not complete under* ϕ.

The proofs are simple and can therefore be omitted.

Let $\overline{\mathbf{D}}$ denote the class of all valued fields which are quotient fields of rings in **D**. The following result classifies the Archimedean and non-Archimedean members of $\overline{\mathbf{D}}$:

Proposition 1. *Let* $(k,\phi) \in \overline{\mathbf{D}}$. ϕ *is Archimedean if, and only if,* $k|k'$ *is a purely algebraic extension for every subfield* k' *of* k. ϕ *is non-Archimedean if, and only if, there exists a subfield* k' *of* k *such that the extension* $k|k'$ *is transcendental.*

Proof. Obviously it suffices to prove the second assertion only. Suppose there exists a subfield k' of k and $z \in k$ such that z is transcendental over k'. The subfield $k'(z)$ has only non-Archimedean valuations. Thus the restriction of ϕ to $k'(z)$ is non-Archimedean and therefore ϕ itself.

Conversely, let ϕ be non-Archimedean. Let (R,ϕ) denote the ring of (k,ϕ). Since ϕ is non-Archimedean, there exist proper subfields F of k such that $\phi(F^*) \subset [\delta(\phi), \frac{1}{\delta(\phi)}]$ (at least the prime field is such an F). Here $F^* := F-\{0\}$ and $[a,b]$, with a and b in R, denotes an interval. For fixed F suppose $k|F$ were purely algebraic. Then, for every α in k, there exists a positive integer n and

$0 \neq P := \sum_{i=0}^{n} a_i X^i \in F[X]$, $a_n = 1$, such that $P(\alpha) = 0$. Then

$$(\phi(\alpha))^n = \phi(a_0 + a_1\alpha + \dots + a_{n-1}\alpha^{n-1})$$

$$\leqslant \max_{0 \leqslant j \leqslant n-1} \phi(a_j)(\phi(\alpha))^j$$

$$\leqslant \frac{1}{\delta(\phi)} \max_{0 \leqslant j \leqslant n-1} (\phi(\alpha))^j .$$

This shows $\phi(\alpha) \leqslant \frac{1}{\delta(\phi)}$. Since α was arbitrary, we have the above inequality especially for all $a \in R$, a contradiction.

Remark. Proposition 1 shows that $(k,\phi) \in \overline{\mathbf{D}}$ with ϕ Archimedian, iff k is a purely algebraic extension of **Q**, whereas ϕ is non-Archimedian, iff k is an algebraic extension of a rational function field in arbitrarily many variables over a field F with $\delta(\phi) \leqslant \phi(a) \leqslant \frac{1}{\delta(\phi)}$ for all $a \in F$, $a \neq 0$.

Arguments analogous to those used in [Sie.4, Sec.2] immediately show

Lemma 2. *Let* $(k,\phi) \in \overline{\mathbf{D}}$ *and let* k^{ϕ} *denote the completion of* k *under* ϕ. *Then* $k^{\phi} \notin \overline{\mathbf{D}}$, *thus* k *cannot be complete.*

Remark. This shows that all local fields are contained in the class of all rings R with real multiplicative valuation ϕ satisfying $\delta(\phi) = 0$. $\overline{\mathbf{D}}$ contains all global fields.

3. Transcendence results.

Let $(k,\phi) \in \overline{\mathbf{D}}$ be fixed. From the view point of transcendence theory it is sufficient to consider as groundfields

$k = \mathbf{Q}$, iff ϕ is Archimedian

$k = F(S)$, iff ϕ is non-Archimedian.

Here F denotes a subfield of k with $\delta(\phi) \leqslant \phi(a) \leqslant \frac{1}{\delta(\phi)}$ for all $0 \neq a \in F$, and S is a non-empty transcendence basis of k over F (see Prop. 1, proof). The Archimedian case is classically well-known. Therefore we will consider non-Archimedian ϕ only. Hence, for the rest of this paper we can make the

GENERAL ASSUMPTIONS: $R = F[S]$, $k = F(S)$, $\phi = |\ |_\infty$.

We need some additional notations. Let $(K,\hat{\phi})$ denote an algebraically closed, complete extension of k. Let $\mathbf{R}_+$ ($\mathbf{N}$ resp.) denote the set of all positive real numbers (positive rational integers) and let $\mathbf{R}_{+,o} := R_+ \cup \{0\}$ ($\mathbf{N}_o := \mathbf{N} \cup \{0\}$). Let $\mathbf{L}^K := \{ \sum_{i \in \mathbf{Z}} a_i X^i : a_i \in K\}$ denote the K[X]-module of formal Laurent series with coefficients in K and let K[[X]] denote the integral domain of formal power series over K. For $t \in \mathbf{R}_{+,o}$ let

$$\mathbf{L}_t^K = \begin{cases} K[[X]] & , \quad \text{iff } t = 0 \\ \{f \in \mathbf{L}^K : \lim\limits_{|i| \to \infty} \phi(a_i)t^i = 0 \} , & \text{iff } t \neq 0. \end{cases}$$

and let $P_t^K := \mathbf{L}_t^K \cap K[[X]]$. Futhermore, for every $f \in \mathbf{L}^K$ and fixed $t \in \mathbf{R}_{+,o}$ define

$$\|f\|_t := \begin{cases} \phi(a_o) & , \quad \text{iff } t = 0 \\ \sup\{\phi(a_i)t^i : i \in \mathbf{Z}\} , & \text{iff } t \neq 0 . \end{cases}$$

Let A(K) (T(K) resp.) denote the set of all algebraic (transcendental) elements of K over k. For every $\alpha \in A(K)$ let $f_\alpha \in k[X]$ denote the minimal polynomial of α, $\deg(\alpha) := \deg(f_\alpha)$ denote the degree of α, and $D_\alpha := \{x \in R : \alpha x \in I(R,K)\}$ denote the denominator ideal of α, where I(R,K) is the integral closure of R in K. We have $D_\alpha = (d(\alpha))$ for $0 \neq d(\alpha) \in R$ uniquely determined up to units. The $d(\alpha)$ is called the denominator of α. Let $\alpha \in A(K)$ with $\nu := \deg(\alpha)$. Let $\alpha_1 := \alpha, \alpha_2, \ldots, \alpha_\nu$ denote the conjugates of α. Then

$\lceil\alpha\rceil := \max_{1 \leq i \leq \nu} \hat{\phi}(\alpha_i)$ denotes the house of α,

and

$$s(\alpha) := \begin{cases} 0 & , \quad \text{iff } \alpha = 0 \\ \max\{\log \lceil\alpha\rceil, \log\phi(d(\alpha))\} , & \text{iff } \alpha \neq 0 \end{cases}$$

denotes the size of α. We note that $s(\alpha)$ is invariant under changes of $d(\alpha)$.

Finally, for arbitrary field extensions $F_2|F_1$ let $\mathrm{tdeg}_{F_1}F_2$ be the transcendence degree of F_2 over F_1.

We can now state the transcendence theorems mentioned above. Their complete proofs for char(k) = 0 can be found in Sie.[3]. It is not difficult to see that, after suitable technical adjustments, they also hold for char(k) > 0. In order to illustrate our approach we will outline the proof of Theorem 1 in Section 5.

Theorem 1. (Schneider's method). *Let* $k'|k$ *be a finite, separable extension. Let* $\ell \in \mathbf{N}$, $r \in \hat{\phi}(K) \cap \mathbf{R}_+$ *and* $f_1, \ldots, f_\ell \in P_r^K$ *be algebraically independent over* K. *Let* $r' \in \mathbf{R}_+$ *with* $r' < r$. *Let* $(\alpha_n)_{n \in \mathbf{N}}$ *denote a sequence of elements in*

$$U'_{r'}(0) := \{z \in K : \hat{\phi}(z) \leqslant r'\}$$

such that $f_i(\alpha_n) \in k'$ *for all* $1 \leqslant i \leqslant \ell$ *and* $n \in \mathbf{N}$. *Then*

$$\sum_{i=1}^{\ell} \limsup_{T \quad \mathbf{N}} \frac{1}{\log T} \log\Big(\max_{1 \leqslant j \leqslant T} s(f_i(\alpha_j))\Big) \geqslant \ell - 1.$$

For technical reasons we will state the applications only for char(k) = 0. Then the exponential and logarithmn functions are defined via the usual power series expansions for $z \in K$ such that $\hat{\phi}(z) < 1$ and $\hat{\phi}(z-1) < 1$ respectively.

Corollary 1. (Theorem of Gelfond-Schneider). *Let* char(k) = 0. *Let* $\alpha, \beta \in K$ *with* $\beta \notin \mathbf{Q}$, $0 < \hat{\phi}(\alpha-1) < 1$ *and* $\hat{\phi}(\beta)\hat{\phi}(\alpha-1) < 1$. *Let* $\alpha^\beta := \exp(\beta\log(\alpha))$. *Then* $\mathrm{tdeg}_k\, k(\alpha,\beta,\alpha^\beta) \geqslant 1$.

Proof. Let $k' := k(\alpha,\beta,\alpha^\beta)$, $r' := \max\{1, \hat{\phi}(\beta)\}$ and $\tilde{r} := \big(\hat{\phi}(\log\alpha)\big)^{-1}$. Obviously $\tilde{r} > r'$. Let $r \in \hat{\phi}(K) \cap [r',r]$ such that $f_1, f_2 \in P_{r'}^K$, where $f_1 := X$, $f_2 := \exp(X\log(\alpha))$. It is not difficult to see that f_1, f_2 are algebraically independent. Since $\beta \notin \mathbf{Q}$ the elements of the sequence

$$(\mu_1 + \mu_2\beta)_{(\mu_1,\mu_2) \in \mathbf{N}^2}$$

are all distinct and in $U'_{r'}(0)$.

Suppose that $\mathrm{tdeg}_k k' = 0$. Then $k'|k$ is a finite extension such that $f_1(\mu_1 + \mu_2\beta)$, $f_2(\mu_1 + \mu_2\beta) \in k'$ for all $(\mu_1,\mu_2) \in \mathbf{N}^2$. For $T \in \mathbf{N}$ let

$$x_T^{(1)} := \max_{1 \leqslant \mu_1,\mu_2 \leqslant T} s(\mu_1 + \mu_2\beta)$$

and

$$x_T^{(2)} := \max_{1 \leqslant \mu_1,\mu_2 \leqslant T} s(\alpha^{\mu_1}(\alpha^\beta)^{\mu_2}) .$$

We find $x_T^{(1)} \leqslant c_1$ and $x_T^{(2)} \leqslant c_2 T$ with suitable constants depending only on k'. This together with Theorem 1 shows

$$1 \leqslant \overline{\lim_{T\to\infty}} \frac{\log x_T^{(1)}}{2 \log T} + \overline{\lim_{T\to\infty}} \frac{\log x_T^{(2)}}{2 \log T} \leqslant \frac{1}{2} ,$$

a contradiction.

In a similar way one proves

Corollary 2. *Let* $m, n \in \mathbf{N}$ *be such that* $mn > m+n$. *Let* $\{u_1,\ldots, u_m\}$, $\{v_1, \ldots, v_n\}$ *denote* $\mathbf{Q}$*-linearly independent subsets of* K *such that* $\hat{\phi}(u_i v_j) < 1$ *for all* $1 \leqslant i \leqslant m$, $1 \leqslant j \leqslant n$. *Then*

$$\mathrm{tdeg}_k\, k(\exp(u_1v_1),\ldots,\exp(u_mv_n)) \geqslant 1.$$

Remark. The simplest cases are $(m,n) = (2,3)$ and $(3,2)$ ("Theorem of six exponentials").

From Corollary 2 we deduce the following

Corollary 3. *Let* $\alpha \in A(K)$, $0 < \hat{\phi}(\alpha-1) < 1$. *Let* $\beta \in K$ *such that* $\beta \in A(K)$ *implies* $\deg(\beta) \geqslant 3$. *Let* $\hat{\phi}(\beta^\nu \log(\alpha)) < 1$ *for* $0 \leqslant \nu \leqslant 3$. *Then*

$$\mathrm{tdeg}_k\, k(\alpha^\beta,\alpha^{\beta^2},\alpha^{\beta^3}) \geqslant 1 .$$

Proof. Take $n = 3$, $m = 2$, $u_1 = \log(\alpha)$, $u_2 = \beta \log(\alpha)$, $v_1 = 1$,

$v_2 = \beta$, $v_3 = \beta^2$ in Corollary 2.

Our second main result is

Theorem 2. (Gelfond's method). *Let* $k'|k$ *be a finite, separable extension. Let* $\ell \in \mathbf{N}$, $\ell \leqslant 2$, *and* $r \in \hat{\phi}(K) \cap \mathbf{R}_+$. *Let* $f_1, \ldots, f_\ell \in P_r^K$ *such that at least two are algebraically independent over* K. *Let* $r' \in \mathbf{R}_+$, $r' < r$. *Let* $(\alpha_n)_{n \in \mathbf{N}}$ *denote a sequence in* $U'_{r'}(0)$ *whose elements are all distinct and such that* $f_i(\alpha_n) \in k'$ *for all* $1 \leqslant i \leqslant \ell$ *and* $n \in \mathbf{N}$. *Let*

$$D\Big(\sum_{n=0}^{\infty} a_n X^n\Big) := \sum_{n=1}^{\infty} n a_n X^{n-1}$$

denote the standard derivative on $K[[X]]$. *Finally let* $D(k'[f_1,\ldots,f_\ell]) \subset k'[f_1,\ldots,f_\ell]$. *Then*

$$(1) \quad \liminf_{n \in \mathbf{N}} \frac{1}{n} \max_{1\leqslant j\leqslant \ell} \max_{1\leqslant i\leqslant n} s\big(f_j(\alpha_i)\big) > 0 .$$

If D *operates on the* k'*-vector space* $k' + k'f_1 + \ldots + k'f_\ell$, *then*

$$(2) \quad \liminf_{n \in \mathbf{N}} \frac{1}{n} \max_{1\leqslant j\leqslant \ell} \max_{1\leqslant i\leqslant n} s(f_j(\alpha_i)) = +\infty .$$

Theorem 2 provides an alternative proof for Corollary 1. In addition we have

Corollary 4. (Theorem of Hermite-Lindemann). *Let* $\mathrm{char}(k) = 0$. *Let* $\alpha \in K$ *be such that* $0 < \hat{\phi}(\alpha) < 1$. *Then* $\mathrm{tdeg}_k\, k(\alpha,\exp(\alpha)) \geqslant 1$.

Proof. Let $k' := k(\alpha,\exp(\alpha))$, $r' := \hat{\phi}(\alpha)$. Let $r \in \hat{\phi}(K) \cap\,]r',1[$ be such that $f_1 := X$ and $f_2 := \exp(X)$ are in P_r^K. Then f_1, f_2 are algebraically independent over K. For all $n \in \mathbf{N}$, let $\alpha_n := n\alpha \in U'_{r'}(0)$. D operates on $k' + k'f_1 + k'f_2$. We have $s(\alpha_n) \leqslant ns(\alpha)$ and $s(\exp(\alpha_n)) \leqslant ns(\exp(\alpha))$ for all $n \in \mathbf{N}$ and therefore there exists a $C \in \mathbf{R}_+$ such that

$$A := \limsup_{n \in \mathbf{N}} \frac{1}{n} \max\Big\{\max_{1\leqslant j\leqslant n} s(j\alpha),\ \max_{1\leqslant j\leqslant n} s(\exp(j\alpha))\Big\} \leqslant C .$$

If $\mathrm{tdeg}_k\, k' = 0$ then by Theorem 2, $A = +\infty$, a contradiction.

For any $\alpha \in A(K)$ we call $H(\alpha)$ the height of α, which is defined as the maximum of the $\hat{\phi}$-values of the coeffieients of the minimal polynomial of α over R.

Theorem 3. (Baker's method). *Let* char(k) = 0. *Let* $n \in \mathbf{N}_0$, $d \in \mathbf{N}$, $A, B \in \mathbf{R}_+$, $A \geq 3$, $B \geq 3$. *Let* $\alpha_1, \ldots, \alpha_n \in A(K)$ *such that* $\hat{\phi}(\alpha_i - 1) < 1$, $\deg(\alpha_i) \leq d$, $H(\alpha_i) \leq A$ *for all* $1 \leq i \leq n$. *There exists an effective constant* $C \in \mathbf{R}_+$, *depending only on* n *and* d, *such that either*

$$\beta_0 + \sum_{j=1}^{n} \beta_j \log(\alpha_j) = 0$$

or

$$\hat{\phi}(\beta_0 + \sum_{j=1}^{n} \beta_j \log(\alpha_j)) > B^{-C(\log(A))^{2n^2+5n+8}}$$

for all $\beta_0, \beta_1, \ldots, \beta_n \in A(K)$ *with* $\deg(\beta_\nu) \leq d$ *and* $H(\beta_\nu) \leq B$ *for* $0 \leq \nu \leq n$.

From this we deduce the following *approximation measure.*

Corollary 5. *Let* char(k) = 0. *Let* $n, d \in \mathbf{N}$, $A \in \mathbf{R}_+$, $A \geq 3$. *Let* $\alpha_1, \ldots, \alpha_n, \beta_0, \beta_1, \ldots, \beta_n \in A(K)$ *with* $0 < \hat{\phi}(\alpha_i - 1) < 1$ *and* $\hat{\phi}(\beta_i)\hat{\phi}(\alpha_i - 1) < 1$ *for* $1 \leq i \leq n$. *Suppose either* $0 < \hat{\phi}(\beta_0) < 1$, *or* $1, \beta_1, \ldots, \beta_n$ *are* $\mathbf{Q}$*-linearly independent. Let* $e^{\beta_0} = \exp(\beta_0)$ *and* $\alpha_i^{\beta_i} := \exp(\beta_i \log(\alpha_i))$ *for* $1 \leq i \leq n$. *Then there exists an effective positive real constant* C, *depending only on* $n, d, \beta_0, \beta_1, \ldots, \beta_n, \alpha_1, \ldots, \alpha_n$ *such that for all* $\eta \in A(K)$ *with* $\deg(\eta) \leq d$ *and* $H(\eta) \leq A$

$$(e^{\beta_0} \alpha_1^{\beta_1} \cdots \alpha_n^{\beta_n} - \eta) > e^{-C(\log(A))^{2n^2+9n+15}}.$$

With the usual method, suitably adjusted for our purposes, one obtains the following transcendence measures for arbitrary characteristic.

Theorem 4. *Let* $b \in R$, $\phi(b) > 1$, *be fixed. Let* $(c_n)_{n \in \mathbf{N}}$ *denote a sequence of integers such that* $c_n \neq 0$ *infinitely often. Let*

$$\alpha := c_0 + \sum_{n=1}^{\infty} c_n b^{-n!} .$$

Then, for every polynomial $P \in R[X]$ *of degree* $D \leqslant 1$ *and height* H *one has*

$$\log\left(\hat{\phi}(P(\alpha))\right) \geqslant -51(D^{D-1} + DH\log^2(2H)) .$$

Finally let us note that, using the same methods as in [1] and [2], we can prove

Theorem 5. (Schanuel's conjecture). *Let* $(k,\phi) \in \overline{\mathbf{D}}$, ϕ *non-Archimedian,* $\mathrm{char}(k) = 0$. *Let* $n \in \mathbf{N}$ *and* $\alpha_1, \ldots, \alpha_n \in K$ *be* $\mathbf{Q}$-*linearly independent. Let* $\hat{\phi}(\alpha_i) < 1$ *for all* $1 \leqslant i \leqslant n$. *Then*

$$\mathrm{tdeg}_k\ k(\alpha_1,\ldots,\alpha_n,\exp(\alpha_1),\ldots,\exp(\alpha_n)) \geqslant n .$$

4. Auxiliary results.

For the proof of Theorem 1 we will need the following lemmas.

Lemma 3. (Fundamental inequality). *Let* $\alpha \in A(K)$, $\alpha \neq 0$. *Then*

$$\log(\hat{\phi}(\alpha)) \geqslant -\ 2\deg(\alpha)s(\alpha) .$$

Proof. See Sie.[3], Chapter 1.

Lemma 4. (Siegel's lemma). *Let* $k'|k$ *be a finite, separable extension. Let* $m, n \in \mathbf{N}$, $m > n$. *Let* $a_{ij} \in I(R,k')$, $1 \leqslant i \leqslant n$, $1 \leqslant j \leqslant m$. *Let* $S \in \mathbf{R}_+$, $S \geqslant 1$, *be such that* $\max_{i,j} \overline{|a_{ij}|} \leqslant S$. *Then the system*

$$\sum_{j=1}^{m} a_{ij}X_j = 0, \qquad 1 \leqslant i \leqslant n,$$

has a non-zero solution $(z_1,\dots,z_m) \in I(R,k')^m$ *such that*

$$\overline{|z_j|} \leq c_1(c_2S)^{\frac{n}{m-n}} \quad , \; 1 \leq j \leq m ,$$

where the constants c_1, $c_2 \in \mathbf{R}_+$ *depend only on* k'.

Proof. see [3, Chapter 1].

From non-Archimedian analysis we need the following two results.

Lemma 5. *Let* $r \in \mathbf{R}_+$ *and* $f \in P_r^K$, $f \neq 0$ *and put* $f = \sum_{n=0}^{\infty} a_n X^n$. *Then* f *has at most finitely many, namely* $d^+(f,r) := \max\{j \in \mathbf{N}_0 \mid \hat{\phi}(a_j)r^j = \|f\|_r\}$ *zeros in* $U_r^{\prime}(0)$ *(counted with multiplicities).*

Lemma 6. *Let* $r \in \mathbf{R}_+$, $r' \in \mathbf{R}_{+,0}$, $r' \leq r$. *Let* $0 \neq f \in P_r^K$ *have* h *zeros in* $U'_{r'}(0)$, $h \in \mathbf{N}_0$. *For* $\ell \in \mathbf{N}_0$ *let* $f^{(\ell)}$ *denote the* ℓ*-th formal derivative of the power series* f. *Then*

$$\|f^{(\ell)}\|_{r'} \leq \left(\frac{r'}{r}\right)^h r'^{-\ell} \|f\|_r .$$

5. Proof of Theorem 1.

Let $T \in \mathbf{N}$ be sufficiently large. Put $L = [k':k]$. The assertion is trivial if there exists $i_o \in \{1,\dots,\ell\}$ such that

$$\limsup_{T \in \mathbf{N}} \frac{1}{\log(T)} \log\left(\max_{1\leq j\leq T} s(f_{i_o}(\alpha_j))\right) = +\infty .$$

If for all $1 \leq i \leq \ell$

$$\limsup_{T \in \mathbf{N}} \frac{1}{\log(T)} \log\left(\max_{1\leq j\leq T} s(f_i(\alpha_j))\right) < +\infty ,$$

then we show: if $\rho_1, \dots, \rho_\ell \in \mathbf{R}_+$ are such that

$$\max_{1\leq j\leq T} s(f_i(\alpha_j)) \leq \tilde{T}^{\rho_i} \qquad \text{for all } \tilde{T} \geq T,\ 1 \leq i \leq \ell,$$

then $\sum_{i=1}^{\ell} \rho_i \geq \ell - 1$. A simple argument shows that we can restrict to the case $\max_{1 \leq i \leq \ell} \rho_i < \rho + \frac{1}{\ell}$, where $\rho := \frac{1}{\ell} \sum_{i=1}^{\ell} \rho_i$. Let $\varepsilon := \rho + \frac{1}{\ell}$.

Step 1. We construct an auxiliary function.

Let $G_i = [2T^{\varepsilon-\rho_i}]$, $1 \leq i \leq \ell$. We show that there exists a polynomial

$$P := \sum_{\lambda_1=0}^{G_1} \cdots \sum_{\lambda_\ell=0}^{G_\ell} p(\lambda_1,\ldots,\lambda_\ell)\, X_1^{\lambda_1} \ldots X_\ell^{\lambda_\ell},$$

not the zero-polynomial and with coefficients in $I(R,k')$, such that all $\overline{|p(\lambda_1,\ldots,\lambda_\ell)|} \leq \exp(c_4 T^{\varepsilon})$ and such that $F := P(f_1,\ldots,f_\ell) \in P_r^K$ vanishes for all α_j, $1 \leq j \leq T$.

From the last condition we obtain a system of T linear forms with coefficients $(f_1(\alpha_j))^{\lambda_1} \ldots (f_\ell(\alpha_j))^{\lambda_\ell}$ in k' in the $\prod_{i=1}^{\ell} (G_i+1) > 2^{\ell} T > T$ unknowns $p(\lambda_1,\ldots,\lambda_\ell)$. Let

$$\Theta_{ij} := d(f_i(\alpha_j)), \quad 1 \leq i \leq \ell, \quad 1 \leq j \leq T,$$

$$E_{\underline{\lambda},j} := \prod_{i=1}^{\ell} \Theta_{ij}^{G_i} \left(f_i(\alpha_j)\right)^{\lambda_i} \in I(R,k').$$

Here $\underline{\lambda} := (\lambda_1,\ldots,\lambda_\ell)$. Then the system

$$F(\alpha_j) = 0, \quad 1 \leq j \leq T, \tag{*}$$

is equivalent to the system

$$\sum_{\underline{\lambda}} p(\underline{\lambda})\, E_{\underline{\lambda},j} = 0, \quad 1 \leq j \leq T, \tag{**}$$

which has coefficients in $I(R,k')$ satisfying

$$\overline{|E_{\underline{\lambda},j}|} \leq \exp\Big(2 \sum_{i=1}^{\ell} G_i \max_{1 \leq \nu \leq T} s(f_i(\alpha_\nu))\Big)$$

$$\leq \exp(4\ell T^{\varepsilon}) =: S.$$

Applying Lemma 4 with $m := \prod_{i=1}^{\ell} (G_i+1)$, $n := T$ and S as above, we obtain $p(\underline{\lambda}) \in I(R,k')$, not all zero, which solve (*) <=> (**) and are such that

$$\overline{|p(\underline{\lambda})|} \leq c_1 c_2 S \leq \exp(c_4 T^{\varepsilon}).$$

Step 2. We construct a suitable non-zero element in k'.

Since $f_1,\ldots,f_\ell$ are K-algebraically independent, F is not the zero function. From Lemma 5 we know that F has only finitely many zeros in $U'_{r'}(0)$. Therefore there is at least one $F(\alpha_j) \neq 0$. Let $T^* := \min\{j \in \mathbf{N} : F(\alpha_j) \neq 0\}$. Step 1 shows that $T^* \geq T+1$. Now define $\eta_0 = F(\alpha_{T^*})$. By construction η_0 is a non-zero element of k'.

Step 3. We estimate η_0 from below.

Using Lemma 3 we obtain $\hat{\phi}(\eta_0) \leq \exp(-2Ls(\eta_0))$. The size of η_0 can be estimated by

$$s(\eta_0) \leq \log \overline{|P|} + \sum_{i=1}^{\ell} G_i \, s(f_i(\alpha_{T^*})),$$

where P denotes the maximum of the houses of the coefficients of P. We have

$$\log \overline{|P|} < T^{\varepsilon} \qquad \text{(from step 1)}$$

and

$$s(f_i(\alpha_{T^*})) \leq \max_{1 \leq \nu \leq T^*} s(f_i(\alpha_\nu)) \leq (T^*)^{\rho_i}, \qquad 1 \leq i \leq \ell .$$

Noting that $T^* \geq T+1$ we obtain

$$s(\eta_0) \ll (T^*)^{\varepsilon} .$$

Thus, for a suitable constant $c_5 \in \mathbf{R}_+$

$$\hat{\phi}(\eta_0) \geq \exp(-c5(T^*)^{\varepsilon}). \tag{1}$$

Step 4. We estimate η_0 from above.

Apply Lemma 6 with $f = F$, $h = T^* - 1$ and $\ell = 0$. We obtain

$$\hat{\phi}(\eta 0) \leq \|F\|_{\hat{\phi}(\alpha_{T^*})} \leq \|F\|_{r'} \leq (r^*)^{-T^*+1} \|F\|_r,$$

where $r^* := r/r' > 1$. Since $\|f_1\|_r, \ldots, \|f_\ell\| \in \mathbf{R}_+$ there exist constants $c^{(i)} \in \mathbf{R}_+$ depending only on f_i, $1 \leq i \leq \ell$, such that

$$\begin{aligned} \|F\|_r &= \| \sum_{\underline{\lambda}} p((\underline{\lambda}) f_1^{\lambda_1} \ldots f_\ell^{\lambda_\ell} \|_r \\ &\leq \max_{\lambda} \hat{\phi}(p(\lambda)) \|f_1\|_r \ldots \|f_\ell\|_r \\ &\leq \exp\left(c_4 T^{\varepsilon} + \sum_{i=1}^{\ell} c^{(i)} G_i\right) \\ &\leq \exp\left(c_6 (T^*)^{\varepsilon}\right) \end{aligned}$$

for some constant $c_6 \in \mathbf{R}_+$. Thus

$$\hat{\phi}(\eta 0) \leq \exp\left(-c_7 T^* + c_8 (T^*)^{\varepsilon}\right), \tag{2}$$

for suitable positive real constants c_7, c_8.

Now T^* is large since T is large. Thus the corresponding inequalities (1) and (2) give $\varepsilon \geq 1$, from which our assertion follows.

References.

[1] Ax, J. "On Schanuel's conjectures," *Ann. of Math.* **93** (1971), 252-268.

[2] Coleman, R. F. "On a stronger version of the Schanuel-Ax theorem," *Am. J. Math.* **102** (1979), 595-624.

[3] Sieburg, H. B. *Transzendenz und algebraische Unabhängigkeit in einer Klasse nicht-Archimedisch bewerteter Körper der Charakteristik Null*. Thesis, Köln 1983.

[4] Sieburg, H. B. "Algebraically independent values of Liouville-von Neumann series over QV-fields." To appear *Arch. Math.* 1984.

H. B. Sieburg
Stanford University and The Salk Institute for
Stanford, CA 94305 Biological Studies,
U.S.A. P.O.Box 85800
San Diego, CA 92168

Progress in Mathematics

1 GROSS. Quadratic Forms in Infinite-Dimensional Vector Spaces
2 PHAM. Singularités des Systèmes Différentiels de Gauss-Manin
3 OKONEK/SCHNEIDER/SPINDLER. Vector Bundles on Complex Projective Spaces
4 AUPETIT. Complex Approximation, Proceedings, Quebec, Canada, July 3-8, 1978
5 HELGASON. The Radon Transform
6 LION/VERGNE. The Weil Representation, Maslov Index and Theta Series
7 HIRSCHOWITZ. Vector Bundles and Differential Equations Proceedings, Nice, France, June 12-17, 1979
8 GUCKENHEIMER/MOSER/NEWHOUSE. Dynamical Systems, C.I.M.E. Lectures, Bressanone, Italy, June, 1978
9 SPRINGER. Linear Algebraic Groups
10 KATOK. Ergodic Theory and Dynamical Systems I
11 BALSLEV. 18th Scandinavian Congress of Mathematicians, Aarhus, Denmark, 1980
12 BERTIN. Séminaire de Théorie des Nombres, Paris 1979-80
13 HELGASON. Topics in Harmonic Analysis on Homogeneous Spaces
14 HANO/MARIMOTO/MURAKAMI/OKAMOTO/OZEKI. Manifolds and Lie Groups: Papers in Honor of Yozo Matsushima
15 VOGAN. Representations of Real Reductive Lie Groups
16 GRIFFITHS/MORGAN. Rational Homotopy Theory and Differential Forms
17 VOVSI. Triangular Products of Group Representations and Their Applications
18 FRESNEL/VAN DER PUT. Géometrie Analytique Rigide et Applications
19 ODA. Periods of Hilbert Modular Surfaces
20 STEVENS. Arithmetic on Modular Curves
21 KATOK. Ergodic Theory and Dynamical Systems II
22 BERTIN. Séminaire de Théorie des Nombres, Paris 1980-81
23 WEIL. Adeles and Algebraic Groups
24 LE BARZ/HERVIER. Enumerative Geometry and Classical Algebraic Geometry
25 GRIFFITHS. Exterior Differential Systems and the Calculus of Variations
26 KOBLITZ. Number Theory Related to Fermat's Last Theorem
27 BROCKETT/MILLMAN/SUSSMAN. Differential Geometric Control Theory
28 MUMFORD. Tata Lectures on Theta I
29 FRIEDMAN/MORRISON. Birational Geometry of Degenerations
30 YANO/KON. CR Submanifolds of Kaehlerian and Sasakian Manifolds
31 BERTRAND/WALDSCHMIDT. Approximations Diophantiennes et Nombres Transcendants
32 BOOKS/GRAY/REINHART. Differential Geometry
33 ZUILY. Uniqueness and Non-Uniqueness in the Cauchy Problem
34 KASHIWARA. Systems of Microdifferential Equations
35 ARTIN/TATE. Arithmetic and Geometry: Papers Dedicated to I.R. Shafarevich on the Occasion of His Sixtieth Birthday, Vol. I
36 ARTIN/TATE. Arithmetic and Geometry: Papers Dedicated to I.R. Shafarevich on the Occasion of His Sixtieth Birthday, Vol. II
37 DE MONVEL. Mathématique et Physique
38 BERTIN. Séminaire de Théorie des Nombres, Paris 1981-82
39 UENO. Classification of Algebraic and Analytic Manifolds
40 TROMBI. Representation Theory of Reductive Groups

41 STANELY. Combinatories and Commutative Algebra
42 JOUANOLOU. Théorèmes de Bertini et Applications
43 MUMFORD. Tata Lectures on Theta II
44 KAC. Infinite Dimensional Lie Algebras
45 BISMUT. Large Deviations and the Malliavin Calculus
46 SATAKE/MORITA. Automorphic Forms of Several Variables Taniguchi Symposium, Katata, 1983
47 TATE. Les Conjectures de Stark sur les Fonctions L d'Artin en s = 0
48 FRÖHLICH. Classgroups and Hermitian Modules
49 SCHLICHTKRULL. Hyperfunctions and Harmonic Analysis on Symmetric Spaces
50 BOREL, ET AL. Intersection Cohomology
51 BERTIN/GOLDSTEIN. Séminaire de Théoire des Nombres. Paris 1982–83
52 GASQUI/GOLDSCHMIDT. Déformations Infinitesimales des Structures Conformes Plates
53 LAURENT. Théorie de la Deuxième Microlocalisation dans le Domaine Complèxe
54 VERDIER/LE POTIER. Module des Fibres Stables sur les Courbes Algébriques Notes de l'Ecole Normale Supérieure, Printemps, 1983
55 EICHLER/ZAGIER. The Theory of Jacobi Forms
56 SHIFFMAN/SOMMESE. Vanishing Theorems on Complex Manifolds
57 RIESEL. Prime Numbers and Computer Methods for Factorization
58 HELFFER/NOURRIGAT. Hypoellipticité Maximale pour des Opérateurs Polynomes de Champs de Vecteurs
59 GOLDSTEIN. Séminarie de Théorie des Nombres, Paris 1983–84
60 PROCESI. Geometry Today: Giornate Di Geometria, Roma. 1984
61 BALLMANN/GROMOV/SCHROEDER. Manifolds of Nonpositive Curvature
62 GUILLOU/MARIN. A la Recherche de la Topologie Perdue
63 GOLDSTEIN. Séminaire de Théorie des Nombres, Paris 1984–85
64 MYUNG. Malcev-Admissible Algebras
65 GRUBB. Functional Calculus of Pseudo-Differential Boundary Problems
66 CASSOU-NOGUÈS/TAYLOR. Elliptic Functions and Rings and Integers
67 HOWE. Discrete Groups in Geometry and Analysis: Papers in Honor of G.D. Mostow on His Sixtieth Birthday
68 ROBERT. Antour de L'Approximation Semi-Classique
69 FARAUT/HARZALLAH. Deux Cours d'Analyse
70 ADOLPHSON/CONREY/GOSH/YAGER. Number Theory and Diophantine Problems: Proceedings of a Conference at Oklahoma State University